Quantum Imaging

Mikhail I. Kolobov
Editor

Quantum Imaging

Mikhail Kolobov
Laboratoire PhLAM
Université de Lille 1
Cité Scientifique, Bât. P5
F-59655 Villeneuve d'Ascq Cedex
FRANCE

Library of Congress Control Number: 2006924113

ISBN 10: 0-387-33818-7 e-ISBN 0-387-33988-4
ISBN 13: 978-0387-33818-7

Printed on acid-free paper.

Printed in Singapore. (TB/KYO)

9 8 7 6 5 4 3 2 1

springer.com

Preface

Optical imaging techniques are nowadays used not only by the scientific community but also in our daily life due to an extraordinary development of digital photo and video cameras, the last generation of cellular telephones, and so on.

Until very recently optical imaging has been ignoring the quantum nature of light. The reason for this is very simple: usually the intensity of light used in optical imaging is so high that quantum fluctuations inherently present in the light are completely negligible. However, rapid technological progress often brings us new challenges. For example, charge-coupled devices (CCD) are currently widely used to capture digital images. In a quest for improvement of the quality and resolution of captured still images and movies, current optical sensor technologies are struggling to reduce the pixel size (or to increase the number of pixels per unit area). As the pixel size decreases, the amount of light illuminating each individual pixel decreases as well. At low light levels the quantum nature of light will inevitably manifest itself in the appearance of photons or shot noise that will ultimately limit the quality of the detected image. This example gives us an idea of how the quantum nature of light puts ultimate performance limits on optical imaging.

This is just one example from the new area of quantum imaging that has as one of its goals investigation of such ultimate performance limits in optical imaging imposed by the quantum nature of light. Quantum imaging uses the latest achievements in quantum optics that allow us to tailor the spatial distribution of quantum fluctuations in the transverse area of light beams and to reduce these quantum fluctuations below the shot-noise limit. This reduction of spatial quantum fluctuations brings new opportunities for improving the performance in recording, storage, and read-out of optical images beyond the limits set by the spatial shot noise. Moreover, quantum imaging offers numerous exciting opportunities in the area of quantum information due to an intrinsic parallelism of optical image processing. This feature of quantum imaging allows us to greatly increase the information capacity of several quantum information protocols such as, for instance, quantum teleportation and quantum dense coding.

The area of quantum imaging has been pioneered by the teams in Russia (St. Petersburg, Moscow), Europe (Como, Paris, Lille), and the United States

(Boston, Evanston). This book presents the latest results in the area of quantum imaging that have been obtained in the framework of the European project QUANTIM ("Quantum Imaging") funded by the European Community. I would like to thank all the contributing authors because without their work this book could not have been written.

Lille, October 2005 *Mikhail Kolobov*

Contents

List of Contributors

Morten Bache
Optical Fiber Group
Research Center COM
Technical University of Denmark
DK-2800 Lyngby
Denmark

Hans A. Bachor
Department of Physics
Australian National University
Building 38
Canberra ACT 0200
Australia

Stephen M. Barnett
Department of Physics
University of Strathclyde
107 Rottenrow
Glasgow G4 0NG
United Kingdom

Enrico Brambilla
INFM
Università dell'Insubria
Via Valleggio 11
22100 Como
Italy

Pere Colet
Instituto Mediterraneo des Estudios Avanzados
Campus Universitat de les Illes Balears
E-07122 Palma de Mallorca
Spain

Claude Fabre
Laboratoire Kastler Brossel
Université Pierre et Marie Curie
Case 74
4 Place Jussieu
75252 Paris Cedex 05
France

Alessandra Gatti
INFM
Università dell'Insubria
Via Valleggio 11
22100 Como
Italy

Adrian Jacobo
Instituto Mediterraneo des Estudios Avanzados
Campus Universitat de les Illes Balears
E-07122 Palma de Mallorca
Spain

Ottavia Jedrkiewicz
INFM
Università dell'Insubria
Via Valleggio 11
22100 Como
Italy

John Jeffers
Department of Physics
University of Strathclyde
107 Rottenrow
Glasgow G4 0NG
United Kingdom

Mikhail I. Kolobov
Laboratoire PhLAM
Université de Lille 1
Cité Scientifique, Bât P5
59655 Villeneuve d'Ascq Cedex
France

Ping Koy Lam
Department of Physics
Australian National University
Building 38
Canberra ACT 0200
Australia

Eric Lantz
Laboratoire d'Optique
P. M. Duffieux
Université de Franche Comté
Route de Gray
25030 Besançon Cedex
France

Luigi A. Lugiato
INFM
Università dell'Insubria
Via Valleggio 11
22100 Como
Italy

Agnès Maître
Laboratoire Kastler Brossel
Université Pierre et Marie Curie
Case 74
4 Place Jussieu
75252 Paris Cedex 05
France

Gian-Luca Oppo
Department of Physics
University of Strathclyde
107 Rottenrow
Glasgow G4 0NG
United Kingdom

Maxi San Miguel
Instituto Mediterraneo des Estudios Avanzados
Campus Universitat de les Illes Balears
E-07122 Palma de Mallorca
Spain

Pierre Scotto
Instituto Mediterraneo des Estudios Avanzados
Campus Universitat de les Illes Balears
E-07122 Palma de Mallorca
Spain

Ivan V. Sokolov
V. A. Fock Physics Institute
Saint Petersburg State University
198504 Ul'yanovskaya 1
Saint Petersburg
Russian Federation

Nicolas Treps
Laboratoire Kastler Brossel
Université Pierre et Marie Curie
Case 74
4 Place Jussieu
75252 Paris Cedex 05
France

Roberta Zambrini
Department of Physics
University of Strathclyde
107 Rottenrow
Glasgow G4 0NG
United Kingdom

1 Quantum Imaging with Continuous Variables

Luigi A. Lugiato, Alessandra Gatti, and Enrico Brambilla

INFM, Dipartimento di Fisica e Matematica, Universitá dell'Insubria, Via Valleggio 11, 22100, Como, Italy luigi.lugiato@uninsubria.it

1.1 Introduction

A significant fraction of the research activities in the field of quantum imaging concerns optical Parametric Down-Conversion (PDC). Some basic features of this phenomenon will be described in this first chapter of the volume, which deals with the multiphoton regime of the signal–idler field, that one meets in Optical Parametric Amplifiers (OPA) with medium or high gain or in Optical Parametric Oscillators (OPO). In this case, the behavior of the system is naturally described in terms of continuous variables such as field intensity or field quadratures. On the other hand, in the (very) low gain regime of the OPA, one detects coincidences between signal and idler photons and a significant part of the literature on quantum imaging deals with this case, as illustrated in the review article [1].

The first part of this chapter, which is based on the tutorial delivered by one of us (L. A. L.) in the Cargèse workshop,[1] will introduce some key concepts in the continuous variable description, such as squeezing in quadratures and in photon number difference, or entanglement between quadratures, and the basic connection between this entanglement and squeezing. This will be done with the help of two paradigmatic models, one including a single radiation mode and the other with two modes. The second part of this chapter will be devoted to the spatially multimode configuration that one meets in OPAs and in degenerate OPOs. We will discuss the topic of spatially multimode squeezing or local squeezing and of spatial correlations, with the related near-field/far-field interest for quantum imaging, namely the detection of weak amplitude or phase objects duality in type-I OPA/OPO. Then we will turn our attention to two subjects of direct detection beyond the standard quantum limit and the image amplification by parametric down-conversion. The results illustrated in this chapter have been obtained prior to QUANTIM, whereas new results are included in Chapter 2 and Chapter 5.

[1] "Imaging at the Limits," ESF/PESC Exploratory Workshop, *Cargèse (Corsica), France, 5-11 September 2004.*

1.2 The Concepts of Squeezing and of Entanglement with Continuous Variables, and Their Intrinsic Connection

1.2.1 Prototype Model I

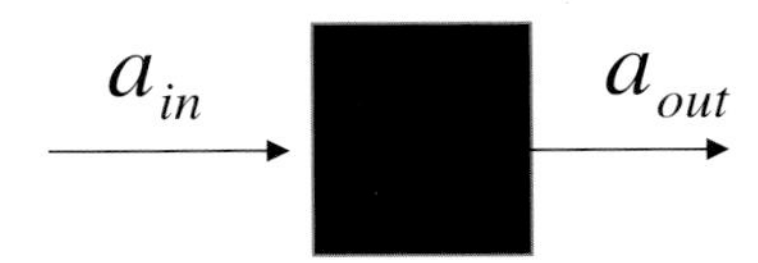

Fig. 1.1. "Input-output box" for an OPA (or an OPO below threshold) in the degenerate single-mode configuration.

Let us consider the "black box" in Fig. 1.1, with an input mode and an output mode associated with annihilation and creation operators $\hat{a}_{\rm in}$ and $\hat{a}_{\rm in}^{\dagger}$ and $\hat{a}_{\rm out}$, $\hat{a}_{\rm out}^{\dagger}$, respectively, with

$$\left[\hat{a}_{\rm in}, \hat{a}_{\rm in}^{\dagger}\right] = 1 , \qquad \left[\hat{a}_{\rm out}, \hat{a}_{\rm out}^{\dagger}\right] = 1 . \tag{1.1}$$

Let us assume the input-output relation

$$\hat{a}_{\rm out} = U\hat{a}_{\rm in} + V\hat{a}_{\rm in}^{\dagger}, \tag{1.2}$$

with coefficients U and V obeying the condition

$$|U|^2 - |V|^2 = 1, \tag{1.3}$$

which ensures the unitarity of transformation (1.2). In the following we take for definiteness

$$U = \cosh g , \qquad V = \sinh g . \tag{1.4}$$

A concrete realization of (1.2) is given, for example, by a degenerate OPA (or OPO below threshold) in the single-mode configuration.

Case 1

If $\hat{a}_{\rm in}$ is in a coherent state $\mid \alpha\rangle$, so that the mean value of $\hat{a}_{\rm in}$ is $\langle\hat{a}_{\rm in}\rangle = \alpha$, one has from (1.2)

$$\langle\hat{a}_{\rm out}\rangle = U\alpha + V\alpha^* . \tag{1.5}$$

Hence the system behaves as a phase-sensitive amplifier/deamplifier; for example, if α is real one has amplification:

$$|\langle\hat{a}_{\rm out}\rangle|^2 = |U+V|^2\,|\alpha|^2 = \mathrm{e}^{2g}\,|\langle\hat{a}_{\rm in}\rangle|^2 , \tag{1.6}$$

whereas if α is imaginary one has deamplification:

$$|\langle\hat{a}_{\rm out}\rangle|^2 = |U-V|^2\,|\alpha|^2 = \mathrm{e}^{-2g}\,|\langle\hat{a}_{\rm in}\rangle|^2 . \tag{1.7}$$

Case 2

Let us focus, instead, on the case that $\hat{a}_{\rm in}$ is in the vacuum state $\mid 0\rangle$. If we consider the quadrature components of the input and output modes

$$\hat{X}_{\rm in} = \frac{\hat{a}_{in} + \hat{a}^{\dagger}_{\rm in}}{2} \,, \qquad \hat{Y}_{\rm in} = \frac{\hat{a}_{\rm in} - \hat{a}^{\dagger}_{\rm in}}{2{\rm i}} \,, \tag{1.8}$$

$$\hat{X}_{\rm out} = \frac{\hat{a}_{out} + \hat{a}^{\dagger}_{\rm out}}{2} \,, \qquad \hat{Y}_{\rm out} = \frac{\hat{a}_{out} - \hat{a}^{\dagger}_{\rm out}}{2{\rm i}} \,, \tag{1.9}$$

the input-output relation (1.2) can be rephrased in the following form,

$$\hat{X}_{\rm out} = {\rm e}^{g} \hat{X}_{\rm in} \,, \qquad \hat{Y}_{\rm out} = {\rm e}^{-g} \hat{Y}_{\rm in} \,, \tag{1.10}$$

hence the quadrature component $\hat{X}$ is amplified whereas the quadrature component $\hat{Y}$ is deamplified, and from Fig. 1.2 one sees that the input vacuum state is transformed into a squeezed vacuum state, with squeezing in the quadrature component $\hat{Y}$. By varying the phase of the coefficients U and V with respect to the choice (1.4), the squeezing can be produced in an arbitrary quadrature component $\hat{X}_\theta = 1/2(\hat{a}{\rm e}^{-{\rm i}\theta} + \hat{a}^{\dagger}{\rm e}^{{\rm i}\theta})$, for any value of θ.

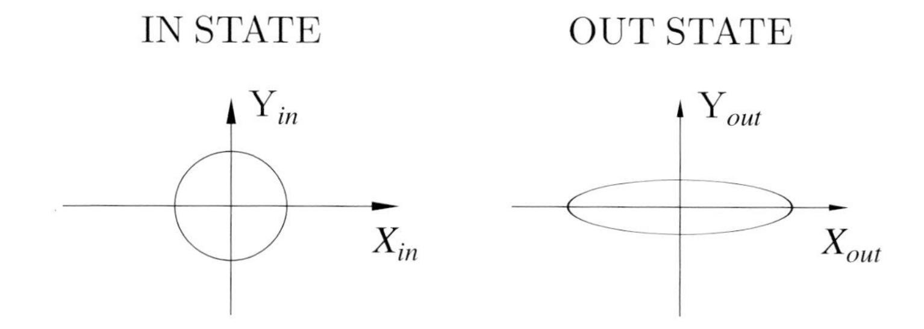

Fig. 1.2. The input-output relation (1.2) with (1.4) transforms the vacuum state into a squeezed vacuum state, with squeezing in the quadrature component $\hat{Y}$.

1.2.2 Prototype Model II

Let us now consider the black box in Fig. 1.3, with two input and two output modes such that

$$\left[\hat{a}_{i,{\rm in}}, \hat{a}^{\dagger}_{j,{\rm in}}\right] = \delta_{ij} \,, \qquad \left[\hat{a}_{i,{\rm out}}, \hat{a}^{\dagger}_{j,{\rm out}}\right] = \delta_{ij} \qquad (i,j=1,2) \,, \tag{1.11}$$

and the input-output relations

$$\begin{aligned} \hat{a}_{1\text{out}} &= U_1 \hat{a}_{1\text{in}} + V_1 \hat{a}^\dagger_{2\text{in}} \,, \\ \hat{a}_{2\text{out}} &= U_2 \hat{a}_{2\text{in}} + V_2 \hat{a}^\dagger_{1\text{in}} \,, \end{aligned} \tag{1.12}$$

with the unitarity condition

$$|U_i|^2 - |V_i|^2 = 1 \,, \qquad U_1 V_2 = U_2 V_1 \,. \tag{1.13}$$

In the following we take for definiteness

$$U_1 = U_2 = U = \cosh g \,, \qquad V_1 = V_2 = V = \sinh g \,. \tag{1.14}$$

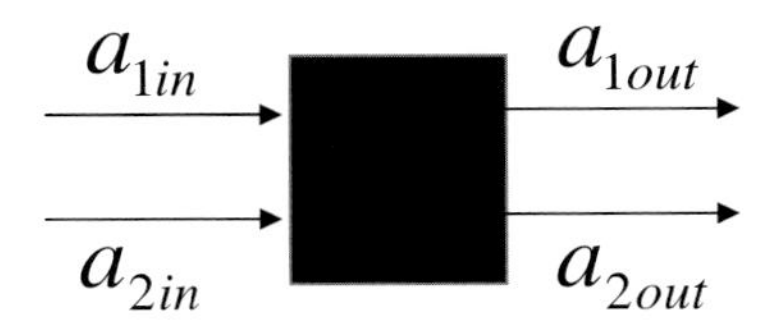

Fig. 1.3. Input-output box for an OPA (or an OPO below threshold) in the nondegenerate two-mode configuration.

A realization of (1.14) is given by a nondegenerate OPA (or OPO below threshold) in the two-mode regime.

Case 1

Let us consider the case when the mode $\hat{a}_{1\text{in}}$ is in a coherent state $|\,\alpha\rangle$, whereas the mode $\hat{a}_{2\text{in}}$ is in the vacuum state $|\,0\rangle$. One obtains from (1.12)

$$\langle \hat{a}_{1\text{out}} \rangle = U_1 \alpha, \qquad \langle \hat{a}_{2\text{out}} \rangle = V_2 \alpha^*, \tag{1.15}$$

so that

$$|\langle \hat{a}_{1\text{out}} \rangle|^2 = \left(\cosh^2 g\right) |\alpha|^2 = \left(\cosh^2 g\right) |\langle \hat{a}_{1\text{in}} \rangle|^2 \,, \tag{1.16}$$

$$|\langle \hat{a}_{2\text{out}} \rangle|^2 = \left(\sinh^2 g\right) |\alpha|^2 = \left(\sinh^2 g\right) |\langle \hat{a}_{2\text{in}} \rangle|^2 \,. \tag{1.17}$$

Hence mode 1 is amplified in a phase-insensitive way, whereas mode 2 is generated from the vacuum, and in the large-gain limit, $g \to \infty$, it becomes equally as strong as mode 1.

Case 1′

Let us assume that both modes $\hat{a}_{1\text{in}}$ and $\hat{a}_{2\text{in}}$ are in the same coherent state $\mid \alpha\rangle$. In this case one has from (1.12), (1.14),

$$\langle \hat{a}_{1\text{out}} \rangle = \langle \hat{a}_{2\text{out}} \rangle = \cosh g\alpha + \sinh g\alpha^* , \tag{1.18}$$

so that as in (1.5), one has a phase-sensitive amplification/deamplification. One can prove that, in general, phase-insensitive amplification degrades the signal-to-noise ratio at least by a factor 2, whereas phase-sensitive amplification can preserve the signal-to-noise ratio (noiseless amplification) [1].

Case 2

Let us consider the case when both $\hat{a}_{1\text{in}}$ and $\hat{a}_{2\text{in}}$ are in the vacuum state. The most interesting situation is in the limit of large g, in which $U_i \approx V_i \approx \mathrm{e}^g/2$ $(i = 1, 2)$, so that by indicating $U = \mathrm{e}^g/2$ relations (1.12) reduce to

$$\begin{aligned} \hat{a}_{1\text{out}} &= U\hat{a}_{1\text{in}} + U\hat{a}^\dagger_{2\text{in}} , \\ \hat{a}_{2\text{out}} &= U\hat{a}_{2\text{in}} + U\hat{a}^\dagger_{1\text{in}} , \end{aligned} \tag{1.19}$$

hence by introducing the quadrature components of the input-output modes

$$\begin{aligned} \hat{X}_{j,\text{in}} &= \frac{\hat{a}_{j,\text{in}} + \hat{a}^\dagger_{j,\text{in}}}{2} , \quad \hat{Y}_{j,\text{in}} = \frac{\hat{a}_{j,\text{in}} - \hat{a}^\dagger_{j,\text{in}}}{2\mathrm{i}} , \quad j = 1, 2, \\ \hat{X}_{j,\text{out}} &= \frac{\hat{a}_{j,\text{out}} + \hat{a}^\dagger_{j,\text{out}}}{2} , \quad \hat{Y}_{j,\text{out}} = \frac{\hat{a}_{j,\text{out}} - \hat{a}^\dagger_{j,\text{out}}}{2\mathrm{i}} \quad j = 1, 2, \end{aligned} \tag{1.20}$$

one obtains the relations

$$\hat{X}_{2\text{out}} = \hat{X}_{1\text{out}} , \qquad \hat{Y}_{2\text{out}} = -\hat{Y}_{1\text{out}} . \tag{1.21}$$

Therefore, if one measures, for example, $\hat{X}_{1\text{out}}$ and $\hat{Y}_{1\text{out}}$, one can immediately infer the values of $\hat{X}_{2\text{out}}$ and $\hat{Y}_{2\text{out}}$. This is precisely the phenomenon of *quantum entanglement*, and this is completely identical to the original Einstein–Podolsky–Rosen paradox [2], which was formulated for the position x and momentum p of two particles. This formulation of the EPR paradox for continuous variables $\hat{X}$ and $\hat{Y}$ (quadrature components) of two radiation modes was introduced in [3] taking into account the uncertainties in the measurements of $\hat{X}$ and $\hat{Y}$. This was experimentally verified in [4].

1.3 Intrinsic Relation Between Squeezing and Entanglement

In this section we show that there is a basic connection between the two paradigmatic models just discussed, which amounts to an intrinsic relation between entanglement and squeezing.

Let us consider a 50/50 beamsplitter (Fig. 1.4). One demonstrates that:

— If $\hat{a}_1$ and $\hat{a}_2$ are EPR entangled beams (in the sense defined before) then the beam $\hat{b}_1$ is squeezed in the $\hat{Y}$ quadrature and the beam $\hat{b}_2$ is squeezed in the $\hat{X}$ quadrature.

— And vice versa.

Proof

Let us consider the input-output relations of the beamsplitter

$$\hat{b}_1 = \frac{\hat{a}_1 + \hat{a}_2}{\sqrt{2}} , \qquad \hat{b}_2 = \frac{\hat{a}_2 - \hat{a}_1}{\sqrt{2}} . \tag{1.22}$$

Next, let us assume that $\hat{a}_1$ and $\hat{a}_2$ are the entangled output modes $\hat{a}_{1\text{out}}$ and $\hat{a}_{2\text{out}}$ of model II, so that

$$\begin{aligned} \hat{a}_1 &= U\hat{a}_{1\text{in}} + V\hat{a}_{2\text{in}}^\dagger , \\ \hat{a}_2 &= U\hat{a}_{2\text{in}} + V\hat{a}_{1\text{in}}^\dagger , \end{aligned} \tag{1.23}$$

where $\hat{a}_{1\text{in}}$ and $\hat{a}_{2\text{in}}$ are in the vacuum state. By inserting (1.23) into (1.22) we obtain

$$\hat{b}_1 = U\hat{f}_1 + V\hat{f}_1^\dagger, \tag{1.24}$$

$$\hat{b}_2 = U\hat{f}_2 - V\hat{f}_2^\dagger, \tag{1.25}$$

where modes $\hat{f}_1$ and $\hat{f}_2$ are defined as

$$\hat{f}_1 = \frac{\hat{a}_{1\text{in}} + \hat{a}_{2\text{in}}}{\sqrt{2}} , \qquad \hat{f}_2 = \frac{\hat{a}_{2\text{in}} - \hat{a}_{1\text{in}}}{\sqrt{2}} . \tag{1.26}$$

Because $\hat{a}_1$ and $\hat{a}_2$ are in the vacuum state, the same is true for $\hat{f}_1$ and $\hat{f}_2$. Now one notes immediately that (1.24) is identical to the prototype model I (1.2), hence we can conclude that $\hat{b}_1$ is squeezed with respect to the $\hat{Y}$ quadrature. On the other hand, one sees that (1.25) has the same form of

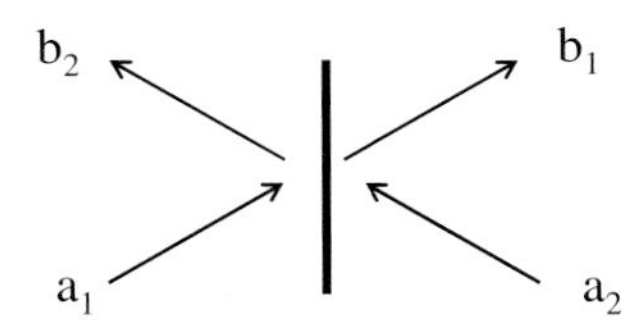

Fig. 1.4. A 50/50 beamsplitter converts modes a_1 and a_2 into modes b_1 and b_2.

model I (1.2) except that V is replaced by $-V$. One can easily prove that this feature implies that $\hat{b}_2$ is squeezed with respect to the $\hat{X}$ quadrature.

1.4 Spatially Multimode Parametric Down-Conversion: Some Topics in Quantum Imaging

1.4.1 Spatially Multimode Versus Single-Mode Squeezing: Optical Parametric Down-Conversion of Type-I

In almost all literature on squeezing one considers single-mode squeezing. If one wants to detect a good level of squeezing, the local oscillator must be matched to the squeezed spatial mode and, in addition, it is necessary to detect the whole beam. If one detects only part of the beam, the squeezing is immediately degraded, because a portion of a mode necessarily involves higher-order modes, in which squeezing is absent. What we can call *local squeezing* (i.e., squeezing in small regions of the transverse plane) can be obtained only in the presence of *spatially multimode squeezing*, (i.e., squeezing in a band of spatial modes). This has been predicted by Sokolov and Kolobov for a traveling-wave optical parametric amplifier (OPA) [6, 7] and by our group for an optical parametric oscillator (OPO) [8, 9]. Let us dwell a moment, for

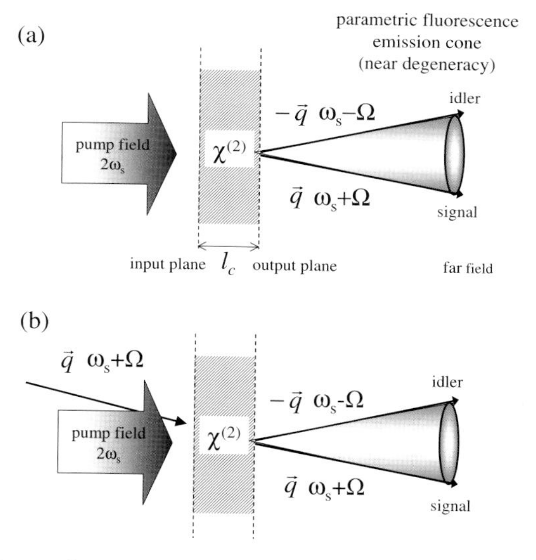

Fig. 1.5. (a) Scheme for parametric down-conversion of type-I. (b) Parametric amplification of a plane-wave; $\boldsymbol{q}$ is the component of the wave-vector in the plane orthogonal to the direction of propagation of the pump.

example, on the case of the OPA of type-I (Fig. 1.5a), in which one has a slab of $\chi^{(2)}$ material that is pumped by a coherent plane wave of frequency $2\omega_s$. A fraction of the pump photons is down-converted into signal–idler photon pairs, which are distributed over a broad band of temporal frequencies around the degenerate frequency ω_s. For each fixed temporal frequency, the photon pairs are distributed over a band of spatial frequencies labeled by the transverse component $\boldsymbol{q}$ of the wave-vector.

If, in addition to the pump field, we inject a coherent plane wave with frequency $\omega_s + \Omega$ and the transverse wave-vector $\boldsymbol{q}$ (Fig. 1.5b), in the output we have a signal wave that corresponds to an amplified version of the input wave, and for this reason the system is called an optical parametric amplifier. Because of the pairwise emission of photons, there is also an idler wave that, close to degeneracy, is symmetrical with respect to the signal wave. Referring to the case in which only the pump is injected, two regimes can be distinguished. One is that of pure spontaneous parametric down-conversion, as in the case of a very thin crystal. In this case coincidences between partners of single photon pairs are detected. The other is that of dominant stimulated parametric down-conversion, in which a large number of photon pairs at a time is detected. In this chapter we will focus on the second case, whereas the first case will be considered in Chapter 2.

On a more formal ground, let us consider the down-converted signal–idler field emitted close to the degenerate frequency ω_s. Let us denote by $\hat{a}_{\rm in}(\boldsymbol{x}, t)$ the signal–idler complex amplitude envelope operator at the input endface of the crystal slab, and by $\hat{a}_{\rm out}(\boldsymbol{x}, t)$ the envelope operator at the output endface; t indicates time and $\boldsymbol{x} \equiv (x, y)$ is the coordinate vector in the endfaces. We expand $\hat{a}_{\rm in}$ and $\hat{a}_{\rm out}$ in Fourier modes in space and time:

$$\hat{a}_{\rm in}(\boldsymbol{x}, t) = \int {\rm d}\boldsymbol{q} \int {\rm d}\Omega \; \hat{a}_{\rm in}(\boldsymbol{q}, \Omega) {\rm e}^{{\rm i}\boldsymbol{q}\cdot\boldsymbol{x} - {\rm i}\Omega t} \,, \tag{1.27}$$

$$\hat{a}_{\rm out}(\boldsymbol{x}, t) = \int {\rm d}\boldsymbol{q} \int {\rm d}\Omega \; \hat{a}_{\rm out}(\boldsymbol{q}, \Omega) {\rm e}^{{\rm i}\boldsymbol{q}\cdot\boldsymbol{x} - {\rm i}\Omega t} \,. \tag{1.28}$$

One can demonstrate that, in the linear regime of an undepleted pump, the following input-output relations hold [9],

$$\hat{a}_{\rm out}(\boldsymbol{q}, \Omega) = U(\boldsymbol{q}, \Omega)\hat{a}_{\rm in}(\boldsymbol{q}, \Omega) + V(\boldsymbol{q}, \Omega)\hat{a}^{\dagger}_{\rm in}(-\boldsymbol{q}, -\Omega) \,, \tag{1.29}$$

$$\hat{a}_{\rm out}(-\boldsymbol{q}, -\Omega) = U(-\boldsymbol{q}, -\Omega)\hat{a}_{\rm in}(-\boldsymbol{q}, -\Omega) + V(-\boldsymbol{q}, -\Omega)\hat{a}^{\dagger}_{\rm in}(\boldsymbol{q}, \Omega) \,, \tag{1.30}$$

where the expressions of $U(\boldsymbol{q}, \Omega)$ and $V(\boldsymbol{q}, \Omega)$ are given in [9]. We can note immediately that, for each fixed $\boldsymbol{q}$, Ω, Eqs. (1.29) have the same form of the prototype model II (1.12). Hence the results of Section 1.2.2 hold for this case; for example, Fig. 1.5b corresponds to case 1 of Section 1.2.2. The case of parametric down-conversion of type-II will be considered in Chapter 5.

1.4.2 Near-Field/Far-Field Duality in Type-I OPAs

We want to illustrate the key spatial quantum properties of the field emitted by an OPA of type-I, in the linear regime of negligible pump depletion, or by an OPO below threshold.

In the near field (see Fig. 1.6) one has the phenomenon of spatially multimode squeezing or local squeezing discussed in Section 1.4.1. A good level of squeezing is found, provided the region that is detected has a linear size not smaller than the inverse of the spatial bandwidth of emission in the Fourier plane. If, on the other hand, one looks at the far field (which can be reached, typically, by using a lens as shown in Fig. 1.6) one finds the phenomenon of *spatial entanglement* between small regions located symmetrically with respect to the center. Precisely, if one considers two symmetrical pixels 1 and 2 (Fig. 1.7a), the intensity fluctuations in the two pixels are very well correlated or, equivalently, the fluctuations in the intensity difference between the two pixels are very much below the shot-noise level [10, 11]. Precisely,

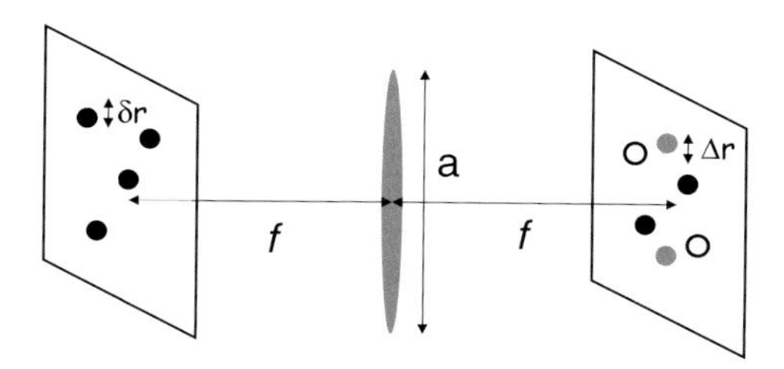

Fig. 1.6. Illustration of the near-field far-field duality; f is the focal plane of the lens. Not shown is the pump field of frequency $2\omega_s$ and the nonlinear slab.

let us consider the number of photons $\hat{N}_1$ and $\hat{N}_2$ detected in pixel 1 and 2, respectively, and the associated fluctuations $\delta\hat{N}_i = \hat{N}_i - \langle\hat{N}_i\rangle$ (i $= 1, 2$). For symmetry reasons one has that $\langle\hat{N}_1\rangle = \langle\hat{N}_2\rangle$, $\langle(\delta\hat{N}_1)^2\rangle = \langle(\delta\hat{N}_2)^2\rangle$. In the limit of the plane-wave pump, the photon number difference $\hat{N}_- = \hat{N}_1 - \hat{N}_2$ turns out to be fluctuationless [10]; that is,

$$\langle(\delta\hat{N}_-)^2\rangle = 0 \,. \tag{1.31}$$

Basically, one has that $\hat{N}_1 = \hat{N}_2$; that is, by measuring $\hat{N}_1$ one can infer the value of $\hat{N}_2$ (entanglement). This result expresses in the most emphatic way the emission of signal and idler photons in pairs, and follows from the perfect correlation between the photons in 1 and 2. As a matter of fact one has

$$\langle(\delta\hat{N}_-)^2\rangle = \langle(\delta\hat{N}_1)^2\rangle + \langle(\delta\hat{N}_2)^2\rangle - 2\langle\delta\hat{N}_1\delta\hat{N}_2\rangle \,; \tag{1.32}$$

because $\langle(\delta\hat{N}_1)^2\rangle = \langle(\delta\hat{N}_2)^2\rangle$ one has from (1.31) that the normalized correlation

$$C \equiv \frac{\langle\delta\hat{N}_1\delta\hat{N}_2\rangle}{\sqrt{\langle(\delta\hat{N}_1)^2\rangle\langle(\delta\hat{N}_2)^2\rangle}} = 1 \,, \tag{1.33}$$

which means perfect correlation. For the realistic case of a Gaussian pump, the fluctuations of $\hat{N}_-$ are below the shot-noise level; [10] that is

$$\langle(\delta\hat{N}_-)^2\rangle < \langle\hat{N}_+\rangle = \langle\hat{N}_1\rangle + \langle\hat{N}_2\rangle \,. \tag{1.34}$$

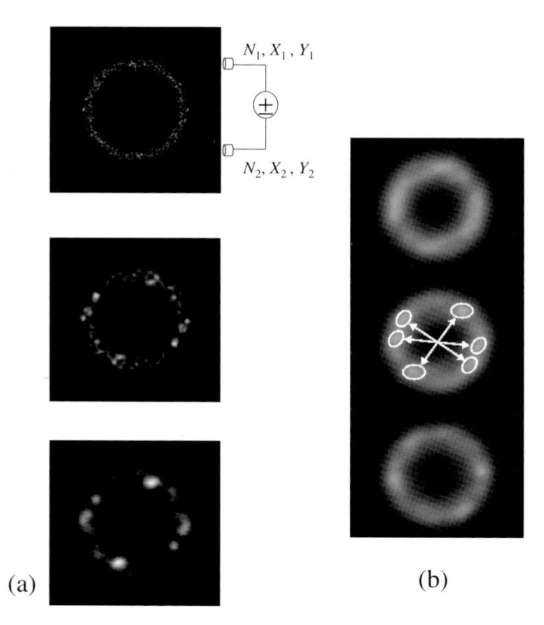

Fig. 1.7. Intensity distribution in the far field for a single shot of the pulsed pump field. (a) Numerical simulations. The waist of the pump beam is 1000 μm, 300 μm, 150 μm in the three frames from the top to the bottom, respectively. N_i, X_i, Y_i, (i $= 1, 2$) denote the photon numbers and the quadrature component measured in the two pixels 1 and 2, respectively. (b) Experimental observation by Devaux and Lantz at University of Besançon (see [13]).

Because this phenomenon arises for any pair of symmetrical pixels, we call it spatial entanglement. The same effect occurs also for quadrature components, because in the two pixels the fluctuations of the quadrature component $\hat{X}$ are almost exactly correlated, and those of the quadrature component $\hat{Y}$ are almost exactly anticorrelated [12]. The case of perfect correlation/anticorrelation occurs in the limit of the plane-wave pump, in which relations (1.21) hold.

The minimum size of the symmetrical small regions, among which one finds spatial entanglement, is determined by the finite aperture of optical elements, and is given, in the paraxial approximation, by $\lambda f/a$, where λ is the wavelength, f is the focal length of the lens and a is the aperture of optical elements (e.g., the lens aperture; Fig. 1.6). In a more realistic model of the OPA, the finite waist of the pump field must be taken into account. In this case the minimum size of the regions where entanglement is detectable in the far field is determined by the pump waist.

The spatial entanglement of intensity fluctuations in the far field is quite evident even in single shots (the pump field is typically pulsed). Figure 1.7a shows a numerical simulation in a case of noncollinear phase matching at the degenerate frequency. One observes the presence of symmetrical intensity peaks, which become broader and broader as one reduces the waist of the pump field. A similar situation is observed in an experiment performed using a LBO crystal [13] (see Fig. 1.7b). We observe finally that the near-field/far-field duality (i.e., squeezing in the near field, entanglement in the far field) can be understood on the basis of the intrinsic connection between squeezing and quantum entanglement. The spatial entanglement in the far field arises from the correlation between the modes $\hat{a}_{\text{out}}(\boldsymbol{q}) \sim \exp[\mathrm{i}\boldsymbol{q} \cdot \boldsymbol{x}]$ and $\hat{a}_{\text{out}}(\boldsymbol{q}) \sim \exp[-\mathrm{i}\boldsymbol{q} \cdot \boldsymbol{x}]$, which in the far field gives rise to two separated and opposite spots in the transverse plane.

On the other hand, in the near field there is no squeezing in modes $\hat{a}(\boldsymbol{q})$ and $\hat{a}(-\boldsymbol{q})$ separately, whereas there is large squeezing in the combination modes $\hat{b}(\boldsymbol{q}) = (\hat{a}(\boldsymbol{q}) + \hat{a}(-\boldsymbol{q}))/\sqrt{2}$ and $\hat{b}(-\boldsymbol{q}) = (\hat{a}(\boldsymbol{q}) - \hat{a}(-\boldsymbol{q}))/\sqrt{2}$, which annihilate photons in spatial modes $\sim\cos(\boldsymbol{q} \cdot \boldsymbol{x})$ and $\sim\sin(\boldsymbol{q} \cdot \boldsymbol{x})$. In the near field it is possible to observe this squeezing by using a local oscillator with a $\cos(\boldsymbol{q} \cdot \boldsymbol{x})$ or $\sin(\boldsymbol{q} \cdot \boldsymbol{x})$ spatial configuration [8]. One notices immediately that the relation between modes $\hat{a}(\boldsymbol{q})$, $\hat{a}(-\boldsymbol{q})$ and modes $\hat{b}(\boldsymbol{q})$, $\hat{b}(-\boldsymbol{q})$ coincides with Eq. (1.22), which, as we have seen, transforms entangled beams into squeezed beams, and vice versa.

1.4.3 Detection of Weak Amplitude or Phase Objects Beyond the Standard Quantum Limit

Let us consider first the case of a weak amplitude object that is located, say, in the signal part of the field emitted by an OPA (Fig. 1.8). Both signal and idler are very noisy and therefore, in the case of a large photon number, if the object is weak and we detect only the signal field, the signal-to-noise ratio for the object is low. But, because of the spatial entanglement, the

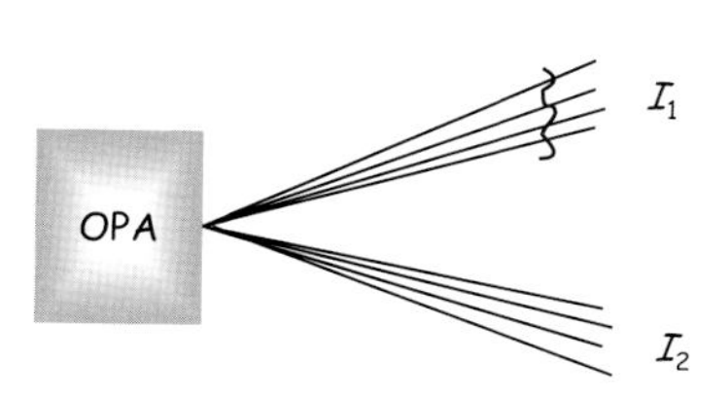

Fig. 1.8. Detection of a weak amplitude object by measuring the intensity difference $I_1 - I_2$.

fluctuations in the intensity difference between the signal and idler are small. Hence if we detect the intensity difference, the signal-to-noise ratio for the object becomes much better. This scheme is the generalization to the spatially multimode configuration of a single-mode scheme utilized to detect a weak absorption [14]. Next, let us pass to the case of a weak phase object in which

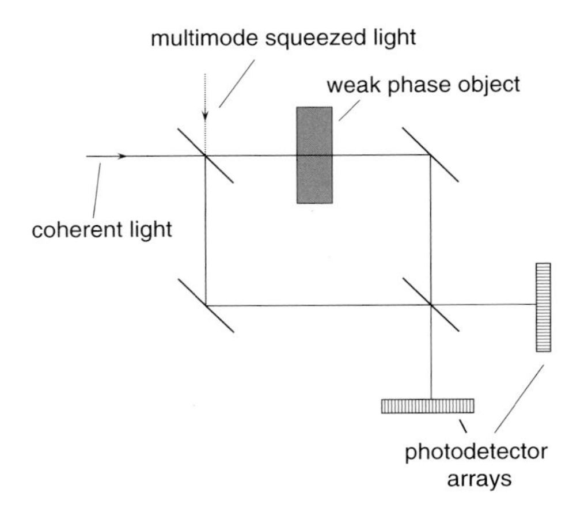

Fig. 1.9. Detection of a weak phase object.

one can exploit, instead, the property of spatially multimode squeezing. The configuration is the standard one of a Mach–Zehnder interferometer in which, as is well known, one can detect a small phase shift with a sensitivity beyond the standard quantum limit by injecting a squeezed beam in the port through which usually normal vacuum enters. If we have a weak phase image (Fig. 1.9) we can obtain the same result by injecting spatially multimode squeezed light [15].

1.4.4 Image Amplification by Parametric Down-Conversion (Type-I)

Let us come back to the configuration of Fig. 1.5b. Let us assume that now, instead of a plane wave at frequency close to ω_s, we inject a coherent monochromatic image (Fig. 1.10) of frequency ω_s. Parametric image amplification has been extensively studied from a classical viewpoint (see, e.g., [13]). A basic point in Fig. 1.10 is that, if the image is injected off axis, one obtains in the output a signal image that represents an amplified version of the input image, and also a symmetrical idler image. An interesting situation arises if one has, in addition to the amplifier, a pair of lenses located at focal distances with respect to the object plane, to the amplifier, and to the image plane (Fig. 1.11). As was shown by our group [11,12,16], in the limit of large amplification the two output images can be considered twins of each other even from a quantum-mechanical viewpoint. As a matter of fact, they do not only display the same intensity distribution but also the same local quantum fluctuations. Precisely, let us consider two symmetrical pixels in the two images (Fig. 1.12). It turns out that the intensity fluctuations in the two pixels are identical, that is, exactly correlated/synchronized. On the other hand, the

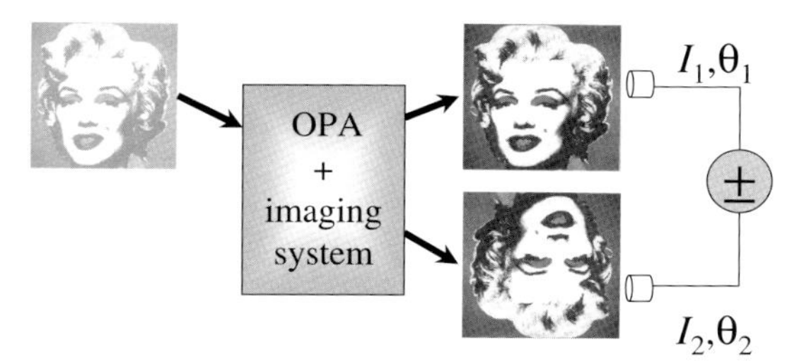

Fig. 1.10. Off-axis injection of an image and generation of twin entangled images.

phase fluctuations are exactly anticorrelated. So in this way, from one image one obtains twin images in a state of spatial entanglement that also involves the quadrature components X and Y, as already described in the case of parametric fluorescence without any signal injection in Section 1.4.2. There

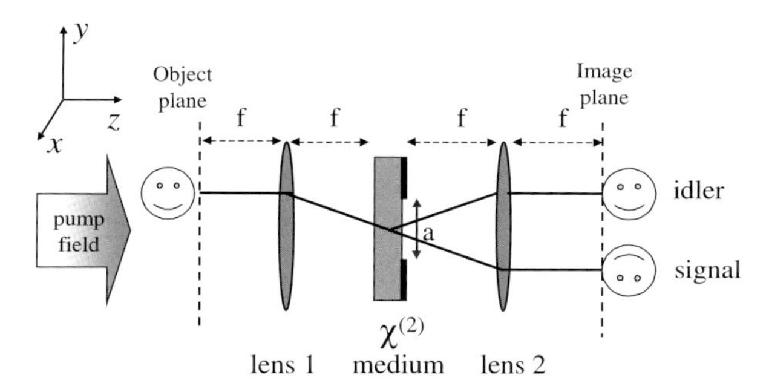

Fig. 1.11. A possible scheme of the parametric optical image amplifier.

is, however, a negative point that concerns the signal-to-noise ratio. When the input image is injected off axis, this mechanism of amplification is phase-insensitive and therefore, as is well known, it adds 3 dB of quantum noise in

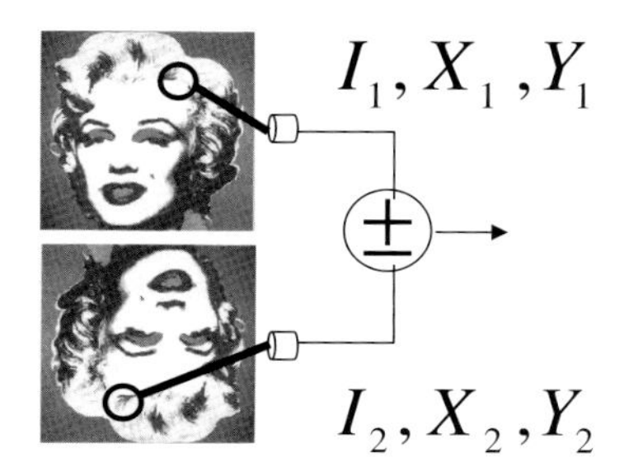

Fig. 1.12. The spatial entanglement between the two output images concerns intensity and phase fluctuations, and also the fluctuations of quadrature components.

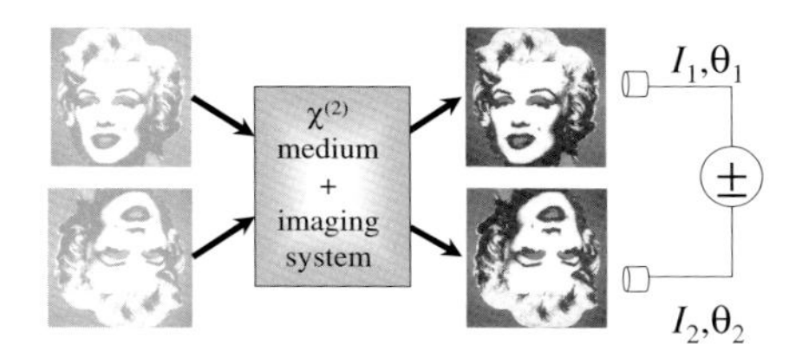

Fig. 1.13. Symmetrical injection of an image.

the output [17]. In order to have noiseless amplification (i.e., amplification that preserves the signal-to-noise ratio) one must inject two coherent images symmetrically (Fig. 1.13) [16]. In this case one has in the input two identical, but uncorrelated images, and in the output two amplified images in a state of spatial quantum entanglement. One can prove that this symmetrical configuration is phase-sensitive (compare with Case 1′ of Section 1.2.2) and, in fact, as shown in [18, 19], the amplification can become noiseless. A few years ago there was a landmark experiment by Kumar and collaborators [20] that demonstrated the noiseless amplification of a simple test pattern. This experiment has a configuration which differs from that of Fig. 1.11 because it does not include the two lenses. A complete theory for this experimental configuration is given in [21].

References

1. L. A. Lugiato, A. Gatti, and E. Brambilla, J. Opt. B: Quant. and Semiclass. Opt. **4**, S176-S183 (2002).
2. D. F. Walls and G. Milburn, *Quantum Optics* (Springer Verlag, Berlin, 1994); M. O. Scully and M. S. Zubairy, *Quantum Optics* (Cambridge University Press, Cambridge, 1997); W. Schleich, *Quantum Optics in Phase Space* (Wiley, New York, 2001); S. M. Barnett and P. M. Radmore, *Methods in Theoretical Quantum Optics* (Clarendon, Oxford, 1997).
3. A. Einstein, B. Podolsky, and N. Rosen, Phys. Rev. **47**, 777 (1935).
4. M. D. Reid and P. D. Drummond, Phys. Rev. Lett. **60**, 2731 (1988); M. D. Reid, Phys. Rev. A **40**, 913 (1989).
5. Z. Y. Ou, S. F. Pereira, H. J. Kimble, and K. C. Peng, Phys. Rev. Lett. **68**, 3663 (1992); Z. Y. Ou, S. F. Pereira and H. J. Kimble, Appl. Phys. B **55**, 265 (1992).
6. M. I. Kolobov and I. V. Sokolov, Sov. Phys JETP **69**, 1097 (1989); Phys. Lett. A **140**, 101 (1989); Europhys. Lett. **15**, 271 (1991).
7. M. I. Kolobov, Rev. Mod. Phys. **71**, 1539 (1999).
8. L. A. Lugiato and A. Gatti, Phys. Rev. Lett. **70**, 3868 (1993); A. Gatti and L. A. Lugiato, Phys. Rev. A **52**, 1675 (1995); L. A. Lugiato, M. Brambilla, and A. Gatti, Optical Pattern Formation, in *Advances in Atomic, Molecular and Optical Physics*, vol. 40, 229 (Academic, Boston, 1999).
9. L. A. Lugiato and Ph. Grangier, J. Opt. Soc. Am. B **14**, 225 (1997); K. I. Petsas, A. Gatti, L. A. Lugiato, and C. Fabre, Eur. Phys. J. D **12**, 501 (2003).

10. E. Brambilla, A. Gatti, L. A. Lugiato, and M. I. Kolobov, Eur. Phys. J. D **15**, 127 (2001).
11. A. Gatti, E. Brambilla, L. A. Lugiato, and M. I. Kolobov, Phys. Rev. Lett. **83**, 1763 (1999).
12. P. Navez, E. Brambilla, A. Gatti, and L. A. Lugiato, Phys. Rev. A **65**, 013813 (2001).
13. F. Devaux and E. Lantz, Eur. Phys. J. D **8**, 117 (2000); E. Lantz and F. Devaux, Eur. Phys. J. D **17**, 93 (2001).
14. P. H. Souto Ribeiro, C. Schwob, A. Maître, and C. Fabre, Opt. Lett. **22**, 1893 (1997).
15. M. I. Kolobov and P. Kumar, Opt. Lett. **18**, 849 (1993).
16. A. Gatti, E. Brambilla, L. A. Lugiato, and M. I. Kolobov, J. Opt. B: Quantum Semiclass. Opt. **2**, 196 (2000).
17. C. M. Caves, Phys. Rev. D **26**, 1817 (1982).
18. M. I. Kolobov and L. A. Lugiato, Phys. Rev. A **52**, 4930 (1995).
19. I. V. Sokolov, M. I. Kolobov, and L. A. Lugiato, Phys. Rev. A **60**, 2420 (1998).
20. S.-K. Choi, M. Vasilyev and P. Kumar, Phys. Rev. Lett. **83**, 1938 (1999).
21. K. Wang, G. Yang, A. Gatti and L. A. Lugiato J. Opt. B: Quantum Semiclass. Opt. **5**, S535 (2003).

2 Spatial Entanglement in Optical Parametric Down-Conversion

Alessandra Gatti, Enrico Brambilla, Ottavia Jedrkiewicz, and Luigi A. Lugiato

INFM, Dipartimento di Fisica e Matematica, Universitá dell'Insubria, Via Valleggio 11, 22100, Como, Italy luigi.lugiato@uninsubria.it

2.1 Introduction

This chapter and Chapter 5 constitute a direct continuation of Chapter 1, with a description of the developments in the continuous variable approach to quantum imaging, achieved in the framework of QUANTIM. Whereas Chapter 2 treats the traveling-wave configuration for PDC, the discussion of Chapter 5 focuses on the cavity configuration of optical parametric oscillators. The structure of this chapter is as follows. Section 2.2 presents the theoretical picture of spatial entanglement in type-II materials. Contrary to most literature on PDC, the emphasis is on the high-gain regime. Section 2.3 describes the experimental observations of those theoretical predictions outlined in Section 2.2, which concern the quantum correlations in the far field, with an associated phenomenon of photon antibunching in space. In the final Section 2.4 we illustrate the multiphoton, multimode polarization entanglement in the high-gain PDC. Instead of considering the standard polarization entanglement in single pairs of signal–idler photons, we focus on collective polarization entanglement properties at the macroscopic level.

2.2 Simultaneous Near-Field and Far-Field Spatial Quantum Correlation in the High-Gain Regime of Type-II Parametric Down-Conversion

This section is based on [1], and we refer the reader to this article for more details about these results.

2.2.1 Propagation Equations for the Signal–Idler Fields and Input-Output Relations

We decompose the electric field in the superposition of three quasi-monochromatic wavepackets (denoted with E_0, E_1, and E_2) of central frequencies ω_0, ω_1, and ω_2, corresponding to the pump, the signal, and the idler fields, respectively. These frequencies are taken to satisfy the energy conservation condition $\omega_1 + \omega_2 = \omega_0$. Assuming that the mean direction of propagation is the z-direction, and denoting with $\boldsymbol{x} = (x, y)$ the coordinate vector in the transverse plane, we can write

$$E_j(z, \boldsymbol{x}, t) \propto A_j(z, \boldsymbol{x}, t)\, \mathrm{e}^{\mathrm{i}k_j z - \mathrm{i}\omega_j t} + c.c. \qquad (j = 0, 1, 2)\ , \tag{2.1}$$

where $k_j = n_j \omega_j / c$ is the wave number of wave j at the carrier frequency along the z-axis; for an extraordinary wave the refractive index n_j depends on the propagation direction, a property leading to spatial walk-off.

In a single-pass configuration with crystal length on the order of a few millimeters, the pump depletion due to down-conversion is indeed a small entity, unless extremely high intensity laser sources are used. We shall therefore work within the parametric approximation, which treats the pump as a known classical field that propagates linearly inside the crystal, while the down-converted fields are quantized. Making the formal substitution for the signal–idler (S–I) field envelopes $A_j(z, \boldsymbol{x}, t) \to \hat{a}_j(z, \boldsymbol{x}, t)$, $(j = 1, 2)$, we impose the following commutation rules at equal z [2],

$$\begin{aligned} &\left[\hat{a}_i(z, \boldsymbol{x}, t), \hat{a}_j^\dagger(z, \boldsymbol{x}\,', t')\right] = \delta_{ij}\ \delta(\boldsymbol{x} - \boldsymbol{x}\,')\delta(t - t')\ , \\ &[\hat{a}_i(z, \boldsymbol{x}, t), \hat{a}_j(z, \boldsymbol{x}\,', t')] = 0 \qquad (i, j = 1, 2)\ , \end{aligned} \tag{2.2}$$

valid within the framework of the paraxial and quasi-monochromatic approximations. With this definition

$$\hat{I}_j(z, \boldsymbol{x}, t) = \hat{a}_j^\dagger(z, \boldsymbol{x}, t)\hat{a}_j(z, \boldsymbol{x}, t) \qquad (j = 1, 2), \tag{2.3}$$

is the photon flux density operator associated with the wave j: its expectation value gives the mean number of photons crossing a region of unit area in the transverse plane. In the linear regime the field operators obey the same equations as the corresponding classical quantities. For our purposes, it is useful to introduce the Fourier transforms of the field envelopes with respect to time and to the transverse plane coordinates:

$$\hat{a}_j(z, \boldsymbol{q}, \Omega) = \int \frac{\mathrm{d}\boldsymbol{x}}{2\pi} \int \frac{\mathrm{d}t}{\sqrt{2\pi}} \hat{a}_j(z, \boldsymbol{x}, t) \mathrm{e}^{-\mathrm{i}\boldsymbol{q}\cdot\boldsymbol{x} + \mathrm{i}\Omega t} \qquad (j = 1, 2)\ . \tag{2.4}$$

A similar definition holds also for the Fourier component $A_0(z, \boldsymbol{q}, \Omega)$ of the classical pump field envelope. The propagation equations then take the form

$$\begin{aligned} \frac{\partial \hat{a}_j(z, \boldsymbol{q}, \Omega)}{\partial z} &= \mathrm{i}\delta_j(\boldsymbol{q}, \Omega)\hat{a}_j(z, \boldsymbol{q}, \Omega) \\ &\quad + \sigma \mathrm{e}^{-\mathrm{i}\Delta_0 z} \int \frac{\mathrm{d}\boldsymbol{q}\,'}{2\pi} \int \frac{\mathrm{d}\Omega'}{\sqrt{2\pi}} A_0(z, \boldsymbol{q} - \boldsymbol{q}', \Omega - \Omega') \hat{a}_l^\dagger(z, -\boldsymbol{q}', -\Omega') \\ &\quad (j, l = 1, 2;\ j \neq l)\ , \end{aligned} \tag{2.5}$$

where the coupling constant σ is proportional to the effective second-order susceptibility $\chi^{(2)}_{eff}$ characterizing the down-conversion process, and

$$\delta_j(\boldsymbol{q},\Omega) = k'_j\Omega + \frac{1}{2}k''_j\Omega^2 + \varrho_j q_y - \frac{1}{2k_j}(q_x^2+q_y^2) , \qquad (j=1,2), \tag{2.6}$$

is the quadratic expansion of $k_{jz}(\omega_j+\Omega,\boldsymbol{q}) - k_j$ around $\boldsymbol{q}=0, \Omega=0$, and $k_{jz}(\omega_j+\Omega,\boldsymbol{q}) = \sqrt{k_j^2(\omega_j+\Omega,\boldsymbol{q}) - q^2}$ denotes the z-component of the k-vector associated with the $(\boldsymbol{q},\Omega)_j$ plane-wave mode. In particular the walk-off angle ϱ_j can be identified as $\partial k_j/\partial q_y$ calculated for $\boldsymbol{q}=0, \Omega=0$. A more detailed derivation can be found in [1–4].

Equations (2.5) contain the convolution integral in Fourier space of the S/I field envelope with the pump field envelope. Within the undepleted pump approximation, the latter can be expressed as

$$A_0(z,\boldsymbol{q},\Omega) = \mathrm{e}^{\mathrm{i}\delta_0(\boldsymbol{q},\Omega)z} A_0(z=0,\boldsymbol{q},\Omega) , \tag{2.7}$$

$$\delta_0(\boldsymbol{q},\Omega) = k'_0\Omega + \frac{1}{2}k''_0\Omega^2 + \varrho_0 q_y - \frac{1}{2k_0}(q_x^2+q_y^2) , \qquad (j=1,2), \tag{2.8}$$

the $z=0$ plane being taken at the input face of the slab. In the following we shall assume that the pump pulse has a Gaussian profile both in space and time, of beam waist w_0 and time duration τ_0 at $z=0$:

$$A_0(z=0,\boldsymbol{x},t) = (2\pi)^{3/2} A_p \mathrm{e}^{-(x^2+y^2)/w_0^2} \mathrm{e}^{-t^2/\tau_0^2} . \tag{2.9}$$

In Fourier space we have then the expression

$$A_0(z=0,\boldsymbol{q},\Omega) = 2\sqrt{2}\frac{A_p}{\delta q_0^2 \delta\omega_0} \mathrm{e}^{-(q_x^2+q_y^2)/\delta q_0^2} \mathrm{e}^{-\Omega^2/\delta\omega_0^2} , \tag{2.10}$$

where

$$\delta q_0 = 2/w_0 , \qquad \delta\omega_0 = 2/\tau_0, \tag{2.11}$$

denote the bandwidths of the pump in the spatial frequency domain and in the temporal frequency domain, respectively.

Let us now consider the limit of the stationary and plane-wave pump approximation (PWPA), in which w_0 and τ_0 tend to infinity and

$$A_0(z,\boldsymbol{q},\Omega) \to (2\pi)^{3/2} A_p \,\delta(\boldsymbol{q})\delta(\Omega) . \tag{2.12}$$

Under this condition Eqs. (2.5) couple only pairs of phase-conjugated modes $(\boldsymbol{q},\Omega)_1$ and $(-\boldsymbol{q},-\Omega)_2$ and can be solved analytically. The unitary input-output transformations relating the field operators at the output face of the slab of thickness l_c, $\hat{a}_{j,\mathrm{out}}(\boldsymbol{q},\Omega) \equiv \hat{a}_j(z=l_c,\boldsymbol{q},\Omega)$, to those at the input face, $\hat{a}_{j,\mathrm{in}}(\boldsymbol{q},\Omega) \equiv \hat{a}_j(z=0,\boldsymbol{q},\Omega)$, take the following form.

$$\begin{aligned} \hat{a}_{1\mathrm{out}}(\boldsymbol{q},\Omega) &= U_1(\boldsymbol{q},\Omega)\hat{a}_{1\mathrm{in}}(\boldsymbol{q},\Omega) + V_1(\boldsymbol{q},\Omega)\hat{a}^\dagger_{2\mathrm{in}}(-\boldsymbol{q},-\Omega) , \\ \hat{a}_{2\mathrm{out}}(\boldsymbol{q},\Omega) &= U_2(\boldsymbol{q},\Omega)\hat{a}_{2\mathrm{in}}(\boldsymbol{q},\Omega) + V_2(\boldsymbol{q},\Omega)\hat{a}^\dagger_{1\mathrm{in}}(-\boldsymbol{q},-\Omega) , \end{aligned} \tag{2.13}$$

with

$$U_1(\boldsymbol{q},\Omega) = \exp\left[\mathrm{i}\frac{\delta_1(\boldsymbol{q},\Omega)-\delta_2(-\boldsymbol{q},-\Omega)-\Delta_0}{2}l_c\right] \times \left[\cosh(\Gamma(\boldsymbol{q},\Omega)l_c)+\mathrm{i}\frac{\Delta(\boldsymbol{q},\Omega)}{2\Gamma(\boldsymbol{q},\Omega)}\sinh(\Gamma(\boldsymbol{q},\Omega)l_c)\right], \tag{2.14}$$

$$V_1(\boldsymbol{q},\Omega) = \exp\left[\mathrm{i}\frac{\delta_1(\boldsymbol{q},\Omega)-\delta_2(-\boldsymbol{q},-\Omega)-\Delta_0}{2}l_c\right] \times \frac{\sigma_p}{\Gamma(\boldsymbol{q},\Omega)}\sinh(\Gamma(\boldsymbol{q},\Omega)l_c), \tag{2.15}$$

$$U_2(\boldsymbol{q},\Omega) = \exp\left[\mathrm{i}\frac{\delta_2(\boldsymbol{q},\Omega)-\delta_1(-\boldsymbol{q},-\Omega)-\Delta_0}{2}l_c\right] \times \left[\cosh(\Gamma(-\boldsymbol{q},-\Omega)l_c)+\mathrm{i}\frac{\Delta(-\boldsymbol{q},-\Omega)}{2\Gamma(-\boldsymbol{q},-\Omega)}\sinh(\Gamma(-\boldsymbol{q},-\Omega)l_c)\right], \tag{2.16}$$

$$V_2(\boldsymbol{q},\Omega) = \exp\left[\mathrm{i}\frac{\delta_2(\boldsymbol{q},\Omega)-\delta_1(-\boldsymbol{q},-\Omega)-\Delta_0}{2}l_c\right] \times \frac{\sigma_p}{\Gamma(-\boldsymbol{q},-\Omega)}\sinh(\Gamma(-\boldsymbol{q},-\Omega)l_c), \tag{2.17}$$

where $\Delta_0 = k_1 + k_2 - k_0$ is the collinear phase mismatch of the central frequency components, and

$$\Gamma(\boldsymbol{q},\Omega) = \sqrt{\sigma_p^2 - \frac{\Delta(\boldsymbol{q},\Omega)^2}{4}}, \qquad \sigma_p = \sigma A_p, \tag{2.18}$$

$$\Delta(\boldsymbol{q},\Omega) = \Delta_0 + \delta_1(\boldsymbol{q},\Omega) + \delta_2(-\boldsymbol{q},-\Omega) \approx k_{1z}(\boldsymbol{q},\Omega) + k_{2z}(-\boldsymbol{q},-\Omega) - k_0 . \tag{2.19}$$

It is important to note that the gain functions U_j and V_j given by Eq. (2.14) satisfy the following unitarity conditions,

$$|U_j(\boldsymbol{q},\Omega)|^2 - |V_j(\boldsymbol{q},\Omega)|^2 = 1 \qquad (j=1,2) \tag{2.20}$$

$$U_1(\boldsymbol{q},\Omega)V_2(-\boldsymbol{q},-\Omega) = U_2(-\boldsymbol{q},-\Omega)V_1(\boldsymbol{q},\Omega), \tag{2.21}$$

which guarantee the conservation of the free-field commutation relations (2.2) after propagation. In the following we shall consider measurements either in the near-field or in the far-field zones of the nonlinear crystal. In order to simplify the notation we shall omit the explicit dependence of the fields on the z-coordinate: when specification is explicitly needed, the measured quantities will be labeled with π or π', which will denote the near-field and the far-field detection planes, respectively (see the scheme of Fig. 2.1).

In the following, we focus on the frequency-degenerate case $\omega_1 = \omega_2$, $k_1 = k_2 = k$, even if $k_1' \neq k_2'$ (see Eq. (2.6)). In this case, the phase mismatch accumulated during propagation is given, using (2.19) and (2.6), by

$$\Delta(\boldsymbol{q},\Omega)l_c = \Delta_0 l_c + \mathrm{sign}[k_1' - k_2']\frac{\Omega}{\Omega_0} - \varrho_2 q_y - \frac{q_x^2+q_y^2}{q_0^2}, \tag{2.22}$$

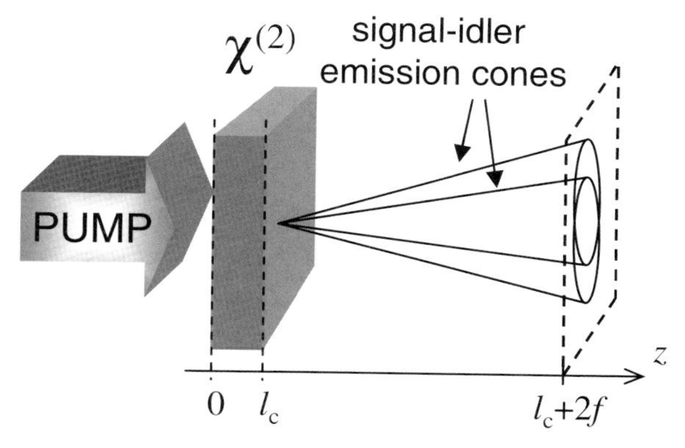

Fig. 2.1. Scheme for the observation of down-conversion in the far-field zone. The lens used to reach the far field (not shown in the figure) is located at $z = l_c + f$.

where we neglected the terms $\propto \Omega^2$, we assumed that the signal wave is ordinarily polarized so that $\varrho_1 = 0$, and we denoted

$$q_0 = \sqrt{\frac{k}{l_c}}\,, \qquad \Omega_0 = \frac{1}{|k_1' - k_2'|l_c}\,. \tag{2.23}$$

The parameters Ω_0 and q_0 are the characteristic bandwidths of parametric down-conversion (PDC) in the spatial frequency and the temporal frequency domain, respectively.

Figure 2.2 illustrates the kind of far-field patterns that can be obtained from a single pump pulse at frequency degeneracy in a type-II crystal. They are obtained by numerical integration of the classical-looking field equations (2.5), with a white input noise that simulates the vacuum fluctuations that trigger the process, as will be described in Section 2.2.3. The pump pulse duration is 1.5 ps and the large waist condition $\delta q_0 \ll q_0$ is fulfilled. The width of the rings is determined by the interval of frequencies of the numerical grid. In the examples shown in Fig. 2.2 the grid acts as a 15 nm box-shaped interference filter. The intensity peaks (white spots in the figure) become broader and broader as the beam waist w_0 is reduced, because their size is on the order of the resolution length imposed by the transverse size of the pump beam waist; that is, $x_{diff} = (\lambda f/2\pi)\delta q_0$. One notes that, whenever one has an intensity peak in the signal ring (Fig. 2.2a,b) or circle (Fig. 2.2c), one has a symmetrical intensity peak in the idler ring or circle. This feature will be confirmed by the experimental results discussed in Section 2.3.

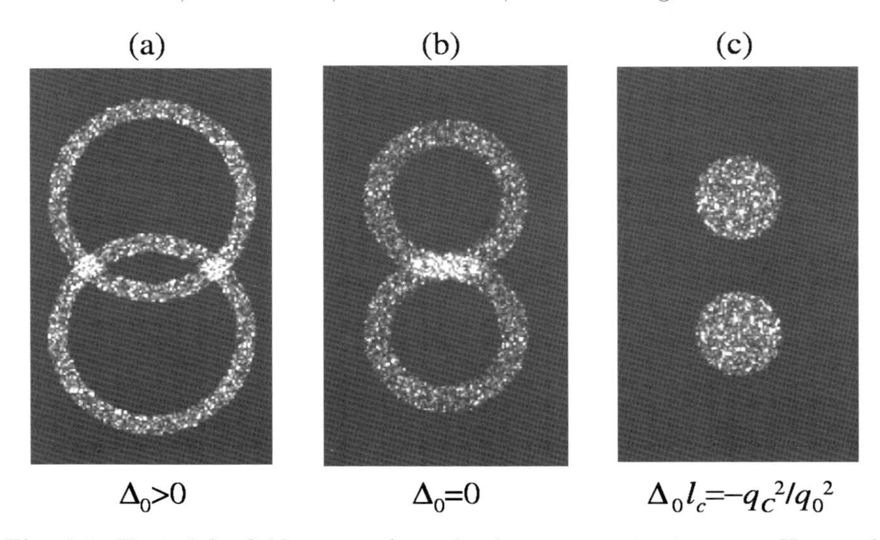

Fig. 2.2. Typical far-field pattern from the down-conversion in a type-II crystal, assuming that the observation is performed at the degenerate frequency (i.e., at $\omega_1 = \omega_2$). They are obtained for decreasing values of the collinear phase-mismatch parameter Δ_0, which makes the radius of the rings shrink to zero. The pump pulse duration is $\tau_0 = 1.5$ ps, the pump beam waist is $w_0 = 664\,\mu$m ($\delta q_0/q_0 = 0.05$), and the parametric gain is $\sigma_p l_c = 4$.

2.2.2 Near- and Far-Field Correlations in the Stationary and Plane-Wave Pump Approximation

Let us consider first the far-field correlations, which find their origin in the conservation of the transverse momentum in the generated photon pairs. Therefore, we consider two pointlike detectors, located symmetrically with respect to the axis of the system. One of them detects the signal photons, the other the idler photons (if necessary, signal and idler photons can be separated with the help of a polarizing beamsplitter because they are polarized in orthogonal directions).

If the detectors are the pixels of a CCD camera, they do not allow any spectral measurement due to the very low resolution power of the device in the time domain. They simply measure the total number of incoming photons down-converted in each single pump shot and the measurement time can be identified with the pump pulse duration. Hence we introduce two operators $\hat{N}_1$ and $\hat{N}_2$ which correspond to the number of photons detected in the two pixels during the whole pump pulse duration, and the associated fluctuations $\delta\hat{N}_i = \hat{N}_i - \langle\hat{N}_i\rangle$, (i = 1, 2). In the PWPA limit, it can be analytically proven [1] that the fluctuations in the photon number difference $\hat{N}_- = \hat{N}_1 - \hat{N}_2$ vanish exactly; that is,

$$\langle(\delta\hat{N}_-)^2\rangle = 0 \; , \tag{2.24}$$

as in the case of type-I PDC (see Eq. (1.31) of Chapter 1). Let us now turn to the near-field correlation. In this case, the separation between the signal and idler photons due to the walk-off is small. Near-field correlations between the signal and the idler photons arise from the fact that they are generated in the same spatial position, hence in this case we consider two pointlike detectors that occupy the same region in the near-field plane. In order to perform the measurement with physically separated pixels, the signal and idler beams can be separated by using a polarizing beamsplitter (see Section 2.2.3). In the near-field case, the result (2.24) holds only asymptotically in the limit $l_c \to 0$ [1] or, more physically, when the transverse area of the detector pixel is larger than the coherence area x_{coh}^2, where

$$x_{coh} \equiv \frac{1}{q_0} = \sqrt{\frac{l_c}{k}} \,. \tag{2.25}$$

This finite correlation length comes from the spread-out of the generated photons due to diffraction, which increases proportionally to the square root of the propagation distance.

We note that if the losses of the detection process are taken into account, the ideal result $\langle(\delta\hat{N}_-)^2\rangle = 0$ must be replaced with

$$\langle(\delta\hat{N}_-)^2\rangle = \eta(1-\eta)\langle\hat{N}_+\rangle \,, \qquad \langle\hat{N}_+\rangle = \langle\hat{N}_1\rangle + \langle\hat{N}_2\rangle, \tag{2.26}$$

η denoting the finite quantum efficiency of the detectors.

In order to understand the spatial entanglement properties of the signal–idler fields, it is convenient to use the Schrödinger picture instead of the Heisenberg picture we utilized up to now. For simplicity, we ignore the time and the frequency variables t and Ω in the remainder of this section. In the Schrödinger picture, the state at the input face of the slab is the vacuum state for all modes; that is,

$$|\psi\rangle_{\text{in}} = \prod_{\boldsymbol{q}} |0,\boldsymbol{q}\rangle_1 |0,\boldsymbol{q}\rangle_2 \,, \tag{2.27}$$

where we indicate by $|n,\boldsymbol{q}\rangle_i$ the Fock state with n photons in mode $\boldsymbol{q}$; $i = 1(2)$ indicates the signal (idler) as usual. On the other hand, in the PWPA the state of the signal–idler field at the output face of slab given by [3, 5],

$$|\psi\rangle_{\text{out}} = \prod_{\boldsymbol{q}} \left\{ \sum_{n=0}^{\infty} c_n(\boldsymbol{q}) |n,\boldsymbol{q}\rangle_1 |n,-\boldsymbol{q}\rangle_2 \right\} \,, \tag{2.28}$$

where $c_n(\boldsymbol{q}) = \{U_1(\boldsymbol{q})V_2(-\boldsymbol{q})\}^n \, |U_1(\boldsymbol{q})|^{-(2n+1)}$. Equation (2.28) is a superposition of states with the same number of photons in the mode $\boldsymbol{q}$ for the signal field and in the mode $(-\boldsymbol{q})$ for the idler field. This expresses in a very emphatic way the momentum entanglement between the signal and idler photons which, in the far field, gives rise to the spatial entanglement between positions $\boldsymbol{x}$ and $(-\boldsymbol{x})$. Equation (2.28) is an eigenstate of $\hat{N}_- = \hat{N}_1 - \hat{N}_2$ with the eigenvalue zero, which explains why N_- is fluctuationless. Two remarks are in order. The first is that, if only one of the two fields is detected (e.g., the signal field) and the idler is disregarded, the output state of the signal field, obtained by tracing away the degrees of freedom of the idler, is described by the density matrix

$$\varrho_{1\mathrm{out}} = \prod_{\boldsymbol{q}} \left\{ \sum_{n=0}^{\infty} |c_n(\boldsymbol{q})|^2 |n, \boldsymbol{q}\rangle_{11}\langle n, \boldsymbol{q}| \right\} . \tag{2.29}$$

It can be verified that

$$|c_n(\boldsymbol{q})|^2 = \frac{\langle \hat{n}(\boldsymbol{q}) \rangle^n}{[1 + \langle \hat{n}(\boldsymbol{q}) \rangle]^{n+1}} , \tag{2.30}$$

where $\langle \hat{n}(\boldsymbol{q}) \rangle$ is the average number of photons in mode $\boldsymbol{q}$, so that the photon statistics of the signal field is thermal for all modes $\boldsymbol{q}$. The same is true for the idler. Hence when $\langle \hat{n}(\boldsymbol{q}) \rangle \gg 1$, both the signal and the idler photon numbers undergo large fluctuations.

The second remark is that almost all literature on PDC is limited to the case $\langle \hat{n}(\boldsymbol{q}) \rangle \ll 1$, in which the state $|\psi\rangle_{\mathrm{out}}$ reduces to

$$\begin{aligned} |\psi\rangle_{\mathrm{out}} \approx & \prod_{\boldsymbol{q}} c_0(\boldsymbol{q}) |0, \boldsymbol{q}\rangle_1 |0, -\boldsymbol{q}\rangle_2 \\ & + \sum_{\boldsymbol{q}} \left\{ c_1(\boldsymbol{q}) |1, \boldsymbol{q}\rangle_1 |1, -\boldsymbol{q}\rangle_2 \times \prod_{\boldsymbol{q}' \neq \boldsymbol{q}} c_0(\boldsymbol{q}') |0, \boldsymbol{q}'\rangle_1 |0, -\boldsymbol{q}'\rangle_2 \right\} . \end{aligned} \tag{2.31}$$

In this case, coincidences of single photon pairs are detected. We focus, instead, on the case that $\langle \hat{n}(\boldsymbol{q}) \rangle$ is not negligible, so that a large number of terms in the expression (2.28) are relevant (macroscopic case).

The position entanglement between signal and idler photons in the near field can be retrieved in the following way. In the short crystal limit, where diffraction and walk-off along the crystal are negligible, the coefficients $U_i(\boldsymbol{q})$ and $V_i(\boldsymbol{q})$ in (2.13) become practically constant with respect to $\boldsymbol{q}$ and can be replaced with their values for $\boldsymbol{q} = 0$. Back-transforming Eqs. (2.13) to the real space $\boldsymbol{x}$, they become

$$\begin{aligned} \hat{a}_{1\mathrm{out}}(\boldsymbol{x}) &= U_1(\boldsymbol{q} = 0) \hat{a}_{1\mathrm{in}}(\boldsymbol{x}) + V_1(\boldsymbol{q} = 0) \hat{a}^\dagger_{2\mathrm{in}}(\boldsymbol{x}) , \\ \hat{a}_{2\mathrm{out}}(\boldsymbol{x}) &= U_2(\boldsymbol{q} = 0) \hat{a}_{2\mathrm{in}}(\boldsymbol{x}) + V_2(\boldsymbol{q} = 0) \hat{a}^\dagger_{1\mathrm{in}}(\boldsymbol{x}) , \end{aligned} \tag{2.32}$$

where, as we said, we ignore the frequency variable Ω. The input-output relations (2.32) are local in the position $\boldsymbol{x}$ in the crystal output plane ("near field"), and the corresponding output state reads

$$|\psi\rangle = \prod_{\boldsymbol{x}} \left\{ \sum_{n=0}^{\infty} c_n(\boldsymbol{q} = 0) |n, \boldsymbol{x}\rangle_1 |n, \boldsymbol{x}\rangle_2 \right\} , \tag{2.33}$$

where $|n, \boldsymbol{x}\rangle$ is the Fock state with n photons at point $\boldsymbol{x}$. In this limit, there is ideally a perfect correlation in the number of signal–idler photons at the same near-field position (position entanglement).

2.2.3 Near- and Far-Field Correlations: Numerical Results in the General Case

We now present the results obtained from the numerical model that includes the effects of the finite pump. The quantum averages in which we are interested (i.e., photon number correlations) are evaluated through a stochastic method based on the Wigner representation. With respect to other representations in phase space, the Wigner representation presents the advantage that the c-number stochastic equations equivalent to the equations for the field operators (2.5) do not contain Langevin noise terms (because of linearity and absence of dissipation) and are therefore identical to the classical propagation equations. The statistical character of the quantum fields is therefore wholly contained in the stochastic input field (see e.g., [6] for a more detailed discussion). We generate the input field with the appropriate phase-space probability distribution, that is a Gaussian white noise with zero mean, corresponding to the vacuum state in the Wigner representation [7]. With such an input field, we perform the numerical integration of Eqs. (2.5). We use a split-step algorithm [8] that integrates separately the terms describing linear propagation and the term describing the wave-mixing process: the former are integrated in the Fourier space, the latter in the real space. The obtained output fields are used to evaluate the correlation functions of interest. The procedure must be reiterated a sufficiently large number of times, so that the stochastic averages performed become good approximations to the corresponding quantum expectation values. Furthermore, some corrections are necessary in order to convert them to the desired operator ordering (the Wigner representation yields quantum expectation values of symmetrized operator products).

2.2.4 Far-Field Correlations

We shall consider explicitly the system described in [10]: a 1.5 ps high-intensity laser pulse is injected in a 4 mm long beta barium borate (BBO) crystal cut for type-II phase-matching. In the example we consider the pump oriented at an angle close to 48.2° with respect to the crystal axis and PDC is observed around the degenerate wavelength $\lambda_1 = \lambda_2 = 704\,\mathrm{nm}$ with a 10 nm interference filter. We investigated the momentum correlation that can be observed in the far-field plane π', by considering two symmetrical detection areas. The variance of $\hat{N}_-$ normalized to the shot noise is plotted in Fig. 2.3 as a function of the detector size d, normalized to the spatial scale of the photon number distribution in the far-field plane $x_{coh} = (\lambda f/2\pi)q_0$, for different values of the pump-beam waist. Fluctuations are well below shot noise

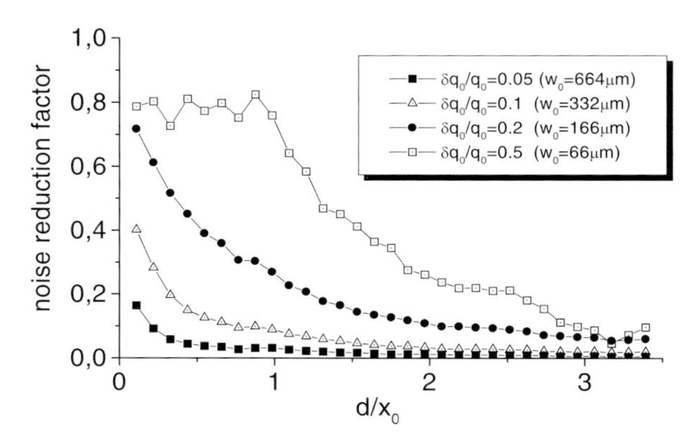

Fig. 2.3. Far-field correlation: the ratio $\langle(\delta\hat{N}_-)^2\rangle_{\pi'}/\langle\hat{N}_+\rangle_{\pi'}$ is plotted as a function of the detector size d for increasing value of the ratio $\delta q_0/q_0$. The parametric gain is $\sigma_p l_c = 4$, the negative value of the collinear phase mismatch is $\Delta_0 l_c = -q_C^2/q_0^2 = -74.4$, and the corresponding field intensity pattern is of the kind shown in Fig. 2.2c.

only if the detection size d is larger than the characteristic resolution length of the system $x_{diff} = (\lambda f)/(2\pi)\delta q_0$, that is, for $d/x_{coh} > \delta q_0/q_0$.

2.2.5 Near-Field Correlations

A main advantage of the type-II configuration lies in the fact that the signal and the idler fields have different polarizations and can therefore be manipulated more easily. In particular, it is possible to measure their mutual correlation in the near field after they have been physically separated by a polarizing beamsplitter, as shown schematically in Fig. 2.4. The lenses L and L' shown in the figure simply perform the $2f-2f$ imaging of the "near-field plane" π onto the two detection planes. For the moment, we only assume that the plane π is located at some coordinate z inside the crystal. The S–I field correlation functions display pronounced peaks for $\boldsymbol{x}' = \boldsymbol{x}$ describing the position entanglement of the S/I photons, which are generated in pairs in the same region of the crystal. The width of the peaks is on the order of the coherence length $x_{coh} = 1/q_0$. In the simulation illustrated in Fig. 2.5, we consider the type-II BBO crystal in the same conditions described before. The near-field coherence length is $x_{coh} = 16.6\,\mu$m. The plot displays the noise reduction factor $\langle(\delta\hat{N}_-)^2\rangle_{\pi}/\langle\hat{N}_+\rangle_{\pi}$, evaluated numerically as a function of the 1D detector size d normalized to x_{coh}. If the near-field observation plane π coincides with the output face of the crystal at $z = l_c$ (triangles), we see that the fluctuations are significantly reduced only when d is about 15 times larger than x_{coh}. The improved result (black squares) has been obtained by imaging onto the detection planes of a plane inside the crystal at $z = l_c - \Delta z$, rather than the crystal output face (see Fig. 2.4). Furthermore, the arrays of

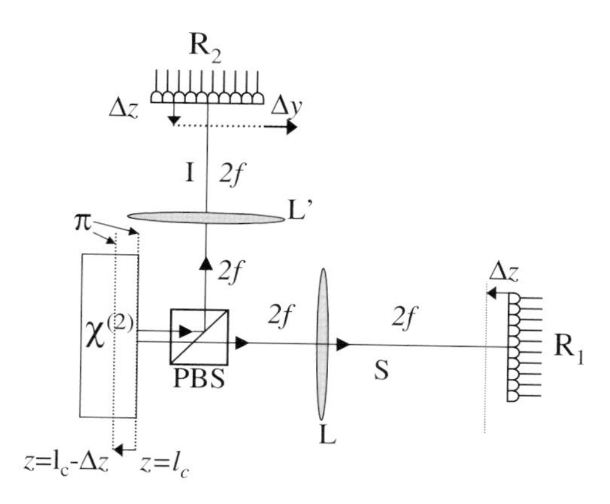

Fig. 2.4. Detection scheme to measure spatial correlations in the near field. A polarizing beamsplitter (PBS) separates the S/I beams. Their near fields, at the plane $\pi : z = l_c - \Delta z$, are imaged by two lenses (L and L′) onto the pixel detectors R_1 and R_2, which are in the plane conjugate to plane π; Δz and Δy indicate the spatial shifts applied to the optical devices that are necessary to optimize the measurement.

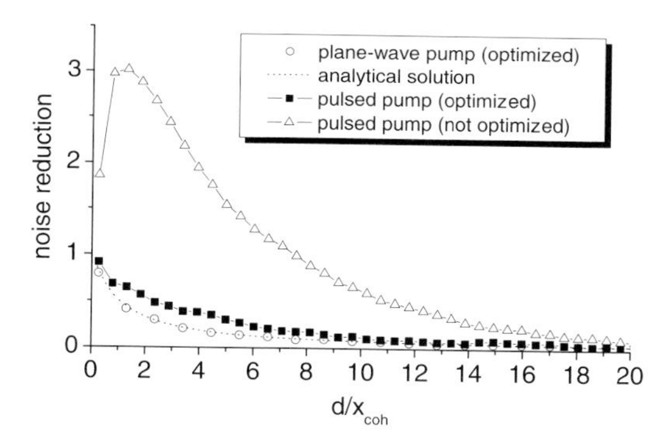

Fig. 2.5. Near-field correlations: the ratio $\langle(\delta\hat{N}_-)^2\rangle_\pi/\langle\hat{N}_+\rangle_\pi$ is plotted as a function of the detector size. The parameters of the pulsed Gaussian pump are $w_0 = 332\mu\text{m}$ $(\delta q_0/q_0 = 0.1)$ and $\tau_0 = 1.5\,\text{ps}$ $(\delta\omega_0/\Omega_0 = 1.14)$; the gain is $\sigma_p l_c = 3$ and $q_R = 0$. The simulations performed applying diffraction and walk-off compensation (squares) are well below the one performed without optimization (white triangle). The dashed line corresponds to the analytical solution obtained in the PWPA.

pixel detectors in the signal and idler arms are shifted with respect to each other by a distance Δy in the transverse direction of walk-off (see Fig. 2.4). This optimization procedure, which takes carefully into account the effects of walk-off and diffraction along the crystal, is described in detail in [1].

2.3 Detection of Sub-Shot-Noise Spatial Correlations in High-Gain Parametric Down-Conversion

In this section we describe the first experimental observation of spatial entanglement in the far field, as theoretically and numerically described in Sections 2.2.2 and 2.2.3. The experiment utilizes a high-efficiency scientific CCD camera with the pixel area on the order of the characteristic resolution area (or coherence area in the far field) x_{diff}^2, with $x_{diff} = (\lambda f/2\pi)\delta q_0$.

The results of this section are based on Refs. [11,12].

There is now a large literature on spatial effects in the low-gain regime, where photon pairs are detected via coincidence counting [13]. This literature includes several imaging experiments (e.g., ghost interference [14]) that exploit the spatial correlation of twin photons, although recent investigations [15,16] have shown that some of these experiments can be reproduced with classically correlated photons. Genuine spatial quantum effects have been shown in [17], which demonstrates spatial antibunching, and in [18], which reports on realization of an EPR paradox regarding the position-momentum uncertainty relation for photons.

Recent theoretical investigations done for arbitrary gains [1,19] have predicted multimode spatial correlations below shot noise between different portions of the signal and the idler emission cones that correspond to the phase-conjugate modes. This kind of k-vector (i.e., far-field) correlation amounts to the multimode version of the well-known twin-beam effect, that is, a sub-shot-noise correlation between the whole signal and idler beams which was evidenced, for example, in [20] in the medium-gain regime of PDC. These spatial correlations have also been investigated in the low-gain regime in [21,22] with a high-sensitive CCD camera, although it was not possible to determine its quantum nature quantitatively. More recently, the measurement of small displacements beyond the Rayleigh limit [23] and the realization of noiseless amplification of optical images [24] put in evidence the potentiality of such multimode quantum correlations in the large photon number regime for potential applications. The aim of our experiment is to demonstrate the predicted quantum character of the correlation that can be observed in the far field of PDC.

In the following we present a description of the far-field detection of the PDC radiation emitted by a β-barium borate nonlinear crystal pumped by a low-repetition rate (2 Hz) pulsed high-power laser (1 GW–1 ps).

The use of the pulsed high-power laser enables us to tune the PDC to the high-gain regime while keeping a large pump beam size (of the order of $\sim$1 mm). Thanks to the huge number of radiation transverse modes, we can concentrate on a portion of the parametric fluorescence close to the collinear direction and within a narrow frequency bandwidth around the degeneracy. This portion still contains a large ($>$1000) number of pairs of signal/idler correlated phase-conjugate modes, propagating in symmetrical directions with respect to the pump in order to fulfill the phase-matching constraints. In the far field, where the measurement is performed, the couples of modes correspond to pairs of symmetrical spots, which can be considered as independent and equivalent spatial replicas of the same quantum system. Thanks to the very large number of these, the statistical ensemble averaging necessary for the quantum measurement can be solely done over the spatial replicas for each, single, pump-laser pulse. Thus, differently from the experiment in [20], where the statistics was performed over different temporal replicas of the system, here no temporal averages over successive laser shots are considered. In our experiment the single-shot measurements reveal sub-shot-noise spatial correlations for a PDC gain corresponding to detection of up to $\simeq$100 photoelectrons per mode [11].

2.3.1 Detection of the Spatial Features of the Far-Field PDC Radiation by Means of the CCD

Before going to the quantitative investigation of existence of the spatial correlations between the signal and the idler beams, we perform a preliminary characterization of the generated parametric radiation. The type-II $5 \times 7 \times 4\,\text{mm}^3$ BBO nonlinear crystal, operated in the regime of parametric amplification of the vacuum-state fluctuations, is pumped by the third harmonic (352 nm) of a 1 ps pulse from a chirped-pulse amplified Nd:glass laser (TWINKLE, Light Conversion Ltd.). The input and output facets of the crystal are anti-reflection-coated at 352 nm and 704 nm, respectively. The pump beam (vertically polarized (e)) is spatially filtered and collimated to a beam waist characterized by a full width at half maximum (FWHM) of approximately 1 mm at the crystal input facet. The energy of the 352 nm pump pulse can be continuously tuned in the range 0.1–0.4 mJ by means of suitable attenuating filters and by changing the energy of the 1055 nm pump laser pulse, allowing a gain G (representing the intensity amplification factor) in the range $10 \leq G \leq 10^3$. The parametric fluorescence of the horizontally polarized signal (o) and vertically polarized (e) idler modes is emitted over two cones, whose apertures depend on the specific wavelengths (see, e.g., [10]). The BBO crystal ($\theta = 49.05°$, $\phi = 0$) is oriented in order to generate signal and idler radiation cones tangent to the collinear direction at the degenerate wavelength $\omega_s = \omega_i = \omega_p/2$ (s, i, and p referring to signal, idler, and pump, respectively).

A simple far-field detection set-up is initially mounted as shown in Fig. 2.6. A deep-depletion back-illuminated charged coupled device (CCD) camera [25] (Roper Scientific, NTE/CCD-400EHRBG1, with quantum efficiency $\eta \approx 89\%$ at 704 nm) triggered by a pulse from the laser system, is placed in the focal plane of a single large-diameter lens (f = 5 cm), which collects at a distance f the far-field PDC radiation emitted by the BBO. The CCD detection array has 1340×400 pixels, with a pixel size of $20\,\mu\text{m} \times 20\,\mu\text{m}$. The pump-frequency contribution is removed by using a normal incidence high-reflectivity (HR) mirror coated for 352 nm placed after the BBO. By using a 10 nm broad

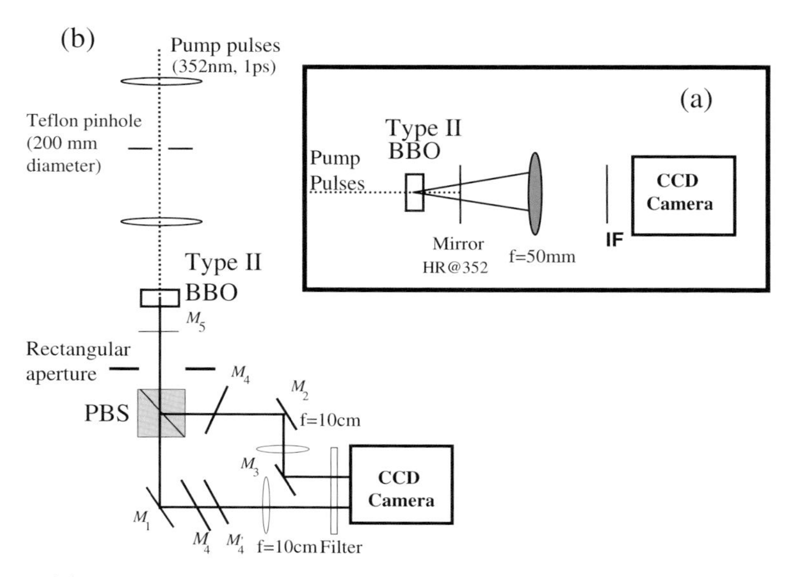

Fig. 2.6. (a) Scheme of the diagnostics for the far-field detection of the degenerate signal and idler ring-type modes. (b) Detailed scheme of the experimental set-up used for the spatial-correlations measurements. The third harmonic of the Nd:Glass laser is used to pump at 352 nm the BBO crystal which is cut for degeneracy at 704 nm ($\theta = 49.05°$, $\phi = 0$).

interferential filter (IF), centered at 704 nm, we are able to visualize the degenerate signal and idler far-field beams emitted in the parametric process. It is worth pointing out that without any spectral filtering, emission occurs on a very wide range of wavelengths and emission angles (see, e.g., [10]). Typical far-field images recorded at degeneracy in a single shot (for 1 ps pump pulse) are shown in Fig. 2.7a,b for two different values of the pump intensity and, in the particular cases illustrated, for two different pump beam sizes. The ring-shaped angular distribution is determined by the phase-matching conditions [10], and the rings at degeneracy are characterized by an angular width of about 8° each. Note that with this set-up the two rings, which are emitted along the vertical direction, are recorded by rotating the CCD by 90° in order to fit the entire ring pattern inside the rectangular chip. The shape of the rings is similar to the numerical Fig. 2.2b. A detailed comparison between theory and experiment as well as a discussion of the photon distribution over the rings, is given in [12].

2.3.2 Experimental Set-Up for Spatial-Correlations Measurements

The existence of spatial correlations already appears from the symmetrical properties of the signal and the idler patterns recorded experimentally and

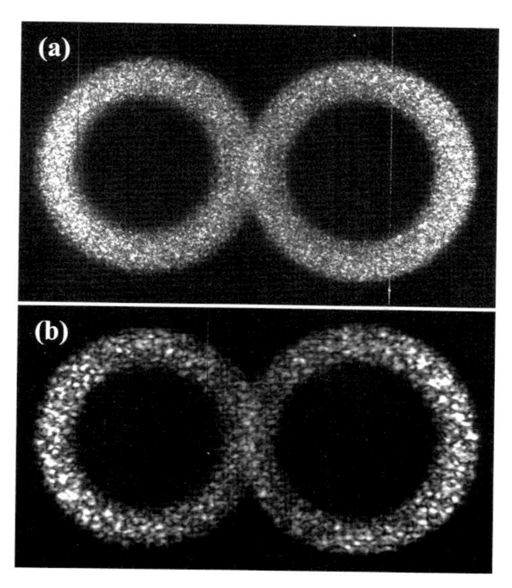

Fig. 2.7. Experimental far-field images of the degenerate signal (left ring) and idler (right ring) beams recorded in single shot by the CCD placed in the focal place of a single lens (f = 50 cm), with pump intensity (a) $I \approx 30$ GW/cm^2, and (b) $I \approx 50$ GW/cm^2, and FWHM pump beam size of 1 mm (a) and 0.4 mm (b) respectively.

shown in Fig. 2.7. However, to investigate and reveal the quantum character of the correlations we use the experimental set-up illustrated in Fig. 2.6b where, with respect to Fig. 2.6a, a different diagnostics configuration is adopted. We consider a pump beam size of 1 mm and we now select the fluorescence around the collinear direction by means of a 5 mm × 8 mm aperture placed 15 cm from the output facet of the BBO. The radiation is then transmitted through a polarizing beamsplitter (PBS) that separates the signal and idler beams. The aperture prevents beam clipping by the PBS and thereby reduces substantially scattered radiation. The beams are finally sent onto two separate regions of the CCD, which is placed in the common focal plane of two lenses (f = 10 cm) used to image the signal and idler far fields. In this set-up the correlation measurements are performed without using any narrowband IFs, because these unavoidably introduce relevant transmission losses reducing the visibility of sub-shot-noise correlations, and could also introduce distortion or even attenuation effects. Here the pump-frequency contribution is removed by using normal incidence (M_5) and at 45° (M_4) high-reflectivity (HR) mirrors coated for 352 nm placed before and after the PBS, respectively, and a low-bandpass color filter (90% transmission around 704 nm) placed in front of the CCD. Note that a second PBS (not shown in the figure) is placed in the arm of the (e) idler beam to remove the residual contribution of ordinary (o) radiation reflected by the first PBS (3%), and a further HR@352 nm mirror (M_4') is placed in the signal arm at a suitable angle in order to balance the unequal transmission of radiation in the two arms. All the optical components (except the color filter) have antireflection coatings at 704 nm. The estimated quantum efficiency of each detection line, which accounts for both the transmission losses and the detector efficiency, is $\eta_{tot} \simeq 75\%$.

It is worth pointing out that prior to the experiment for the detection of quantum spatial correlations, we performed a test of the capabilities of the scientific CCD camera to perform spatially resolved measurements of the photon shot noise. The CCD used for the diagnostic has in fact been calibrated pixel by pixel to compensate for the gain inhomogeneity of the pixels on the CCD chip, allowing the retrieval of the Poissonian statistics of the spatial fluctuations of a uniform enlightening in the full range of the camera dynamics [26]. Retrieving the shot noise in the CCD full dynamic range using classical sources paved the way to spatially resolved photon noise measurements at the sub-shot-noise level (SNL), and turned out to be a necessary step to demonstrate quantum properties of images by means of a CCD diagnostic.

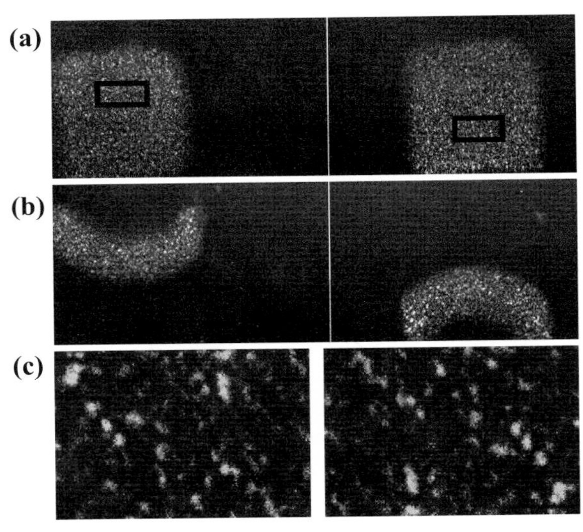

Fig. 2.8. (a) Single-shot far-field image recorded by the CCD for a pump intensity $I \simeq 30$ GW/cm^2. The spatial areas for statistics are delimited by the black boxes selected within the degenerate signal and idler modes, spatially localized from the single-shot image recorded with the 10 nm broad IF (b). (c) Zoom of two symmetrical areas of the signal and the idler far fields.

Figure 2.8a shows a typical far-field image recorded in a single shot in the experimental configuration of Fig. 2.6b, where a fairly broadband radiation (i.e., the one transmitted by the rectangular aperture) is acquired in the signal (left) and idler (right) branches. Selection of the desired temporal and angular bandwidths around the degeneracy is made by inserting in front of the CCD a 10 nm wide IF around 704 nm, allowing us to locate the collinear degeneracy point (see Fig. 2.8b). The data analysis is limited within two rectangular boxes (black frames in Fig. 2.8a) corresponding to an angular bandwidth of 20 mrad $\times$ 8 mrad and to a temporal bandwidth smaller than 10 nm. The selected regions respectively contain 4000 pixels over which the mean number of photons is approximately uniform so that spatial averages are performed over identical replicas. Because the aim of this work is to investigate pixel-pair correlation, and because the size of the CCD pixel approximately corresponds to the physical size of a replica (coherence area in the far field), the ensemble is large enough to perform the desired statistics. A zoom of the selected areas is presented in Fig. 2.8c, where the rather spectacular symmetry of the intensity distribution in the signal and idler branches shows the twin-beam character of the phase-conjugate modes.

2.3.3 Detection of Quantum Spatial Correlations: Spatial Analogue of Photon Antibunching in Time

Each of the signal and idler far-field patterns taken separately looks like a speckle pattern produced by a pseudo-thermal source, such as, for instance, a ground glass illuminated by a laser beam. When this thermal light is split by a macroscopic device such as a beamsplitter, the two resulting beams show a high level of spatial correlation, which is, however, limited by the shot noise. The spatial correlations of the signal and idler beams generated by PDC are, instead, of microscopic origin, and are not limited by the shot noise. The aim of this experiment is just to show the sub-shot-noise nature of the spatial correlation of the PDC beams. We are first interested in the symmetrical pixel-pair correlations, which are evaluated experimentally by measuring the variance σ^2_{s-i} of the PDC photoelectrons (pe) difference $n_s - n_i$ of the signal/idler pixel-pair versus the mean total number of down-converted pe of the pixel-pair. This variance is

$$\sigma^2_{s-i} = \langle (n_s - n_i)^2 \rangle - \langle n_s - n_i \rangle^2 . \tag{2.34}$$

In the experiment these averages are evaluated as spatial averages performed over the set of equivalent symmetrical pixel-pairs contained in the chosen sample regions, on which the mean photon number distribution is nearly uniform (see Fig. 2.8a,c). Each single shot of the laser provides a different ensemble, characterized by its pixel-pair average pe number $\langle n_s + n_i \rangle$, in turn related to the parametric gain. In the experiment, ensembles corresponding to different gains are obtained by varying the pump-pulse energy. We note that the read-out noise of the detector, its dark current, and some unavoidable light scattered from the pump, signal, and idler fields contribute with a nonnegligible background noise to the process. This is taken into account by applying a standard correction procedure (see, e.g., [27]), by subtracting the background fluctuations σ_b^2 from the effectively measured variance $\sigma^2_{(s+b)-(i+b)}$ of the total intensity difference (signal + background) – (idler + background) obtaining $\sigma^2_{s-i} = \sigma^2_{(s+b)-(i+b)} - 2\sigma_b^2$.

Figure 1.9 shows the experimental results where each point is associated with a different laser shot. The data are normalized to the shot-noise level, and their statistical spread accounts for the background correction. Although the noise on the individual signal and idler beams is found to be very high and much greater than their shot-noise level (given by $\langle n_s \rangle$ and $\langle n_i \rangle$, respectively), we observe an evident sub-shot-noise pixel-pair correlation up to the gains characterized by $\langle n_s + n_i \rangle \approx 15 - 20$. Because in that regime the observed transverse size of the coherence areas (i.e., of the modes) is about 2–4 pixels, this approximately corresponds to 100 pe per mode. The maximum level of noise reduction observed experimentally agrees with the theoretical limit (dotted line in Fig. 2.9) determined by the total losses of the system ($\sim 1 - \eta_{tot}$ [1]).

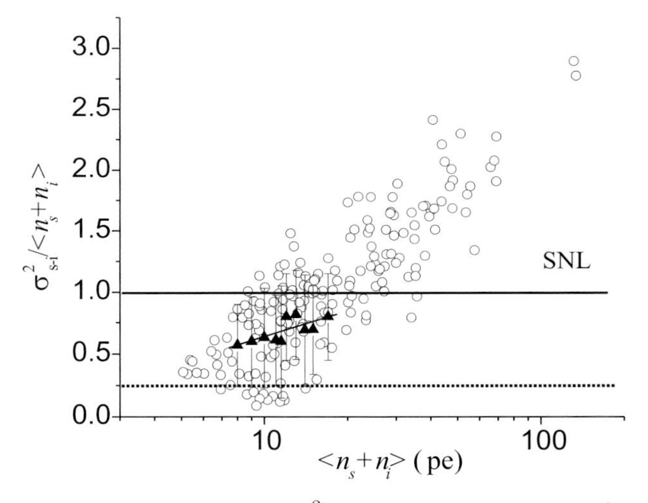

Fig. 2.9. Intensity difference variance σ^2_{s-i} normalized to the SNL $\langle n_s + n_i \rangle$. Each point (white circle) corresponds to a single-shot measurement where the spatial ensemble statistics has been performed over a 100 × 40 pixel region. The triangles (each one obtained by averaging the experimental points corresponding to a certain gain) and their linear fit illustrate the trend of the data in the region between $\langle n_s + n_i \rangle = 8$ and 20.

We can have an idea of the transverse size of the mode by looking at the standard two-dimensional cross-correlation degree,

$$\gamma = \frac{\langle n_s n_i \rangle - \langle n_s \rangle \langle n_i \rangle}{\sqrt{\sigma_s^2 \sigma_i^2}}, \tag{2.35}$$

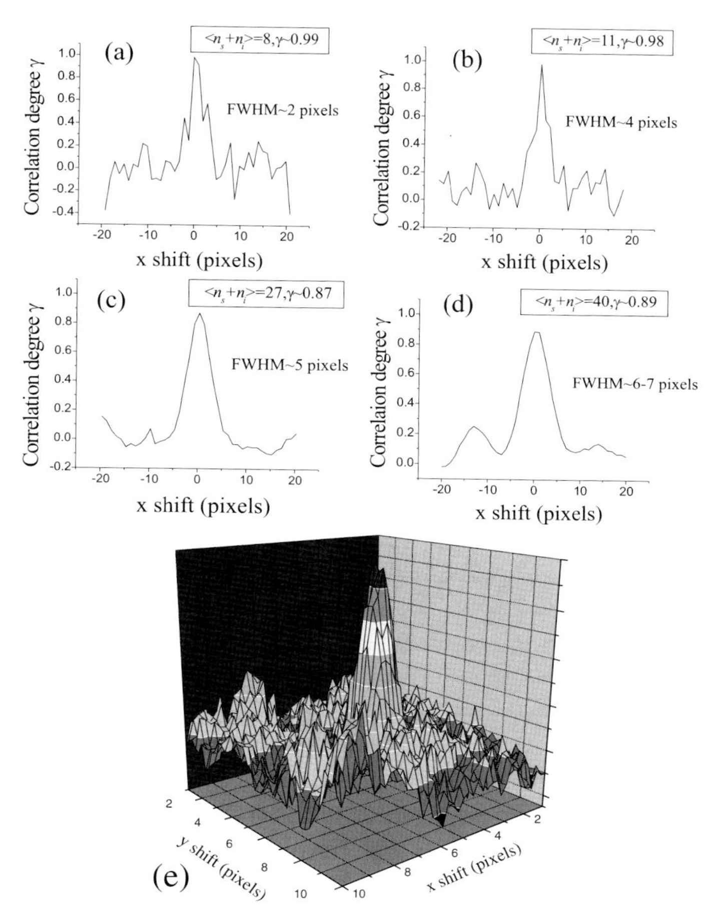

Fig. 2.10. Correlation degree profiles (a)–(d) plotted for four different gain values; (e) displays the full 3D plot of the correlation function surface corresponding to case (a) and obtained as a function of both the horizontal and vertical shifts of the recorded image on the CCD.

between all the angularly symmetrical signal and idler pixels contained within the black boxes (see Fig. 2.8). This can be plotted, for instance, as a function of the horizontal and vertical shifts of the recorded image on the CCD, keeping fixed the position of the boxes. In general $| \gamma | \leq 1$ with $\gamma = 1$ for perfect correlations. The transverse section of the correlation function γ plotted as a function of the horizontal shift x (in pixel units) and obtained from four single-shot images corresponding to different gains is presented in Fig. 2.10a–d. Figure 2.10e shows the full 3D plot of the correlation function for the same gain of Fig. 2.10a. We can notice how the FWHM of the curves increases for increasing gain, clearly reflecting the increment of the speckle size (and thus of the transverse mode size) already observed in Fig. 2.7. As expected, virtually perfect correlation (in our case we have peak values of γ up to $\simeq 0.99$) is obtained for perfect determination (i.e., within one pixel) of the center of symmetry between the signal and the idler regions.

It is interesting to note that the quantum nature of the correlation can also be estimated from the peak value of the correlation degree. As a matter of fact, because

$$\sigma_{s-i}^2 = \langle (n_s - n_i - \langle n_s \rangle - \langle n_i \rangle)^2 \rangle = \sigma_s^2 + \sigma_i^2 - 2(\langle n_s n_i \rangle - \langle n_s \rangle \langle n_i \rangle), \quad (2.36)$$

the sub-shot-noise condition for the intensity difference variance,

$$\sigma_{s-i}^2 < \langle n_s + n_i \rangle, \quad (2.37)$$

can be rephrased in the form

$$\gamma > 1 - \frac{\langle n_{s,i} \rangle}{\sigma_{s,i}^2}, \quad (2.38)$$

if we use (2.35) and assume that

$$\langle n_s \rangle \simeq \langle n_i \rangle \equiv \langle n_{s,i} \rangle \quad \text{and} \quad \sigma_s^2 \simeq \sigma_i^2 \equiv \sigma_{s,i}^2, \quad (2.39)$$

as also confirmed experimentally within a good approximation. We can rewrite (2.38) further by taking into account that, as is well known, the signal and idler beams taken alone display a thermallike statistics. Therefore, $\sigma_s^2 \approx \sigma_i^2 \equiv \sigma_{s,i}^2 \approx \langle n_{s,i} \rangle (1 + \langle n_{s,i} \rangle / M)$ [28], where M is the degeneracy factor representing the number of spatial and temporal modes intercepted by the pixel detectors. In the condition of the experiment the pump duration is slightly longer than the PDC coherence time whereas the pixel area is smaller than the coherence area, so that M is expected to be only slightly larger than unity [27]. Using this, (2.38) becomes

$$\gamma > \frac{\langle n_{s,i} \rangle}{M + \langle n_{s,i} \rangle}. \quad (2.40)$$

Figure 2.11 illustrates the trend of the peak value of γ for different gains (black triangles), extracted from the six images. The dashed curve corresponds to the classical boundary γ_b obtained by interpolation of the function $\gamma_b = 1 - \langle n_{s,i} \rangle / \sigma_{s,i}^2$, calculated for different gains using the values for the mean and variance obtained from the experimental far-field patterns considered. The full line is the theoretical limit obtained from (2.40) by using $M = 2.4$ as a fitting parameter. We thus observe spatial quantum correlations whenever the value of γ lies in the region above the theoretical quantum limit. This limit becomes more demanding as the gain increases. The experimental values obtained are, as expected, compatible with the trend of the data plotted in Fig. 2.9, and highlight a quantum-correlations region up to the values of $\langle n_s + n_i \rangle$ of about 20 pe. For instance, the first three triangles on the left correspond to three images characterized by an intensity difference variance that is clearly below the SNL in Fig. 2.9, and the other three triangles correspond to the images that are characterized by an intensity difference variance above the SNL.

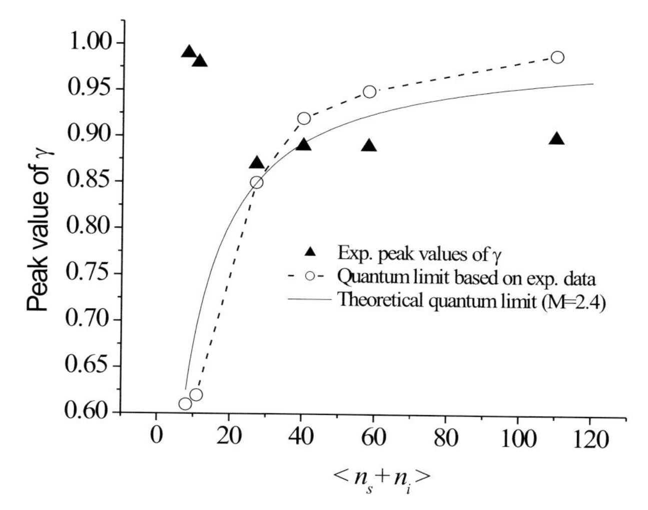

Fig. 2.11. Experimental correlation degree (triangles) measured from six signal/idler far-field images for different values of the gain.

Finally we note that if we multiply Eq. (2.38) by σ^2_{s-i}, with the help of (2.35) and (2.39) we obtain

$$\langle n_s n_i \rangle - \langle n_s \rangle \langle n_i \rangle > \sigma^2_{s,i} - \langle n_{s,i} \rangle; \tag{2.41}$$

that is; by defining

$$\langle \delta n_s \delta n_i \rangle \equiv \langle n_s n_i \rangle - \langle n_s \rangle \langle n_i \rangle, \tag{2.42}$$

and

$$\langle : \delta n^2_{s,i} : \rangle \equiv \sigma^2_{s,i} - \langle n_{s,i} \rangle, \tag{2.43}$$

where the symbol : : indicates the normal ordering, we find the following inequality condition

$$\langle : \delta n_s \delta n_i : \rangle >= \langle \delta n_s \delta n_i \rangle > \langle : \delta n^2_{s,i} : \rangle = [\langle : \delta n^2_s : \rangle \langle : \delta n^2_i : \rangle]^{1/2}, \tag{2.44}$$

equivalent to the sub-shot-noise level condition (2.38), and which states that the cross-correlations between the signal and the idler are larger than the (normally ordered) self-correlation. This corresponds to an apparent violation of the Cauchy–Schwartz inequality. This effect was predicted in [31–33] for the case of an optical parametric oscillator and then generalized [19] to the case of a traveling-wave degenerate optical parametric amplifier. It represents a spatial analogue of the phenomenon of photon antibunching in time and was experimentally demonstrated in the coincidence regime in [17]. Our experiment provides the first evidence of this phenomenon in the high-gain regime.

In our experiment we found a transition from the quantum to the classical regime with increasing gain, similar to the scenario of [20]. Reference [12] includes a discussion of such a transition, as well as of the effects arising from the inaccuracy in the determination of the center of symmetry in the signal–idler field.

2.4 Multiphoton, Multimode Polarization Entanglement in Parametric Down-Conversion

In this section we turn our attention to polarization degrees of freedom of down-converted beams together with the spatial ones. The results reported here are based on [3], where more details can be found.

The quantum properties of light polarization have been widely studied in the regime of single photon counts. In comparison, only recently has there been a rise of interest in the quantum properties of the polarization of macroscopic light beams [34, 35], mainly due to their potential applications in the field of quantum information with continuous variables and to the possibility of mapping the quantum state from light to atomic media [36]. Parametric down-conversion in a type-II crystal probably represents the most well-known source of polarization-entangled photons. Due to spatial walk-off in the crystal, the two emission cones corresponding to the horizontally and vertically polarized beams are slightly displaced, one with respect to the other (see Fig. 2.1), and the far-field intensity distribution has the shape of two rings, whose centers are displaced along the walk-off direction (Fig. 2.2). For certain orientations of the crystal axis, the far-field rings intersect, as, for example, shown in Fig. 2.2b, and the two regions at the ring intersection have a very special role. In the regime of single-photon-pair detection, the polarization of a photon detected in one of these regions is completely undetermined. However, once the polarization of one photon has been measured, the polarization of the other photon, which propagates at the symmetric position, is exactly determined. In other words, when considering photodetection from these regions, the two-photon state can be described as the ideal polarization-entangled state [37]. Photons produced by this process have become an essential ingredient in many recent implementations of quantum information schemes [38, 39].

The question that we address here is whether this microscopic photon polarization entanglement leaves any trace in the regime of high-parametric down-conversion efficiency, where a rather large number of photons are produced, and in which form.

Quantum-optical polarization properties of light are conveniently described within the formalism of Stokes operators, which represent the quantum counterparts of the Stokes vectors of classical optics. These operators obey angular momentumlike commutation rules, and the associated observables are in general noncompatible. Our goal is to study the quantum correlations between the Stokes operators measured from symmetric portions of the far-field beam cross-section. To this end, we consider a measurement of the Stokes operators over a small region $D(\boldsymbol{x})$ centered around the position $\boldsymbol{x}$ in the far-field plane of the down-converted field, and over the detection time T (typically T is much larger than the crystal coherence time):

$$\hat{S}_i(\boldsymbol{x}) = \int_T dt' \int_{D(\boldsymbol{x})} d\boldsymbol{x}' \hat{s}_i(\boldsymbol{x}', t') \ . \tag{2.45}$$

By denoting with $\hat{a}_H, \hat{a}_V$ the field operators of the horizontally and vertically polarized field components,

$$\hat{s}_0(\boldsymbol{x}, t) = \hat{a}_H^\dagger(\boldsymbol{x}, t)\hat{a}_H(\boldsymbol{x}, t) + \hat{a}_V^\dagger(\boldsymbol{x}, t)\hat{a}_V(\boldsymbol{x}, t) \ , \tag{2.46}$$

$$\hat{s}_1(\boldsymbol{x}, t) = \hat{a}_H^\dagger(\boldsymbol{x}, t)\hat{a}_H(\boldsymbol{x}, t) - \hat{a}_V^\dagger(\boldsymbol{x}, t)\hat{a}_V(\boldsymbol{x}, t) \ , \tag{2.47}$$

$$\hat{s}_2(\boldsymbol{x}, t) = \hat{a}_H^\dagger(\boldsymbol{x}, t)\hat{a}_V(\boldsymbol{x}, t) + \hat{a}_V^\dagger(\boldsymbol{x}, t)\hat{a}_H(\boldsymbol{x}, t) \ , \tag{2.48}$$

$$\hat{s}_3(\boldsymbol{x}, t) = -\mathrm{i}\left[\hat{a}_H^\dagger(\boldsymbol{x}, t)\hat{a}_V(\boldsymbol{x}, t) - \hat{a}_V^\dagger(\boldsymbol{x}, t)\hat{a}_H(\boldsymbol{x}, t)\right] . \tag{2.49}$$

The first two operators represent the sum and the difference between the number of horizontal and vertical photons. In the limit of the pump transverse waist much larger than the coherence area of the amplifier and of a pulse duration much longer than the crystal coherence time, the analytical calculations described in Section 2.2.2 show that the number of H and V photons collected from any two symmetric portions of the far-field plane are perfectly correlated observables. This implies ideally perfect correlations, both between $\hat{S}_0(\boldsymbol{x})$, $\hat{S}_0(-\boldsymbol{x})$, and between $\hat{S}_1(\boldsymbol{x})$, $-\hat{S}_1(-\boldsymbol{x})$ for any choice of the position $\boldsymbol{x}$ in the far field [3] (notice that $\hat{S}_0(\boldsymbol{x})$ commutes with $\hat{S}_1(\boldsymbol{x}')$). This result is a direct consequence of the pairwise emission of photons with vertical and horizontal polarizations propagating in symmetric directions, as required by the conservation of the transverse momentum. In the more sophisticated numerical model of Section 2.2.3, the finite width of the pump profile introduces an uncertainty in the directions of propagation of the down-converted photons, so that the correlations in the Stokes operators $\hat{S}_0, \hat{S}_1$ well beyond the shot-noise level are recovered when photons are collected from the regions larger than the resolution area x_{diff}.

Quite different is the situation concerning the other two Stokes operators that involve measurements of the photon number in the oblique and circular polarization basis. Part (b) of Fig. 2.12 shows a typical result for the noise in

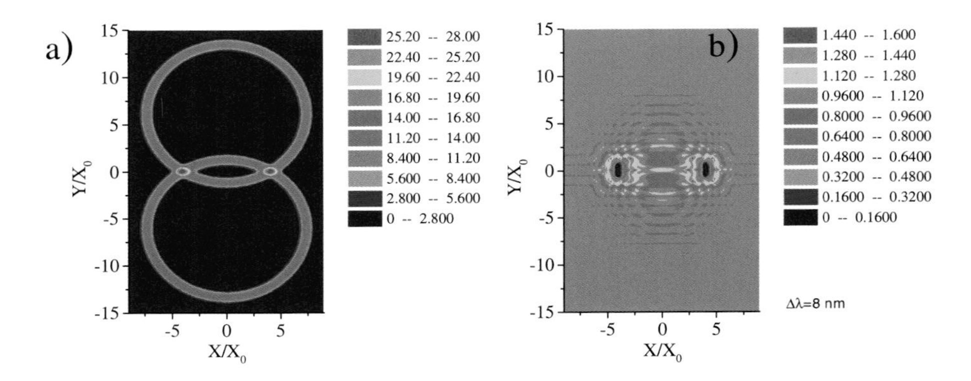

Fig. 2.12. (a) Far-field intensity distribution of the down-converted field. (b) Distribution of the noise in the difference between $\hat{S}_2$ measured from symmetric portions of the beam cross-section, scaled to the shot-noise level. The distribution for $\hat{S}_3$ is identical.

the difference between the Stokes operators measured from small (but larger than x_{diff}) symmetric portions of the far field. Precisely, the figure shows $\langle \left[\hat{S}_2(\boldsymbol{x}) - \hat{S}_2(-\boldsymbol{x})\right]^2 \rangle = \langle \left[\hat{S}_3(\boldsymbol{x}) - \hat{S}_3(-\boldsymbol{x})\right]^2 \rangle$, scaled to the shot-noise level, represented by $\langle \left[\hat{S}_0(\boldsymbol{x}) + \hat{S}_0(-\boldsymbol{x})\right] \rangle$, as a function of the transverse coordinate $\boldsymbol{x} = (x, y)$ scaled to x_0. Parameters in this plot are those of a 2 mm long BBO crystal, cut at 49.6 degrees for degenerate type-II phase matching at 702 nm. For comparison, part (a) of the figure shows the mean intensity distribution in the far field (precisely, the numbers associated with the color scale in (a) represent the number of photons detected over the resolution area x_{diff} and over the crystal coherence time). In plot (b) we see clearly two large blue zones, in correspondence with the intersections of the rings, where the Stokes-operator correlation are almost perfect. Out of these regions, basically no spatial correlation at the quantum level exists for the Stokes operator $\hat{S}_2$ and $\hat{S}_3$.

These results were obtained by exploiting a trick similar to that used in the experiment of [37], in order to partially compensate for the temporal and the spatial walk-off of the down-converted beams. In the regime of single-photon-pair production, the horizontal (ordinary) and vertical (extraordinary) photons can be in principle distinguished because of their different group velocities inside the crystal, and because of their offsets in propagation directions due to walk-off effects inside the crystal. The mere existence of this possibility is detrimental for the entanglement of the state. In the general case (arbitrary number of down-converted photons), it can be shown [3] that the group velocity mismatch and the spatial walk-off are responsible for the appearance of a propagation phase factor that lowers the value of the correlation function between the Stokes operators measured from symmetric regions. In principle this problem can be solved by using an extremely narrow frequency filter, and by performing the measurement over very narrow regions centered around the ring intersections. However, this deteriorates the efficiency of the set-up. Another possibility is to insert a second crystal, after the pump beam has been removed, and after the field polarization has been rotated by 90°. In this way, the slow and fast waves in the first crystal become the fast and the slow waves, respectively, in the second crystal, and the direction of the walk-off is reversed. It can be shown that, differently from the single-photon-pair regime, the correlations are optimized when the length of this second crystal is chosen as $l_c \tanh \sigma/(2\sigma)$, where σ is the linear gain parameter, proportional to the pump amplitude. When this kind of optimization is not possible, our calculations [3] show that similar results can be obtained by a narrowband temporal and spatial filtering, and/or by using crystals that exhibit a smaller amount of walk-off. Figure 2.12 was obtained

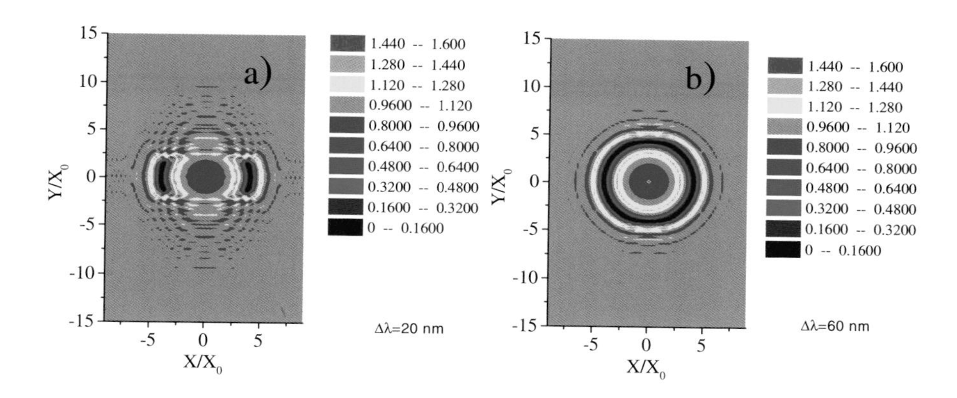

Fig. 2.13. Effect of frequency filtering. Distribution of the noise in the difference between $\hat{S}_2$ ($\hat{S}_3$) measured from symmetric portions of the beam cross-section, scaled to the shot-noise level. In part (a) a frequency filter $\Delta\lambda = 20\,\text{nm}$ wide, centered around the degenerate frequency is used; in (b) $\Delta\lambda = 60\,\text{nm}$.

by using a relatively narrow frequency filter (8 nm). Remarkably, when a broader frequency filter is employed, the regions where the Stokes parameters are correlated stretch to form a ring-shaped region around the pump direction (Fig. 2.13). This kind of shape can be understood by considering the geometry of the down-conversion cones emitted at the various frequencies by a BBO crystal. Figure 2.14 is a polar plot of the phase-matching curves (geometrical loci of the phase-matched modes), with θ being the polar angle from the pump direction of propagation (z-axis) and ϕ the azimuthal angle around z. In this plot the same color identifies the same emission wavelength;

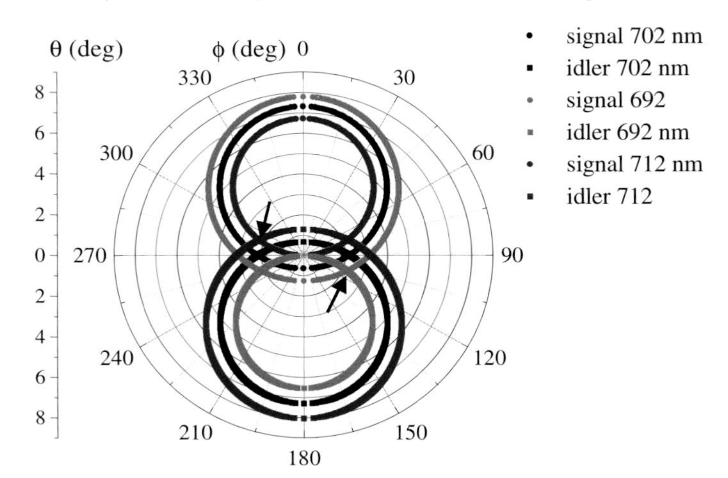

Fig. 2.14. Polar plot of phase-matching curves in a 2 mm BBO crystal cut at 49.6 degrees; θ is the polar angle from the pump direction of propagation, and corresponds to the radial position in the far field; ϕ is the azimuthal angle around the pump. Black curves: $\lambda_{signal} = \lambda_{idler} = 702$ nm; red curves: $\lambda_{signal} = \lambda_{idler} = 692$ nm; blue curves: $\lambda_{signal} = \lambda_{idler} = 712.29$ nm. The signal is an ordinary (slow) wave, and corresponds to the curves in the upper half of the plot. The idler is an extraordinary (fast) wave.

blue/red curves correspond to two conjugate wavelengths, and black curves are the two emission cones at the degenerate wavelength. The horizontally polarized (signal, ordinary) wave emission curves are those in the upper half of the plot. When considering the intersection points of a red circle with a blue circle, which correspond to two conjugate wavelengths, we cannot expect any kind of entanglement, because photons arriving in these positions are clearly distinguishable by their different frequencies. Let us consider, instead, one of the intersections of the two red curves (e.g., the one pointed by the black arrow in the plot). Here horizontally and vertically polarized photons arrive with identical probability, and have the same wavelength. As a consequence, the photon polarization is undetermined, and the light is completely unpolarized. However, each time a horizontal (vertical) photon arrives at this position, a vertical (horizontal) photon, at the conjugate wavelength, will be found at the symmetric position, that corresponds to the intersection of the two blue curves, indicated in the plot by the arrow. Hence, when considering photodetection from the two positions indicated by the arrows we can expect a high degree of polarization entanglement. The same reasoning can be made for any intersection of circles corresponding to the same wavelength (red with red, blue with blue, and black with black). By connecting these points , we can, for example, recognize the geometrical shape of the blue regions in Fig. 2.13a, where a high degree of correlations in all the Stokes parameters exists. By including more frequencies, the ring-shaped region of Fig. 2.13b shows up.

We hence conclude that the polarization entanglement of photon pairs emitted in parametric down-conversion survives in high-gain regimes, where the number of converted photons can be rather large. In this regime it takes the form of nonclassical spatial correlations of all Stokes operators associated with the polarization degrees of freedom. In the regions where the two rings intersect (in a ring-shaped region around the pump direction when a broad frequency filter is employed) all the Stokes operators are highly correlated at the quantum level, realizing in this way a macroscopic polarization entanglement. Although Stokes parameters are extremely noisy (the state is unpolarized), measurement of a Stokes parameter in any polarization basis in one far-field region determines the Stokes parameter collected from the symmetric region, within an uncertainty much below the standard quantum limit.

We believe that this form of entanglement, with its increased complexity in terms of degrees of freedom (photon number, polarization, frequency, and spatial degrees of freedom) can be quite promising for new quantum information schemes.

References

1. E. Brambilla, A. Gatti, M. Bache, and L. A. Lugiato, Phys. Rev. A **69**, 023802 (2004).
2. M. I. Kolobov, Rev. Mod. Phys. **71**, 1539 (1999), and references quoted therein.
3. A. Gatti, R. Zambrini, M. San Miguel, and L. A. Lugiato, Phys. Rev. A **68**, 053807 (2003).
4. P. Scotto and M. San Miguel, Phys. Rev. A **65**, 42811 (2002).
5. A. Gatti, E. Brambilla, and L. A. Lugiato, Phys. Rev. Lett. **90**, 133603 (2003).
6. M. J. Werner, M. G. Raymer, M. Beck, and P. D. Drummond Phys. Rev. A **52**, 4202 (1995); M. J. Werner and P. D. Drummond, Phys. Rev. A **56**, 1508 (1997).
7. In the simulations we made use of a reliable Gaussian random number generator which is discussed in R. Toral, A. Chakrabarti, Comput. Phys. Comm., **74**, 327 (1993).
8. See, e.g., W. Press, B. Flannery, S. Teukolsky, W. Vetterling, *Numerical Recipes* (Cambridge University Press, Cambridge, 1992).
9. B. M. Jost, A. V. Sergienko, A. F. Abouraddy, B. E. A. Saleh, and M. C. Teich, Opt. Expr. **3**, 81 (1998).
10. A. Berzanskis, W. Chinaglia, L. A. Lugiato, K. H. Feller, and P. Di Trapani, Phys. Rev. A **60**, 1626 (1999).
11. O. Jedrkiewicz, E. Brambilla, A. Gatti, M. Bache, L. A. Lugiato, and P. Di Trapani, Phys. Rev. Lett. **93**, 243601 (2004).
12. O. Jedrkiewicz, E. Brambilla, A. Gatti, M. Bache, L. A. Lugiato, and P. Di Trapani, J. Mod. Opt., **53**, 575 (2006).
13. L. A. Lugiato, A. Gatti and E. Brambilla, J. Opt. B: Quant. Semiclass. Opt. **4**, S176 (2002), and references therein.

14. D. V. Strekalov, A. V. Sergienko, D. N. Klyshko, and Y. H. Shih, Phys. Rev. Lett. **74**, 3600 (1995).
15. F. Ferri, D. Magatti, A. Gatti, M. Bache, E. Brambilla, and L. A. Lugiato Phys. Rev. Lett. **94**, 183602 (2005).
16. R. S. Bennink, S. J. Bentley, and R. W. Boyd, Phys. Rev. Lett. **89**, 113601 (2002).
17. W. A. T. Nogueira, S. P. Walborn, S. Padua, and C. H. Monken, Phys. Rev. Lett. **86**, 4009 (2001).
18. J. C. Howell, R. S. Bennink, S. J. Bentley, and R. W. Boyd Phys. Rev. Lett. **92**, 210403 (2004).
19. A. Gatti, E. Brambilla, L. A. Lugiato, and M. I. Kolobov, Phys. Rev. Lett. **83**, 1763 (1999); E. Brambilla, A. Gatti, L. A. Lugiato, and M. I. Kolobov, Eur. Phys. J. D **15**, 117 (2001).
20. O. Aytur and P. Kumar, Phys. Rev. Lett. **65**, 1551 (1990).
21. B. M. Jost, A. V. Sergienko, A. F. Abouraddy, B. E. A. Saleh, and M. C. Teich, Opt. Express **81**, 3 (1998).
22. S. S. R. Oemrawsingh, W. J. van Drunen, E. R. Eliel, and J. P. Woerdman, J. Opt. Soc. Am. B **19**, 2391 (2002).
23. N. Treps, U. Andersen, B. Buchler, P. K. Lam, A. Maître, H.-A. Bachor, and C. Fabre, Phys. Rev. Lett. **88**, 203601, (2002).
24. A. Mosset, F. Devaux, and E. Lantz, Phys. Rev. Lett. **94**, 223603 (2005).
25. J. R. Janesick, *Scientific Charge-Coupled Devices* (SPIE Bellingham, WA, 2001), pp. 204–205; see also http://www.roperscientific.de/theory.html.
26. Y.-K. Jiang, O. Jedrkiewicz, S. Minardi, P. Di Trapani, A. Mosset, E. Lantz, and F. Devaux, Eur. Phys. J. D **22**, 521 (2003).
27. A. Mosset, F. Devaux, G. Fanjoux, and E. Lantz, Eur. Phys. J. D **28**, 447 (2004).
28. J. W. Goodman, *Statistical Optics* (Wiley Classics Library, New York, 2000).
29. S. A. Akhmanov, V. A. Vysloukh, and A. S. Chirkin, *Optics of Femtosecond Laser Pulses* (American Institute of Physics, New York, 1992), p. 151.
30. P. Di Trapani, G. Valiulis, W. Chinaglia, and A. Andreoni, Phys. Rev. Lett. **80**, 265 (1998).
31. I. Marzoli, A. Gatti, and L. A. Lugiato, Phys. Rev. Lett. **78**, 2092 (1997).
32. L. A. Lugiato, A. Gatti, H. Ritsch, I. Marzoli, and G.-L. Oppo, J. Mod. Opt. **44**, 1899 (1997).
33. C. Szwaj, G.-L. Oppo, A. Gatti, and L. A. Lugiato, Eur. Phys. J. D **10**, 433 (2000).
34. W. P. Bowen, R. Schnabel, H.-A. Bachor, and P. K. Lam, Phys. Rev. Lett. **88**, 093601 (2002); W. P. Bowen, N. Treps, R. Schnabel, and P. K. Lam, Phys. Rev. Lett. **89**, 253601 (2002); R. Schnabel, W. P. Bowen, N. Treps, T. C. Ralph, H.-A. Bachor, and P. K. Lam, Phys. Rev. A**67** , 012316 (2003).
35. N. Korolkova, G. Leuchs, R. Loudon, T. C. Ralph, and C. Silberhorn, Phys. Rev. A **65** , 052306 (2002); e-print quant-ph/0108098.
36. J. Hald, J. L. Sorensen, C. Schori, and E. S. Polzik, Phys. Rev. Lett. **83**, 1319 (1999).
37. P. G. Kwiat, K. Mattle, H. Weinfurter, A. Zeilinger, A. V. Sergienko, and Y. Shih, Phys. Rev. Lett. **75**, 4337 (1995).

38. D. Bouwmeester, J.-W. Pan, K. Mattle, M. Eibl, H. Weinfurter and A. Zeilinger, Nature (London) **390**, 575 (1997); D. Boschi, S. Branca, F. De Martini, L. Hardy, and S. Popescu, Phys. Rev. Lett. **80**, 1121 (1998).
39. A. V. Sergienko, M. Atatüre, Z. Walton, G. Jaeger, B. E. A. Saleh, and M. C. Teich, Phys. Rev. A **60**, R2622 (1999); T. Jennewein, C. Simon, G. Weihs, H. Weinfurter, and A. Zeilinger, Phys. Rev. Lett. **84**, 4729 (2000).

3 Quantum Imaging in the Continuous-Wave Regime Using Degenerate Optical Cavities

Agnès Maître, Nicolas Treps, and Claude Fabre

Laboratoire Kastler Brossel, Université Pierre et Marie Curie, case 74, 75252 Paris cedex 05, France.

3.1 Introduction

Images carry a lot of information, contained in the local intensity of the millions of pixels composing the image. From a quantum point of view, they must be described by a quantum state that spans over as many transverse modes as pixels. Quantum imaging is therefore the archetype of highly multimode quantum optics.

Over the last two decades, theoretical and experimental studies have shown that resonant optical cavities are one of the most efficient devices to generate nonclassical states of light, such as squeezed, correlated, or entangled states. However, these states are actually single-mode nonclassical states, because the optical cavity imposes the transverse variation of the field to be one of its eigenmodes, generally the TEM_{00} mode. In order to produce the multimode nonclassical fields that are needed in quantum imaging, one must use special kinds of cavities: the *degenerate optical cavities*, in which a great number of transverse modes are likely to simultaneously oscillate.

In this chapter, we will give a brief account of the domain of multimode quantum imaging using degenerate optical cavities. We will begin by a study at the classical level of the imaging properties of such cavities, then give the main results of the various theoretical studies that have been conducted to show that these devices were indeed interesting for the generation of local squeezing and spatial quantum correlations and entanglement. We will finally describe some recent experiments aimed at showing these effects.

3.2 Classical Imaging Properties of Degenerate Optical Cavities

3.2.1 Introduction

Optical cavities have been widely used in optics for various reasons: they enhance the intracavity field by a large factor at exact resonance, they enable optical feedback in devices inserted in the cavity (as in optical amplifiers inserted in a laser cavity), and they are able to select well-defined modes of the electromagnetic field, thus simplifying the analysis of the system, for example, in quantum optics. This last property seems to be contradictory to what is needed in optical imaging, namely highly multimode, nonfiltering systems. However, as we will see, cavities with degenerate transverse modes are multimode devices that provide useful field enhancement and feedback for a large set of image shapes, because in such cavities an infinite number of modes resonate at the same cavity length. Degenerate optical cavities should thus allow us to observe interesting quantum imaging effects even at the low intensities produced by usual continuous-wave (c.w.) lasers.

Optical cavities have been used and studied in optics for a long time. Let us quote, among others, the spectroscopic studies with the Fabry–Perot resonator [1,2], and the design of laser cavities [3,4]. They are at the same time filters of temporal frequencies and filters of spatial frequencies. An optical cavity is characterized by an eigenmode basis, that is, a family of modes that superimpose onto themselves after a round-trip inside the cavity. Only fields belonging to an eigen subspace can be transmitted through the cavity at resonance: if this subspace contains a single transverse mode, the optical cavity acts as a perfect spatial filter. It has been actually used as such in many experiments (see, e.g., [5]). In degenerate cavities the eigen subspace has an infinite dimension: the filtering effect is partial, and a part of the information conveyed by the image is transmitted through it. The classical imaging properties of these cavities have been envisioned first in [6], and studied in more detail in [7]. We will just recall here the main properties, which are detailed in this latter paper.

3.2.2 Cavity Round-Trip Transform

We consider a monochromatic electromagnetic field $\boldsymbol{E}(\boldsymbol{r},t)$ at frequency ω propagating along a direction Oz of space within the paraxial approximation. At point $\boldsymbol{r} = (x,y)$ in an arbitrary transverse plane of position z inside the cavity, that will be used as a reference plane, it can be written as

$$\boldsymbol{E}(\boldsymbol{r},z,t) = Re[E(\boldsymbol{r},z)\, e^{-i\omega t}\boldsymbol{u}], \tag{3.1}$$

where $\boldsymbol{u}$ is the polarization unit vector. We will call "complex image", or more simply "image", such a transverse distribution of the field amplitude.

The round-trip propagation of the field in an empty optical cavity is characterized by a linear integral transform that we denote $\mathcal{T}$, more precisely:

$$E_{rt}(\boldsymbol{r}) = e^{ikL}\mathcal{T}[E_{in}(\boldsymbol{r})], \tag{3.2}$$

where $E_{in}(\boldsymbol{r})$ and $E_{rt}(\boldsymbol{r})$ are, respectively, the fields in the reference plane before and after one cavity round-trip, and L is the cavity round-trip optical length. If the cavity is made of lenses, curved mirrors, but has no sharp-edge irises diffracting the beam, $\mathcal{T}$ depends only on the coefficients of the Gauss matrix T_{cav} describing the round-trip geometrical-optics properties of the cavity (see, e.g., Ref. [4], p. 781). All the properties of the cavity can be inferred from T_{cav} and L (with $Det(T_{cav}) = 1$). Eigenmodes of such cavities that are localized around the cavity axis only exist when $|Trace(T_{cav})| < 2$. They are the well-known Hermite–Gauss modes TEM_{mn}, which have the following z-dependence on the cavity axis,

$$E_{mn}(\boldsymbol{r} = \boldsymbol{0}, z) = E_0 \frac{w_0}{w(z)} e^{-i(n+m+1)\phi_G(z)} e^{ikz}, \tag{3.3}$$

where $\phi_G(z)$ is the TEM_{00} mode Gouy phase at position z, given by

$$\phi_G(z) = \tan^{-1}(\frac{z}{z_R}), \tag{3.4}$$

$w(z)$ and w_0 are, respectively, the waist size at position z and position 0, and z_R is the Rayleigh length.

The on-axis phase shift of such a mode over one cavity round-trip is equal to $kL + (m + n + 1)\phi_{Grt}$, where ϕ_{Grt} is the round-trip Gouy phase shift, and will play an important role in the discussion. In the case of a simple linear cavity of length L_c closed by two identical spherical mirrors with the radius of curvature R, it is given by:

$$\phi_{Grt} = \cos^{-1}\left(1 - \frac{L_c}{R}\right). \tag{3.5}$$

In the general case, it is related to the round-trip Gauss matrix T_{cav} by the simple relation [6, 7]:

$$e^{\pm i\phi_{Grt}} = eigenvalues(T_{cav}). \tag{3.6}$$

The cavity eigenmodes are the Hermite–Gauss modes that have a total round-trip phase shift equal to $2p\pi$, with p being an integer. This occurs only for a comb of cavity length values L_{mnp} given by

$$L_{mnp} = \frac{\lambda}{2}\left(p + (n + m + 1)\frac{\phi_{Grt}}{2\pi}\right). \tag{3.7}$$

One finds a first kind of degeneracy: all the modes TEM_{mn} with the same value of $m+n$ resonate at the same cavity length. This is to be related to the cylindrical symmetry of the system: if an image is transmitted through the cavity, the same image rotated around the Oz axis by any angle will also be transmitted. The cavity will be "transverse degenerate" when there exists a larger degeneracy than the one related to this symmetry. This occurs when ϕ_{Grt} is a rational fraction of 2π:

$$\phi_{Grt} = 2\pi \frac{K}{N} \quad [2\pi], \tag{3.8}$$

in which K and N are integers and $0 < K/N < 1$ is an irreducible fraction.

It is well known that the free spectral range for longitudinal modes (p periodicity) is equal to $\lambda/2$. But when one scans the cavity length, one finds equally spaced resonances, distant from $\lambda/2N$: there exist N families of modes within a spectral range, corresponding to different values of $s = m + n + 1$. In terms of the ray optics, Eqs. (3.6) and (3.8) imply that [8],

$$(T_{cav})^N = I, \tag{3.9}$$

where I is the identity 2×2 matrix. This relation means that any incoming ray, whatever its position and tilt, will retrace back onto itself after N round-trips, forming a closed trajectory, or *orbit*.

3.2.3 Image Transmission Through an Optical Cavity

We can now determine the fields that can be transmitted through an optical cavity of length L_{mnp} corresponding to one of its resonance peaks. The input field is imaged on the intracavity reference plane, and we call $E_{in}(\boldsymbol{r})$ the input image defined in this plane. In the same way, the output field is the image of the field distribution in the same intracavity reference plane that we call $E_{\text{out}}(\boldsymbol{r})$. E_{out} results from the interference between all the fields created after $1, 2, \ldots$ trips. If the cavity is not a transverse degenerate one, E_{out} is, as is well known, the fundamental Gaussian mode TEM_{00}, or a combination of TEM_{mn} modes with a fixed value of $s = m + n + 1$, depending on the chosen length L_{mnp}. If the cavity is transverse degenerate with degeneracy order K/N, one has:

$$E_{\text{out}}(\boldsymbol{r}) = \sum_{n=0}^{\infty} \left(r_1 r_2 e^{-2i\pi s K/N} \mathcal{T}_{cav} \right)^n E_{\text{in}}(\boldsymbol{r}), \tag{3.10}$$

where r_1 and r_2 are the amplitude reflection coefficients of the two mirrors limiting the cavity. Using the fact that $\left(e^{-2i\pi s K/N} \mathcal{T}_{cav}\right)^N E_{\text{in}}(\boldsymbol{r}) = E_{\text{in}}(\boldsymbol{r})$, we can rewrite the output field of Eq. (3.10) as

$$E_{\text{out}}(\boldsymbol{r}) = \frac{1}{1 - (r_1 r_2)^N} \sum_{n=0}^{N-1} \left(r_1 r_2 e^{-2i\pi s K/N} \mathcal{T}_{cav} \right)^n E_{\text{in}}(\boldsymbol{r}). \tag{3.11}$$

The field $E_{\text{out}}(\boldsymbol{r})$ given by Eq. (3.11) has an important property: $\mathcal{T}_{cav} E_{\text{out}}(\boldsymbol{r})$ is proportional to $E_{\text{out}}(\boldsymbol{r})$. Consequently if it is used as an input field of the cavity, it is transmitted without distortion through it. It is called for this reason a *self-transform field* for $\mathcal{T}_{cav}$ [9]. Let us also note that $E_{\text{out}}(\boldsymbol{r})$ depends on s ($s = 0, \ldots, N-1$), and therefore is different for the different families of modes.

To simplify the following discussion, we will assume that the cavity mirrors have good reflectivities, so that r_1 and r_2 can be approximated by 1 in all terms of the sum in expression (3.11). We can now give three examples of degenerate cavities.

— The confocal cavity (two identical mirrors separated by a distance equal to their radius of curvature) has a Gauss matrix equal to $-I_2$ when one takes its symmetry plane as the reference plane. It has therefore a round-trip Gouy phase equal to π ($K = 1, N = 2$). The round-trip field transformation $\mathcal{T}_{cav}E(\boldsymbol{r})$ gives the field at the symmetrical point $E(-\boldsymbol{r})$. Hence the transmitted field is the the sum $E(\boldsymbol{r}) + e^{-i\pi s}E(-\boldsymbol{r})$, that is, the odd or even part of the input field, depending on s.

— The hemi-confocal cavity (a plane mirror and a curved mirror separated by a distance equal to half its radius of curvature R) has a Gauss matrix, referenced to the plane mirror, equal to:

$$T_{cav} = \begin{bmatrix} 0 & R/2 \\ -2/R & 0 \end{bmatrix}. \tag{3.12}$$

Two round-trips give, as expected, the matrix of the confocal cavity. The hemi-confocal cavity round-trip Gouy phase is equal to $\pi/2$ ($K = 1, N = 4$). The round-trip transformed field $\mathcal{T}_{cav}E(\boldsymbol{r})$ can be shown to be $\tilde{E}(4\pi\boldsymbol{r}/\lambda R)$, where $\tilde{E}(\boldsymbol{q})$ is the spatial Fourier transform of $E(\boldsymbol{r})$. One finds that the transmitted field is equal to $E(\boldsymbol{r}) + e^{-i\pi s/2}\tilde{E}(4\pi\boldsymbol{r}/\lambda R) + e^{-i\pi s}E(-\boldsymbol{r}) + e^{-3i\pi s/2}\tilde{E}(-4\pi\boldsymbol{r}/\lambda R)$. Depending on $s = 1, 2, 3, 4$, it is a combination of the even or odd part of the input field and of the even or odd part of its Fourier transform.

— The "self-imaging cavity" has a Gauss matrix equal to the identity and a Gouy phase equal to zero: any intracavity ray is retraced onto itself after one round-trip; any field configuration, whatever its shape, is perfectly transmitted, and still benefits from the build-up effect of cavity. It is thus an ideal cavity for studies of cavity-enhanced imaging. Self-imaging cavities have been extensively studied in [6]. They cannot be built using only two spherical mirrors, and can consist, for example, of a plane end-mirror, an intracavity lens, and a curved end-mirror, separated by appropriate distances. They have been used in experiments on generation of optical patterns [10].

In contrast, the planar cavity, limited by two plane mirrors, a configuration widely studied theoretically and used in some experiments with pulsed lasers, is described by the Gauss matrix of a mere length interval of value L, which therefore does not fulfill condition (3.9) for any N value. A planar cavity is not a transverse degenerate cavity. The transmitted fields will be cones of tilted plane waves making angles with Oz that are selected by the cavity length: as is well known, rings will be observed at the output.

In an analogous way, the concentric cavity (two curved mirrors sharing the same center of curvature) is described by a round-trip Gauss matrix, referenced to the symmetry plane of the cavity, equal to that of a diverging thin lens of focal length $-R/2$. None of its powers is equal to the identity, and therefore a concentric cavity is not a transverse degenerate cavity either, though, as the previous one, its round-trip Gouy phase shift is zero. Only one class of rays retraces onto itself in these cavities: the rays parallel to the optical axis for the planar cavity, the rays passing through the cavity center for the concentric cavity. All other light rays diverge after many reflections: they go further and further from the cavity axis in the planar cavity case, and are more and more tilted in the concentric cavity case. In both cases the problem cannot be treated properly within the paraxial approximation, where the Gouy phase is defined.

3.3 Theory of Optical Parametric Oscillation in a Degenerate Cavity

Nonclassical states of light are generated when one inserts a nonlinear element in a resonant or quasi-resonant optical cavity. We will not treat here the cases of the second-harmonic generation, Kerr effect, or four-wave mixing, in spite of their interest. We will rather concentrate on the case of intracavity parametric coupling, as most of the studies in quantum imaging have considered this configuration.

3.3.1 Classical Behavior

In a first step, one needs to know the classical behavior of an Optical Parametric Oscillator (OPO) when a nonlinear crystal is inserted in a degenerate cavity, which makes it possible to simultaneously oscillate on different transverse modes. This problem has been considered, for example, in [11–13], where the intracavity field is expanded on the basis of transverse Gaussian modes:

$$E(\boldsymbol{r},t) = \sqrt{\frac{\hbar\omega}{2n_i\varepsilon_0 V}} \sum_{m,n} \alpha^{m,n}(t) E_{m,n}(\boldsymbol{r}), \tag{3.13}$$

the scaling coefficient in Eq. (3.13) being chosen so that $|\alpha^{m,n}|^2$ is a photon number. The evolution equations of the system appear as coupled dynamical equations for the coefficients $\alpha_s^{m,n}(t)$ of the signal wave and $\alpha_i^{m,n}(t)$ of the idler wave expansion. Let us give here as an example the equations for a doubly resonant, nonfrequency-degenerate OPO in the simple case where the modes involved are only two modes at the signal frequency (e.g., TEM_{00} and TEM_{01}), characterized by amplitudes α_s^0 and α_s^1, and two modes at the idler frequency characterized by amplitudes α_i^0, α_i^1. These equations read:

$$\frac{L}{c}\frac{d\alpha_s^0}{dt} = -\alpha_s^0(\kappa - i\delta_s^0) + \chi_{00}\alpha_p^0\alpha_i^{0*} + \chi_{01}\alpha_p\alpha_i^{1*},$$
$$\frac{L}{c}\frac{d\alpha_i^0}{dt} = -\alpha_i^0(\kappa - i\delta_i^0) + \chi_{00}\alpha_p\alpha_s^{0*} + \chi_{10}\alpha_p\alpha_s^{1*},$$

$$\frac{L}{c}\frac{d\alpha_s^1}{dt} = -\alpha_s^1(\kappa - i\delta_s^1) + \chi_{11}\alpha_p\alpha_i^{1*} + \chi_{10}\alpha_p\alpha_i^{0*}, \tag{3.14}$$
$$\frac{L}{c}\frac{d\alpha_i^1}{dt} = -\alpha_i^1(\kappa - i\delta_i^1) + \chi_{11}\alpha_p\alpha_s^{1*} + \chi_{01}\alpha_p\alpha_s^{0*},$$
$$\alpha_p = \alpha_p^{\text{in}} - \frac{1}{2}\left(\chi_{00}\alpha_s^0\alpha_i^0 + \chi_{11}\alpha_s^1\alpha_i^1 + \chi_{01}\alpha_s^0\alpha_i^1 + \chi_{10}\alpha_s^1\alpha_i^0\right),$$

where α_p is the pump field amplitude in the middle plane of the crystal and α_p^{in} the pump field amplitude at the entrance of the crystal; $\kappa = 1 - r$, where r is the amplitude reflection coefficient of the output mirror, and is the cavity loss coefficient, supposed to be equal for all the signal and the idler modes. The various δ are the cavity detunings for different modes. The value of the coupling coefficients χ_{ij} $(i, j = 0, 1)$ can be found, for example, in [11] for a TEM_{00} pump, together with a general stability analysis of the solutions. When the cavity is not degenerate, the detunings are different for different modes. Only one pair of the signal and the idler mode can resonate at a given length. The OPO oscillates on the single pair of signal–idler modes that has the lowest threshold. The OPO behaves in this case as a homogeneously broadened laser, and the first mode that oscillates depletes the pump and prevents the other modes from oscillating. When the cavity is degenerate, all the involved modes can be simultaneously resonant. One finds that in this case the OPO oscillates on a well-defined linear combination of the transverse modes [11]: the OPO therefore remains in a single-mode configuration for the signal and the idler, but this mode is different from the usual Gaussian modes. When an infinite number of modes are coupled together, the problem is more difficult to solve, but the behavior is roughly the same: the OPO oscillates on a well defined linear combination of all these modes, giving rise to *optical patterns*, which have been the subject of many studies. For example, a frequency-degenerate OPO with a planar cavity which is detuned from resonance with a signal-wave propagating perpendicularly to the mirrors generates above threshold a "roll pattern" [14], that is, a field having the shape of interference fringes in the reference plane situated inside the cavity (called "near-field" pattern). To determine the pattern generated in more complex situations, instead of a modal analysis, a "local" analysis can also be used, consisting of direct calculation of the signal and the idler field slowly varying envelope E($\boldsymbol{r}$,t) in a given intracavity transverse plane, in the paraxial and the mean-field approximation [15]. For example, one finds the following evolution equation for the signal mode $E_s(\boldsymbol{r})$,

$$\frac{L}{c}\frac{\partial}{\partial t}E_s(\boldsymbol{r}, t) = -(\kappa + i\delta_s - i\xi L_s)E_s(\boldsymbol{r}, t) + \chi E_p(\boldsymbol{r}, t)E_i^*(\boldsymbol{r}, t), \tag{3.15}$$

where δ_s is the detuning with respect to the closest TEM_{00} mode, ξ is related to the transverse mode spacing (and equal to $\phi_{Grt}/2\pi$ in our notations, where ϕ_{Grt} is defined in Eq. (3.6), and L_s the diffraction operator:

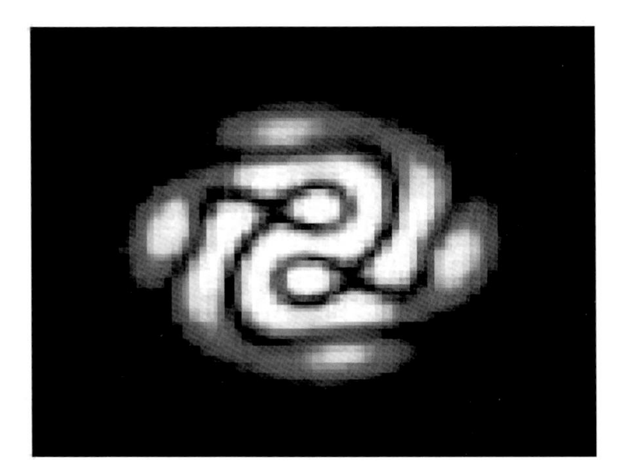

Fig. 3.1. Numerical calculation of the pattern generated by a frequency-degenerate OPO operating in a detuned quasi-confocal cavity ($\delta_s = -1.5\kappa, \xi = 0.25\kappa$), 70 times above threshold, and with a strongly focused pump field ($w_{pump} = 0.25 w_0$).

$$L_s = \frac{w_0^2}{4}\left(\frac{\partial^2}{\partial x^2} + \frac{\partial^2}{\partial y^2}\right) - \frac{x^2 + y^2}{w_0^2} + 1, \tag{3.16}$$

w_0 being the TEM_{00} waist size in the OPO cavity. Equation (3.15), together with similar equations for the idler and the pump modes, can be solved numerically in various configurations. Numerous studies have been devoted to the generation of optical patterns in OPOs [16–24]. They can be, according to the operating conditions, either stationary or evolving in time. Figure 3.1 gives an example of the output field intensity calculated in the case of a quasi-confocal frequency-degenerate OPO pumped well above the threshold. The solution is in this case nonstationary (the movie can be seen online in [13]).

3.3.2 Quantum Properties

It is well known that the phenomenon of the parametric down-conversion, either spontaneous or stimulated by an input wave, gives rise to the signal and the idler photons that are quantum-correlated both in the time domain and in the space domain: if the pump is a plane wave, because of translational invariance in the plane perpendicular to the propagation, the transverse total momentum is exactly conserved in the process, and consequently the directions of emission of the signal and the idler photons are perfectly correlated at the quantum level (see Fig. 3.2). This property has been widely used in the case of spontaneous down-conversion and is at the root of phenomena such as ghost imaging, or spatial pixel-to-pixel correlations, which are described in

Chapters 5 and 2 of this book. Because of the very low efficiency of the parametric down-conversion effect with the presently available nonlinear crystals, one is restricted to work in the photon-counting regime in the case of spontaneous parametric down-conversion, or to use high peak power pulsed lasers with poor pulse-to-pulse stability in order to reach the high-parametric-gain regime. The use of an optical cavity is then of high interest to enhance the efficiency of the parametric effect, so that one can generate highly non-classical fields using as a pump simple and highly stable c.w. solid-state lasers of moderate power. When the parametric crystal is inserted in a cavity, the temporal

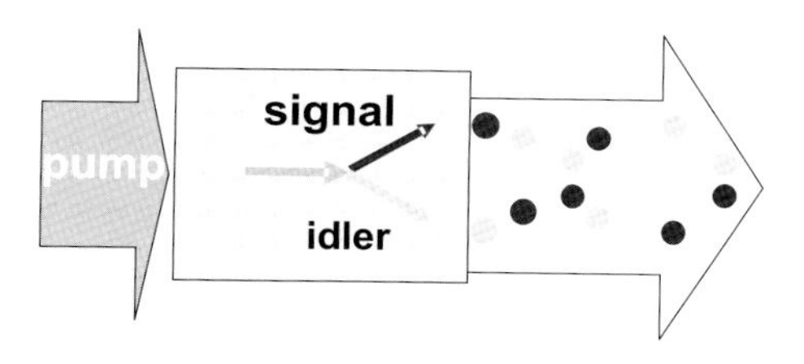

Fig. 3.2. The signal and the idler photons produced by parametric down-conversion are emitted in directions that are perfectly quantum-correlated if the pump is a plane wave.

quantum correlation remains for the Fourier noise temporal frequencies lying within the cavity bandwidth, and the spatial correlations are projected onto the subspace spanned by the transverse modes that are resonant for a given length of the cavity. If this one is a single-mode, all the transverse information is lost. If it is multimode, the situation is more complex to study, and strongly depends on the properties of the cavity.

Let us take first the seemingly simple case where the cavity is simultaneously resonant only for two signal modes and for two idler modes that was outlined previously (Eq. (3.14)). The system is described at the quantum level by four processes where a pump photon is split into twin-photons either in the modes (α_s^0, α_i^0),(α_s^1, α_i^1),(α_s^1, α_i^0), or (α_s^0, α_i^1), and the reverse processes. It corresponds to the interaction Hamiltonian:

$$\hat{H}_{int} \propto \hat{a}_p(\chi_{00}\hat{a}_s^{0\dagger}\hat{a}_i^{0\dagger} + \chi_{11}\hat{a}_s^{1\dagger}\hat{a}_i^{1\dagger} + \chi_{01}\hat{a}_s^{0\dagger}\hat{a}_i^{1\dagger} + \chi_{10}\hat{a}_s^{1\dagger}\hat{a}_i^{0\dagger}) + h.c., \quad (3.17)$$

where, for example, $\hat{a}_s^{0\dagger}$ is the photon creation operator in the mode labeled 0 at the signal frequency. This complex interaction creates a multipartite quantum entanglement between these four modes, which in turn produces quantum spatial correlations between the output signal and idler fields resulting from the superposition of these different modes. When the cavity is simultaneously resonant for an infinite number of transverse modes, the problem is even more complicated. It is generally solved using quantum equations in the Wigner representation [25] analogous to the classical Eq. (3.15), with input quantum-noise terms, to determine the correlation functions between the quantum fluctuations at two different points of a given transverse plane.

The first studies made in the '90s dealt with the "spatial structure of squeezed vacuum" in the regular Degenerate Optical Parametric Amplifier (DOPA), just below threshold. They determine a global property of the system containing spatial information, namely the amount of squeezing as a function of the transverse shape of the local oscillator mode used in the homodyne detection measuring the squeezing [26], in the case of a cavity with planar or spherical mirrors. Later, the spatial spectrum of squeezing was investigated, which yields more detailed information on the spatial distribution of quantum fluctuations in the system [27]. The authors introduced the concept of the *quantum image*, that is, an inhomogeneous field distribution purely generated by the quantum fluctuations. Below the threshold for parametric oscillation, where the near-field distributions are homogeneous both in intensity and phase, appropriate spatial correlation functions anticipate at the level of quantum fluctuations the transverse spatial pattern that appears above the threshold. The concept was later extended to the nonfrequency-degenerate OPO case [28].

Strong quantum correlations were predicted in the DOPA in planar or quasi-planar cavities. They exist in the far field (reference plane at an infinite distance from the cavity) between the quantum intensity fluctuations measured in small areas symmetrical with respect to the cavity optical axis [25,29,30]. Later, it was found that the same device was also able to produce *spatial entanglement*: between the fields measured at symmetrical points in the far-field image, anticorrelations are present in a given field quadrature at the same time as correlations in the conjugate quadrature [31]. This striking property is the starting resource for parallel quantum information processing in images, which represents a promising extension of the regular single-mode EPR entanglement used in various information protocols of quantum information processing. All these properties appear to be robust with respect to the inevitable imperfections of an experimental set-up: slight departure from the planar case and effect of a finite size pump.

The case of an exact confocal cavity pumped by a plane wave pump field turns out to be rather easy to treat. This simplification is related to the fact that, as stated in Section 3.2, the confocal cavity is a true degenerate cavity, which is not the case for the planar cavity. Decomposing the field on the basis of the Laguerre–Gauss modes of a given parity, one finds that in this case there is no cross-coupling between the different Laguerre–Gauss modes $\mathrm{TEM}_{p,\ell}$, so that the system behaves as a superposition of independent degenerate parametric interactions, governed by the Hamiltonian:

$$\hat{H}_{int} \propto \hat{a}_p \sum_{p,\ell} \chi_{p,\ell} (\hat{a}^{\dagger}_{p,\ell})^2 + h.c., \tag{3.18}$$

which is much simpler than the Hamiltonian given by Eq. (3.17). It was predicted that a confocal DOPA pumped by a plane wave produces below threshold a highly multimode squeezed field: strong squeezing can be actually measured whatever the transverse shape of the Local Oscillator field (LO) [32] provided that it is symmetrical with respect to the cavity axis. In particular, using LO with a very small transverse extension, the device produces *local squeezing*, which can be very useful to reduce the quantum noise in applications using pixelized detectors such as CCD cameras. This important result was later extended to more realistic cases, including quasi-confocal cavities, Gaussian pump [33], and long crystals, inside which the diffraction effects can no longer be neglected [7].

There is another very interesting quantum property that has been predicted in these intracavity devices: it is the possibility of noiseless amplification of images, which is the object of Chapter 7 of this book in the case of traveling-wave parametric amplification. The gain reaches very high values when one approaches the oscillation threshold from below, whereas one needs a much higher pump power to get the same gain values in the case of traveling-wave parametric amplification. It has been theoretically shown that a DOPA operating in a planar ring cavity can amplify, in a phase-sensitive way, a small portion of the input image without degradation of the signal-to-noise ratio [35]. If the cavity is confocal, it can simultaneously amplify in a noiseless way all the points of the input image, provided that the input image field is symmetrical with respect to the cavity axis [36].

3.4 Experimental Results

The previous section has shown that many interesting quantum effects are likely to be observed in OPOs contained in degenerate cavities and, more precisely, in confocal cavities. Figure 3.3 gives a general sketch of the experimental set-up used to observe such effects. The parametric crystal, a type-II KTP crystal formed of two optically cemented pieces in order to compensate the walk-off effects, is temperature and angle controlled. It is inserted in a cavity made of two identical spherical mirrors with two different controls of the cavity length: a coarse control, with the help of a micrometric screw, is used to tune the cavity to the confocal point; a fine control, with the help of a PZT stack, is used to put the cavity to exact resonance with the different modes. The pump waist size can be varied, and is usually bigger than the TEM_{00} fundamental cavity mode in order to excite many transverse signal and idler modes. Because the pump is not optimally focused in the cavity, the oscillation threshold is higher than in a single-mode OPO. This is the reason why in most cases a triply resonant configuration is adopted (cavity resonant for signal, idler, and pump fields). The "infrared injection" channel, on the right, is used for experiments on parametric amplification.

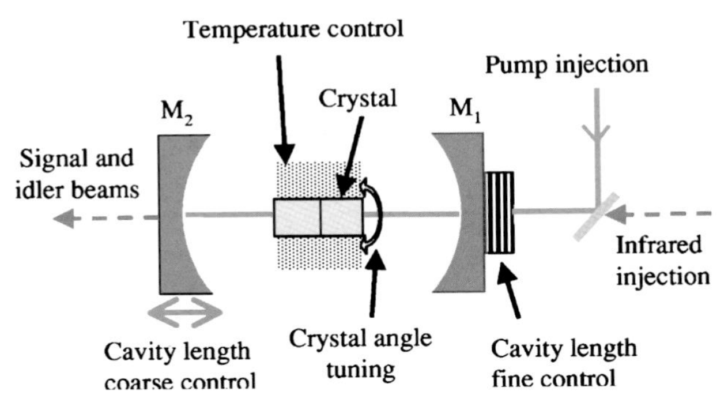

Fig. 3.3. Experimental set-up.

3.4.1 Classical Effects: Observation of Optical Patterns

In a first set of experiments [37, 38], the transverse shape of the fields generated by the system operating above threshold, both in the near-field and in the far-field configuration, was recorded on a CCD camera. The signal and the idler beams, which have different frequencies, were separated by a polarizing beamsplitter. Simple TEM_{00} modes were generated when the cavity length L was larger than L_{conf}, corresponding to exact confocality, by a fraction of a millimeter. Complex patterns could be observed at exact confocality $L = L_{conf}$, but also when $L < L_{conf}$ for higher pump powers. This feature can be explained by a thermal intracavity lensing effect that reduces the actual value of L_{conf} in the presence of an intense pump beam.

Figure 3.4 gives an example of recorded patterns at high pump intensity. One can observe in the center of the generated signal beam, features that are very small compared to the TEM_{00} waist size, and depend on the fine-tuning of the cavity around the resonance length. They require at least 25 Gaussian modes TEM_{pq} in order to be reconstructed. Note that the patterns appearing on the signal and the idler beams are in general different. It is difficult to compare them with the theoretical predictions such as the one given in Fig. 3.1, because many effects present in the experiment are not taken into account in the existing theories: thermal lensing effects, residual walk-off effects in the crystal, diffraction effects in the crystal, and so on. Computer simulations have been nevertheless able to yield patterns with similar features [41]. The same patterns have been also observed later in Ref. [42].

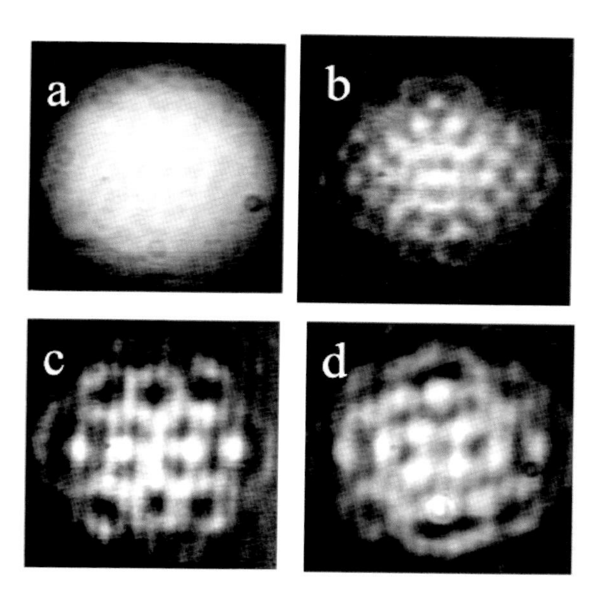

Fig. 3.4. Center of the signal intensity distribution in the far field for a pump power of 360 mW and a coarse length change of (a) L_{conf}, 0.5 mm; (b) L_{conf}, 0.4 mm; (c) and (d): L_{conf}, 1 mm. The cases (c) and (d) correspond to different fine tunings of the cavity.

3.4.2 Observation of Quantum Correlations in Images

The previous investigations showed that it was possible to find experimental conditions where the cavity could be considered as degenerate, meaning that the separation between the different transverse modes could be experimentally adjusted in order to be much smaller than the cavity linewidth, which is a prerequisite for observing quantum spatial effects. A first experiment was devoted to investigation of the spatial distribution of the quantum intensity correlations between the signal and the idler beams [39]. For this purpose an iris of variable diameter was inserted at the output of the OPO (left side of Fig. 3.3). The signal and the idler beams were separated as before by a polarizing beamsplitter after the iris, and the noise N_d of the intensity difference between the two beams was measured while the diameter of the iris was varied. The experimental values of this noise normalized to shot noise are

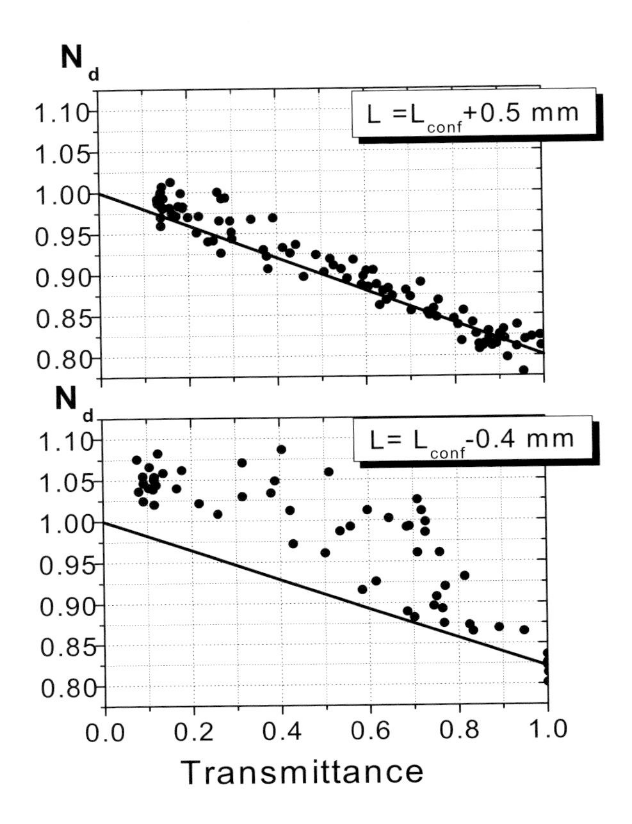

Fig. 3.5. Spatial distribution of the normalized intensity difference noise N_d as a function of the iris transmittance T. Points: values of N_d for a cavity length L beyond the confocality range (upper part) and inside the confocality range (lower part). Straight line: values of N_d that would be obtained with a single-mode beam having the same squeezing when the iris is fully open.

given in Fig. 3.5 as a function of the intensity transmission factor (transmittance) of the iris. A value N_d smaller than unity indicates that there exists a quantum correlation between the corresponding transmitted parts of the signal and the idler beams. The top part of Fig. 3.5 has been recorded for a cavity that was not confocal. The generated signal and idler beams were in this configuration almost perfect TEM_{00} beams. One observes in this case a quantum correlation of 20% ($N_d = 0.8$) between the total intensities of the signal beams (transmittance equal to 1). When the iris is progressively closed, one observes a linear variation of N_d as a function of transmittance: interception of light by the iris is a loss mechanism that destroys the quantum effect just as any other kind of loss. One can describe the nondegenerate

OPO in this case as consisting of the signal and the idler beams composed of time-correlated photons that are spatially randomly distributed in both beams. The bottom part of Fig. 3.5 has been recorded for a cavity that is confocal. The signal and the idler beams now exhibit patterns in the mean field, which are different from each other. In spite of this difference, the two beams are still "twin beams": one indeed measures a quantum correlation of almost 20% between the total intensities of the signal beams. But the behavior now changes when the iris is progressively closed, and the variation of N_d as a function of transmittance is no longer linear. A detailed analysis of the concept of the multimode quantum field [43] shows that this nonlinear behavior is a proof that the generated field cannot be described by a single-mode nonclassical quantum state of light. If the two beams were composed of perfectly spatially correlated photons as in Fig. 3.2, one should observe a value of N_d independent of the transmittance of the iris. This is not at all the case in the present experiment, where it seems that the photons are spatially correlated only in the outer part of the signal and idler beams, and randomly distributed in their central part, a feature that has so far received no theoretical explanation.

The second experiment concerned the frequency-degenerate OPA, below the threshold of oscillation, for which we have seen that many spatial quantum effects are predicted. It studied more precisely the phenomenon of c.w. intracavity parametric amplification of images [40]. It was shown in [44] that parametric amplification using a pump at frequency 2ω, an input signal wave at frequency ω, and a type-II crystal offers interesting possibilities: when the input signal is polarized parallel to the direction of polarization of the signal or the idler, it acts as a phase-insensitive amplifier; when the input signal is polarized at $\pi/4$ from these directions, it acts as a phase-sensitive and, therefore, possibly noiseless amplifier. Furthermore, in the latter configuration at high gain the two projections of the amplified idler wave on the signal and idler polarizations are the "quantum clones" [45]. Noiseless operation of parametric image amplification using powerful pulsed lasers has already been observed in parametric amplification without cavities for temporal fluctuations [46], and more recently for pure spatial fluctuations [47], as described in Chapter 7 of this book. In the experiment described here, using degenerate cavities, we were able to observe large gains, of the order of 23 dB, with a pump power of only 20 mW. Stable phase-sensitive amplification turned out to be difficult to achieve in a regular confocal cavity, which has to be furthermore simultaneously resonant for the pump, the signal, and the idler modes in order to reach a very low oscillation threshold. To solve this problem, a "dual" cavity was used (see Fig. 3.6), for which it is possible to independently tune to resonance the pump mode, by acting on the mirror M_1 position, and the signal and the idler modes, by acting on the mirror M_4 position and the crystal temperature. The fact that the crystal face M_2 was plane prevented us from using a confocal cavity for the signal and the

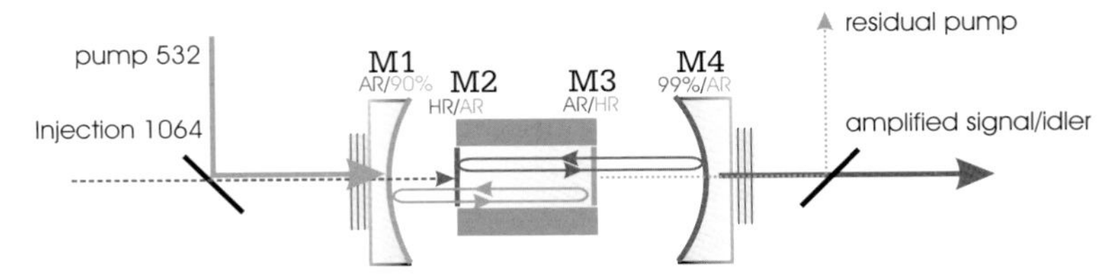

Fig. 3.6. Dual cavity: the pump cavity is limited by the mirror M_1 and the right face M_3 of the crystal; the signal, and the idler cavity are limited by the left face M_2 of the crystal and the mirror M_4, so that they can be independently tuned to resonance.

idler beams. The semi-confocal cavity, another degenerate cavity described in Section 3.3, was used instead. Figure 3.7 shows the experimental results

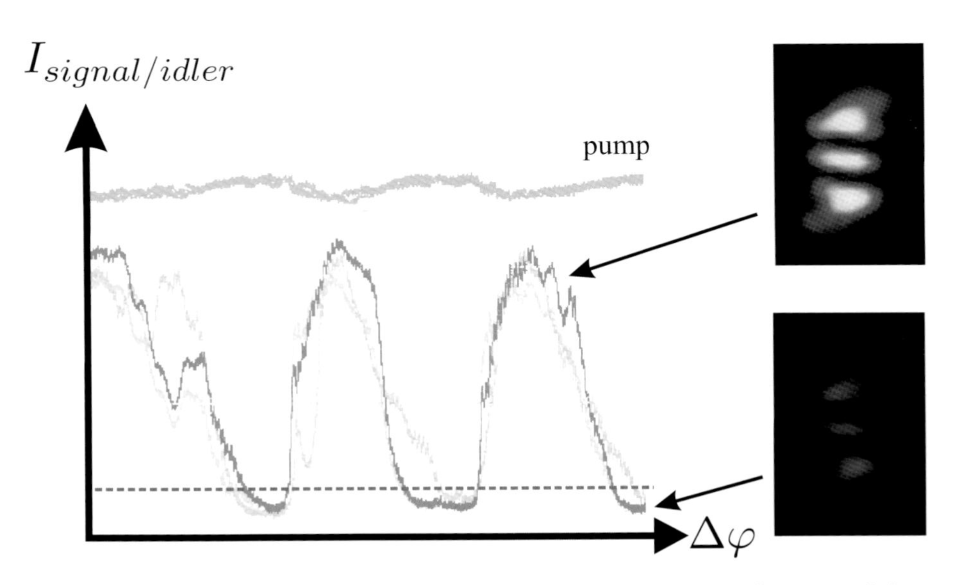

Fig. 3.7. Left side: pump, signal, and idler total intensities as a function of the relative phase between the injected mode and the pump mode. A maximum of signal and idler (amplification) appears when the relative phase is zero, whereas a minimum (deamplification) is present when the relative phase is π. The dashed line is an estimation of the signal and the idler intensity without amplification. Right side: transverse profile of the signal beam at the maximum and minimum amplification.

when a double slit is used as an input image. As explained in Section 3.3, a hemi-confocal cavity transmits the even part of the input image (i.e., the double slit itself), plus the even part of its Fourier transform. This latter part is actually what can be seen more easily in the two rectangles on the right of Fig. 3.7, as it turns out to have more intense central features than the transmitted double slit. The left part of the figure gives the total intensity of the output signal and the idler modes as a function of the phase difference between the pump and the input signal waves. It clearly shows the regime of phase-sensitive amplification of an image, with a maximum gain of 5 dB which is roughly constant over a large area of the input field. The noise figure could not be measured in the present state of the experiment, so that the noiseless character of the amplification remains to be shown. Quantum correlations were proven to be present in the amplified field: noise reduction of 12% below the shot-noise limit was observed between the total intensities of the projections of the output field on the signal and the idler polarizations. The regime of "quantum cloning" of the two polarization components of the amplified image was thus reached in the experiment.

3.5 Conclusion

Pure quantum spatial effects, predicted in OPOs using degenerate optical cavities, have been experimentally demonstrated. The experiments turn out to be rather difficult, as many stringent conditions must be fulfilled and optimized at the same time. This is the reason why, so far, the observed quantum effects are small, and no local squeezing or local quantum entanglement has been observed.

These first c.w. quantum imaging effects are nevertheless encouraging, and will be undoubtedly improved in the years to come. Let us also mention that preliminary experiments seem to show that the "self-imaging cavity," described in Section 3.3, has very promising potential for intracavity quantum imaging. This system will be actively studied in the near future.

References

1. R. Boulouch, Journal de Physique Théorique et Appliquée **2**, 316 (1893).
2. C. Fabry and A. Perot, Comptes-Rendus de l'Académie des Sciences **123**, 802 (1896).
3. A. Schawlow and C. Townes, Phys. Rev. **112**, 1940 (1958)
4. A. Siegman, *Lasers* (University Science Books, Mill Valley, CA, 1986).
5. N. Treps, N. Grosse, W. P. Bowen, C. Fabre, H.-A. Bachor, and P. K. Lam, Science **301**, 940 (2003).
6. J. Arnaud, Appl. Opt. **8**, 189 (1969).
7. S. Gigan, L. Lopez, N. Treps, A. Maître, and C. Fabre, Phys. Rev. A **72**, 023804 (2005).
8. J. Dingjan *Multimode optical resonators and wave chaos*, Ph. D. thesis, Leiden University (2003).
9. A. Lohmann and D. Mendlovic, J. Opt. Soc. Am. A **9**, 2009 (1992).

10. V. B. Taranenko, K. Staliunas, and C. O. Weiss, Phys. Rev. Lett. **81**, 2236 (1998).
11. C. Schwob, P. F. Cohadon, C. Fabre, M. Marte, H. Ritsch, A. Gatti, and L. A. Lugiato, Appl. Phys. B, **66**, 685 (1998).
12. C. Fabre, M. Vaupel, N. Treps, P. F. Cohadon, C. Schwob, and A. Maître, C. R. Acad. Sci. Paris **1** IV, 553 (2000).
13. M. Marte, H. Ritsch, K. I. Petsas, A. Gatti, L. A. Lugiato, C. Fabre, and D. Leduc, Opt., Express **3**, 71 (1998).
14. G. L. Oppo, M. Brambilla, and L. A. Lugiato, Phys. Rev. A **49**, 2028 (1994).
15. G. L. Oppo, M. Brambilla, D. Camesasca, A. Gatti, and L. A. Lugiato, J. Mod. Opt., **41**, 1151 (1994).
16. V. Sanchez-Morcillo, E. Roldan, and G. de Valcarcel, Phys. Rev. A **56**, 3237 (1997).
17. K. Staliunas, Phys. Rev. Lett. **81**, 81 (1998).
18. K. Staliunas and V. Sanchez-Morcillo, Phys. Rev. A **57**, 1454 (1998).
19. M. Santagiustina, P. Colet, M. San Miguel, and D. Walgraef, Phys. Rev. E **58**, 3843 (1998).
20. G. L. Oppo, A. Scroggie, and W. Firth, J. Opt. B: Quantum Semiclass. Opt. **1**, 133 (1999).
21. G. Izus, M. San Miguel, and M. Santagiustina, Opt. Lett. **25**, 1 (2000).
22. G. L. Oppo, A. Scroggie, S. Sinclair, and M. Brambilla, J. Mod. Opt. **47**, 2005 (2000).
23. P. Lodahl, M. Bache, and M. Saffman, Phys. Rev. Lett. **85**, 4506 (2000).
24. H. Ward, M. Ouarzazi, M. Taki, and P. Glorieux, Phys. Rev. E **63**, 016604 (2001).
25. A. Gatti, H. Wiedemann, L. A. Lugiato, I. Marzoli, G. L. Oppo, and S. Barnett, Phys. Rev. A **56**, 877 (1997).
26. L. A. Lugiato and A. Gatti, Phys. Rev. Lett. **70**, 3868 (1993).
27. A. Gatti and L. A. Lugiato, Phys. Rev. A **52**, 1675 (1995).
28. C. Szwaj, G. L. Oppo, A. Gatti, and L. A. Lugiato, Eur. Phys. J. D **10**, 433 (2000).
29. G. Grynberg and L. A. Lugiato, Europhys. Lett. **29**, 678 (1995).
30. I. Marzoli, A. Gatti, and L. A. Lugiato, Phys. Rev. Lett. **78**, 2092 (1997).
31. A. Gatti, L. A. Lugiato, K. I. Petsas, and I. Marzoli, Europhys. Lett. **46**, 461 (1999).
32. P. Grangier and L. A. Lugiato, J. Opt. Soc. Am. B **14**, 225 (1997).
33. K. I. Petsas, A. Gatti, L. A. Lugiato, and C. Fabre, Eur. Phys. J. D **22**, 501 (2003).
34. L. Lopez, S. Gigan, A. Maître, N. Treps, C. Fabre, and A. Gatti, Phys. Rev. A **72**, 3806 (2005).
35. M. I. Kolobov and L. A. Lugiato, Phys. Rev. A **52**, 4930 (1995).
36. S. Mancini, A. Gatti, and L. A. Lugiato, Eur. Phys. J. D **12**, 499 (2000).
37. M. Vaupel, A. Maître, and C. Fabre, Phys. Rev. Lett. **83**, 5278 (1999).
38. S. Ducci, N. Treps, A. Maître, and C. Fabre, Phys. Rev. A **64**, 023803 (2001).
39. M. Martinelli, N. Treps, S. Ducci, S. Gigan, A. Maître, and C. Fabre, Phys. Rev. A **67**, 023808 (2003).
40. S. Gigan, L. Lopez, V. Delaubert, N. Treps, C. Fabre, and A. Maître, J. Mod. Opt., **53**, 809 (2006).
41. M. Le Berre, E. Ressayre, and A. Tallet, Phys. Rev. E **67**, 066207 (2003).

42. M. Lassen, P. Tidemand-Lichtenberg, and P. Buchhave, Phys. Rev. A **72**, 023817 (2005).
43. N. Treps, V. Delaubert, A. Maître, J. M. Courty, and C. Fabre, Phys. Rev. A **71**, 013820 (2005).
44. J. A. Levenson, I. Abram, T. Rivera, P. Fayolle, J. C. Garreau, and P. Grangier, Phys. Rev. Lett. **70**, 267 (1993).
45. J. A. Levenson, I. Abram, T. Rivera, and P. Grangier, J. Opt. Soc. Am. B **10**, 2233 (1993).
46. S.-K. Choi, M. Vasilyev, and P. Kumar, Phys. Rev. Lett. **83**, 1938 (1999).
47. A. Mosset, F. Devaux, and E. Lantz, Phys. Rev. Lett. **94**, 223603 (2005).

4 Quantum Imaging by Synthesis of Multimode Quantum Light

Nicolas Treps[1], Hans A. Bachor[2], Ping Koy Lam[2], and Claude Fabre[1]

[1] Laboratoire Kastler Brossel, Université Pierre et Marie Curie, case 74, 75252 Paris cedex 05, France
[2] Australian Centre for Quantum-Atom Optics, Department of Physics, The Australian National University, Canberra ACT 0200, Australia
treps@spectro.jussieu.fr

4.1 Introduction

An optical image can contain a wealth of information frequently amounting to the equivalence of millions of pixels. From a quantum-mechanical point of view, such images can be described by a quantum state that spans as many transverse modes as pixels [1]. Quantum imaging is therefore the archetype of highly multimode quantum optics. To date, most of the quantum imaging ideas and experiments aim to produce light that is at the same time nonclassical and highly multimode [2, 3]. These fascinating multimode features can be a source of theoretical complications and experimental difficulties.

When dealing with specific applications such as the positioning of a laser beam or the reading of information from a sample by optical means, however, the quantity of information to be extracted from the image can at times be very modest. In many situations, one can safely assume that the image only undergoes changes that are related to the variation of a few parameters, and considerable a priori information about the image is available. We will show that in these situations the highly multimode feature of quantum imaging can be reduced to a tractable problem with only few modes.

When the transverse shape of an unperturbed image is known, each measurement of the image to be performed is associated exactly with one specific transverse mode that is responsible for the noise affecting this measurement. We show that shaping the fluctuations of the incoming light in such a way that they match this transverse mode will allow measurements with precision below the quantum noise limit, thereby increasing the sensitivity in the determination of the relevant parameters. In particular, it will be shown that in order to simultaneously measure N parameters below the shot-noise limit, one needs to use nonclassical multimode light that is the tensor product of $N + 1$ modes, of which N must be in a squeezed state.

The general logic of the above argument evokes a new experimental protocol, which we call the *modal synthesis of spatially nonclassical light*. This protocol requires the coherent interference of carefully chosen transverse mode beams to produce a composite spatially nonclassical beam of light that is suitable for a specific task. In the following sections, we will briefly review the theoretical background, and focus on the experimental techniques associated with modal synthesis quantum imaging. We will also discuss applications concerned with *quantum laser pointing* and the optical read-out of quantum information.

4.2 Quantum Noise in an Arraylike Detection

Let us consider a beam of monochromatic light of frequency ω_0 that is propagating along the z-direction. By using the paraxial and slowly varying envelope approximations, the positive-frequency electric field operator can be expressed as

$$\hat{E}^{(+)}(\boldsymbol{r}, z) = \sqrt{\frac{\hbar\omega_0}{2\epsilon_0 cT}}\hat{A}(\boldsymbol{r}, z), \tag{4.1}$$

where $\boldsymbol{r} = (x, y)$ is the coordinate in the transverse plane, T is the integration time of the detector, and

$$\hat{A}(\boldsymbol{r}, z) = \sum_i \hat{a}_i(z) u_i(\boldsymbol{r}, z). \tag{4.2}$$

In the above expression, an orthonormal transverse mode basis $\{u_i(\boldsymbol{r}, z)\}$ is introduced and $\hat{a}_i(z)$ is the annihilation operator of a photon in the mode $u_i(\boldsymbol{r}, z)$. The field of each mode has to satisfy the free propagation equation in the vacuum [4]. We will henceforth omit the z-dependence of the field mode in the following analysis.

Let us now define what we call a measurement. We consider a beam incident on an array detector, with each pixel of this detector occupying a transverse area D_i. The different pixels do not overlap each other and all light is incident on the detector. Using Eq. (4.2) one finds that the number of photons (and for a detector of unity quantum efficiency the number of electrons) accumulated on each pixel during the integration time T is given by

$$\hat{N}(D_i) = \int_{D_i} \hat{A}^\dagger(\boldsymbol{r})\hat{A}(\boldsymbol{r}) d^2 r. \tag{4.3}$$

Extracting information from an array detector consists of determining, by analogue or digital means, a linear combination of the signals delivered by all pixels. We model the image-processing protocol by introducing a gain σ_i, which can be positive or negative, for each pixel. The resulting signal is then represented by the operator

$$\hat{N}_\sigma = \sum_i \sigma_i \hat{N}(D_i), \tag{4.4}$$

where $\hat{N}_\sigma$ is a quantity that depends both on the image properties and the choice of the gains. The mean value of $\hat{N}_\sigma$ is proportional to the quantity to be measured:the signal. On can show that, as we are performing a single measurement, the modification of the image to which it is sensitive, can be described by a single parameter p (for instance, the value of the displacement in the case of a displacement measurement). We shall call p the *signal* and its variance the *noise* of the measurement. In many practical instances we may wish to perform a *differential measurement* where p is zero (i.e., where the mean value of the signal is zero). In these situations, any common-mode fluctuations, for example, fluctuations in the total intensity of light, will not affect the measurement of small variations of p around zero. The present analysis, however, can be adapted to the cases where the signal is nonzero.

In order to simplify the analysis, we will choose the first mode $u_0(\boldsymbol{r})$ of the transverse mode basis to match the mean electric field transverse distribution when $p = 0$, so that $u_0(\boldsymbol{r}) = 1/\sqrt{N_0}, \langle \hat{A}(\boldsymbol{r})\rangle$, where N_0 is the mean total number of photons in the beam. This implies that the field of all the other modes has a zero mean value, whereas it can still have a variance different from the vacuum fluctuations in the case of a nonclassical mode. The variance of the measurement can then be calculated exactly and one finds that [4]

$$\langle \Delta \hat{N}_\sigma^2 \rangle = f^2 N_0 \langle \Delta \hat{X}_w^{+2} \rangle, \tag{4.5}$$

where f is a normalization factor given by$f^2 = \sum_i \sigma_i^2 \int_{D_i} u_0^*(\boldsymbol{r}) u_0(\boldsymbol{r}) d^2 r$, and $\hat{X}_w^+$ is the quadrature amplitude operator of the transverse detection mode, $w(\boldsymbol{r})$, defined by

$$\forall\, \boldsymbol{r},\ \boldsymbol{r} \in D_i\ \Rightarrow\ w(\boldsymbol{r}) = \frac{1}{f} \sigma_i u_0(\boldsymbol{r}). \tag{4.6}$$

This mode is the mean-field mode u_0 weighted by the gains. For differential measurement (i.e., when $N_\sigma(p = 0) = 0$), the detection mode is orthogonal to u_0. Equation (4.6) deserves some comments: in the case of a coherent field, with u_0 carrying the mean field, all other modes are in the vacuum state. In particular the variance $\langle (\Delta \hat{X}_w^{+2}) \rangle$ is equal to 1. We find that the noise in a measurement performed with a coherent field is proportional to the mean number of photons. Furthermore, we find that in order to reduce this noise, the only possibility is to act on the detection mode. Populating this mode with squeezed vacuum, for example, will reduce the noise of the measurement.

Finally, we note that it is possible to perform simultaneously several measurements of the kind described by Eq. (4.4). Provided that these measurements are "orthogonal" to each other (i.e., lead to linearly independent quantities of N_σ), it is possible to have all these measurement noises be below the standard quantum limit. For each of them we can define the detection modes $w_1, w_2, \ldots$. The noise variances of these signals are thus proportional to the quadrature-amplitude noise in each of the detection modes. To reduce them simultaneously, one needs to fill all these modes with vacuum squeezing in the appropriate quadrature. Therefore, the superposition of N well-defined multimode squeezed states and one coherent state is sufficient to simultaneously improve the measurement of N parameters beyond the standard quantum noise limit. In the next section, we will show how this can be done experimentally.

4.3 Implementing a Sub-Shot-Noise Array Detection

We have shown that the result of a measurement performed on an image with an array detector is affected by the noise whose origin is perfectly identifiable. This noise arises from quantum fluctuations of the detection mode, a transverse mode that is dependent on the image and on the detectors. In order to reduce this noise, it is necessary to use squeezed light with the proper mode shape. To perform such an experiment, one has to interfere with the mean field by squeezed vacuum states in the proper modes with high efficiencies. The light is then incident on a sample under investigation. This produces some variation of the beam parameter(s). Finally, the beam is detected by an array detector with the proper gains.

Some key ingredients are required in order to perform this kind of experiment. First, single-mode vacuum squeezed states from, for example, optical parametric amplifiers operating close to the parametric oscillator threshold are required in order to produce strong and stable squeezing. Second, these squeezed vacuum beams have to be produced or manipulated into the proper transverse mode efficiently. One could think of directly producing the squeezed state using an appropriate cavity. The fragility of squeezed light, however, makes this technique an experimental challenge. The easiest way is to find an efficient method to transform TEM_{00} squeezed beams into the required transverse mode. This can be achieved with many devices, such as a spatial light modulator, holographic masks, or arrays of MEMS (Micro-Electro-Mechanical Systems). The technique used in the experiment presented in this chapter is based on specially fabricated optical half-wave plates that introduce different phase shifts on different parts of a light beam. Finally, the last requisite to our experiment is a method to interfere with and combine all the squeezed beams and the mean field efficiently. Given two optical beams with transverse shapes w_1 and w_2 that are orthogonal, the practical problem is to coherently mix and co-propagate them with minimal losses.

In this chapter, we introduce two transverse mode combination techniques. The first technique depends on the symmetry properties of the constituent modes. Assume that there exists a transverse axis such that, respective to it, w_1 is an even mode and w_2 is odd. An apparatus to mix these two modes can then be a modified Mach–Zehnder interferometer. The modification from the usual Mach–Zehnder interferometer needed, is to make one arm with an even number of reflections, and the other with an odd number of reflections. For any even mode the effect of a reflection is null, whereas for any odd mode each reflection is equivalent to a π phase shift. When the two modes are injected into both input ports of the beamsplitter, the interferometer path-length difference can be adjusted to make both modes exit through the same output port (see Fig. 4.1). However, it is not always possible to

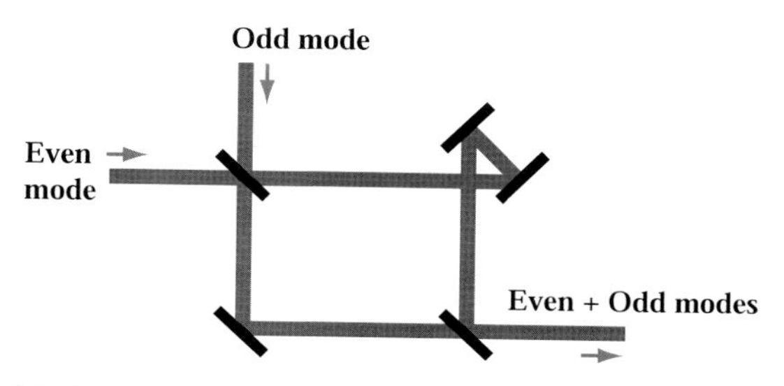

Fig. 4.1. Mode mixing with a modified Mach–Zehnder interferometer.

use such an apparatus and a more general solution is required. Our second technique uses an impedance-matched optical cavity whose resonances are transverse-mode selective. Having a cavity resonant for an arbitrary transverse would be a difficult, but not impossible, task. For a concise discussion, let us assume that the detection mode w_1 is the fundamental TEM_{00} mode and that the cavity has higher-order transverse mode resonances, which are well separated in the resonance frequencies. Let us inject the squeezed TEM_{00} vacuum and adjust the cavity length such that the TEM_{00} mode is in resonance. The squeezed field will then be fully transmitted (at least in the perfect case) through the cavity. All modes w_2 orthogonal to the TEM_{00} mode, on the other hand, will be reflected. A squeezed field in mode w_2 which is incident on the output mirror of the cavity will also be perfectly reflected, and therefore perfectly mixed with the transmitted TEM_{00}. The detailed scheme, in the general case of two orthogonal modes w_1 and w_2, is shown in Fig. 4.2.

4.4 The Quantum Laser Pointer

The first example of applications of the modal synthesis quantum imaging is the so-called *quantum laser pointer*. We consider a laser beam incident on a quadrant detector. This detector allows the measurement of the position of the laser beam in two transverse dimensions with very high accuracies, as shown in Fig. 4.3. For instance, the difference between the sums of two quadrants $a+b$ and $c+d$ gives a signal proportional to the horizontal displacement of the beam. With $\hat{N}_a$, $\hat{N}_b$, $\hat{N}_c$, and $\hat{N}_d$ being the photon numbers delivered by the four quadrants, the horizontal signal is given by $\hat{N} = \hat{N}_a + \hat{N}_b - \hat{N}_c - \hat{N}_d$, which is of the form of Eq.(4.4) with the gain values of ± 1. The detection mode, given by Eq.(4.6) is then reduced to a flipped version of the mean-field mode (due to the minus sign) as shown in Fig. 4.3. This figure represents the

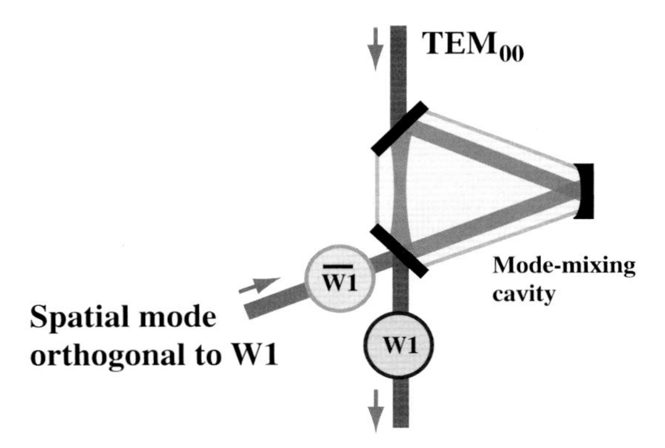

Fig. 4.2. Mixing two modes with an impedance-matched optical cavity. $W1$ is a waveplate that transfer the TEM_{00} into mode w_1 and $\overline{W1}$ does the opposite.

mean-field mode as a TEM_{00} mode, and the detection modes for both measurements are the flipped version of this TEM_{00}. In our experiment [5], we take a laser beam with a Gaussian mode shape and independently prepare two squeezed vacuum beams. These two beams are converted into flipped modes by means of special waveplates. The beams are then coherently mixed using a ring cavity, as described in the previous section. Because small displacement of a light beam can be overwhelmed by classical fluctuations due to vibrations and the air-index fluctuations, the very small amplitude signal (smaller than a nanometer) was chosen to be an oscillation of the beam measured at 4 MHz. This displacement modulation was induced by a mirror mounted on a piezo-electric actuator. Two sets of measurements have been

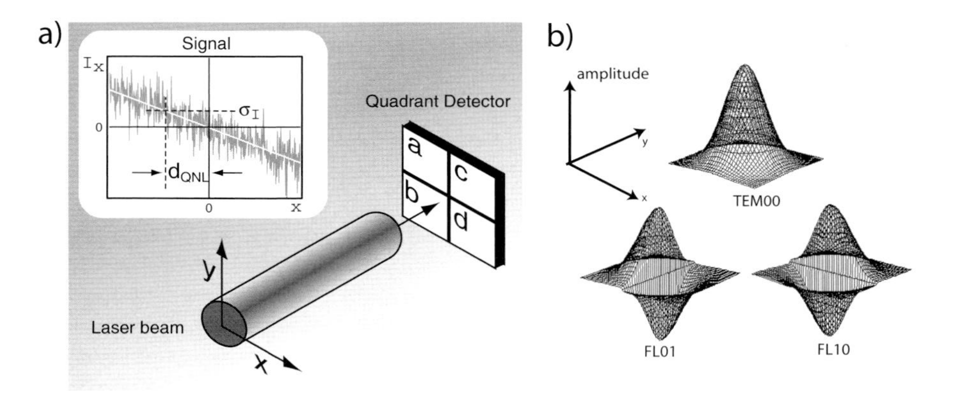

Fig. 4.3. Experimental scheme of the quantum laser pointer.

performed. First, we looked at the fluctuations of the position at the detection frequency of this beam induced by the quantum nature of light. This is displayed in Fig. 4.4. In this graph, each point can be interpreted as an instantaneous measurement of the oscillation amplitude in two dimensions [6]. This plot was recorded with a constant displacement and the fluctuations statistics form a Gaussian distribution around the means value. The left graph shows the two-dimensional noise of a coherent laser beam, whereas the right one shows the fluctuations of the nonclassical multimode beam. Clearly, the pointing noise is reduced in both directions, and we could say that the nonclassical beam propagates in a straighter line than a coherent state beam. Such a spatial noise reduction allows us to improve the sensitivity of the

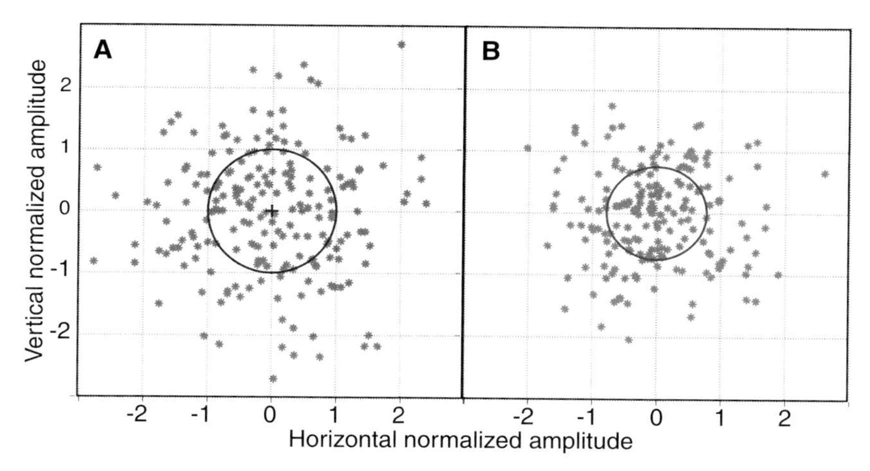

Fig. 4.4. Localization of the quantum laser pointer.

apparatus. In Fig. 4.5 we have introduced a displacement modulation, and increased the amplitude of this oscillation linearly, from 0 to a few angstroms. The results show the signal measured by the quadrant photodetector plotted against the modulation amplitude. We can see that it is not possible to make a distinction between the signal and the noise in the case of a coherent laser beam with small displacement modulations. At larger oscillations the signal and the noise can be distinguished. When squeezed light is used, the signal for a small displacement modulation emerges earlier from the noise floor, thus demonstrating the improvement of the sensitivity of the measurement.

4.5 Optical Read-Out

The second example that we shall consider as an illustration of the theory of the modal synthesis quantum imaging is the issue of the optical read-out.

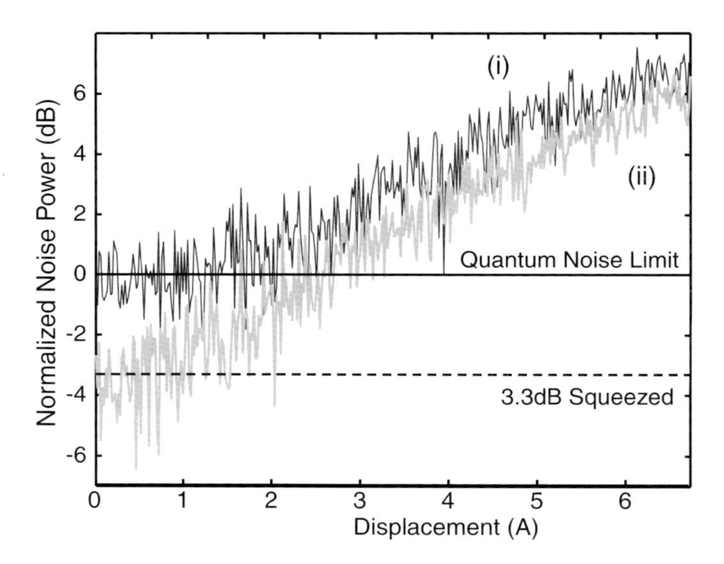

Fig. 4.5. Sensitivity to small displacement measurement is improved with spatial squeezing.

This section is purely theoretical with the aim to demonstrate that it is possible, using mode synthesis, to improve the density of information that can be read by a laser beam. Let us consider as a model system the read-out of information performed on a compact disc. On such a system, information is encoded on the disc by a surface change that induces a π phase shift on the reflected light. The reflected field can look like the flipped mode, introduced in the previous section. Having high spatial frequency components, flipped mode diffracts more than the TEM_{00} mode. A single-pixel detector, placed in the far field of the disk surface, will measure only the central part of the image. The record intensity difference between a flat surface and a surface edge is then used to represent "0" and "1". Only one bit of information can be encoded in a focal area of the laser beam. The density of information is then limited by the wavelength of the light. This is why the CD manufacturers are developing devices that operate with shorter and shorter wavelengths (400 nm for the next generation of CDs). It will be more and more difficult, however, to generate light at the UV region. Alternative ways of storing more than one bit of information per focal area must therefore be seriously investigated.

Having this goal in mind, we propose to extend the use of the super-resolution techniques [7, 8]. Instead of detecting the reflected light with a single-pixel detector, one can use an array detector. In super-resolution methods, having a high number of pixels is very important for fine reconstruction of the field shape. However, in the present case, we know that there are only 2^n possibilities for the shape of the reflected light if we place n bits of information within a focal spot. This matches the assumption of the modal synthesis, where considerable a priori information is available. Hence, it is sufficient to use a finite number of pixels to record the field in order to differentiate among all these possibilities. A sketch of the proposed scheme is

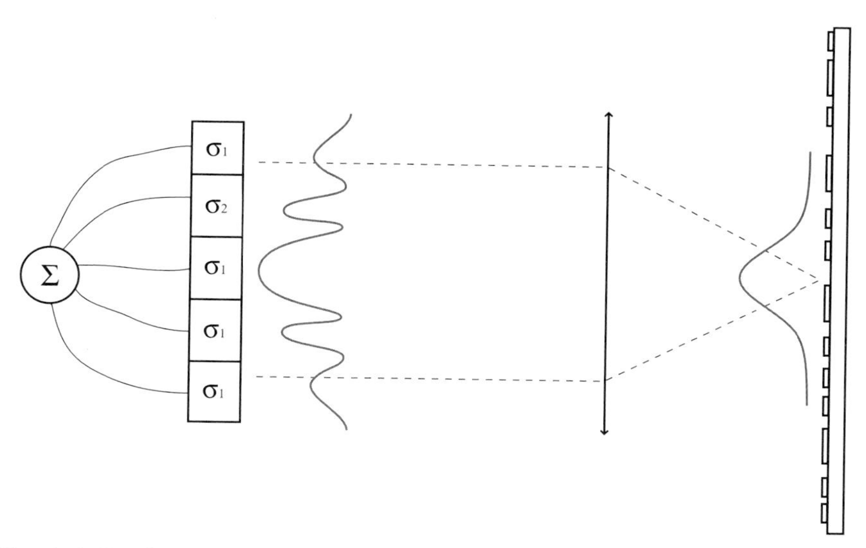

Fig. 4.6. Read-out of a Gaussian beam after interacting with the CD surface and going through the aperture with an array detector.

given in Fig. 4.6. We consider a standard optical disc read-out system, where the input beam is strongly focused on the disc with a high numerical aperture. Several bits of information are placed on the disc within the focal spot. Hence, a complex phase pattern is imprinted onto the beam when it is reflected by the disc. The light is then propagated onto the detector. Due to the high numerical aperture and the small details, the calculation of the field shape after propagation to the detector plane cannot be performed within the paraxial approximation and more sophisticated integration techniques have to be used [9]. Indeed, if we continue with the example of three bits of information per focal point, there are only eight possible field shapes incident on the detector. One then has to design an array detector that allows us, by using appropriate sets of gains on the different pixels, to differentiate efficiently among these eight possibilities. The figure arbitrarily displays a five-pixel detector only as an example.

Each time a measurement is performed, one has to test, in parallel, the eight possible sets of gains that correspond to each sequence of bits. The set that gives the results closest to zero, corresponds to the preferred sequence. However, such a reconstruction technique is very sensitive to noise due to the spatial frequency bandwidth of the optical apparatus. This practical situation is within the framework of the theory presented above. To each of the possible measurements there corresponds a detection mode that is responsible for the noise. This mode is well defined because both the mean field shape and the detector gains are known. Thus it is possible to improve the signal-to-noise ratio of the detection process by injecting squeezed vacuum light with the proper mode shape. Such a system would yield an increase of the density of information encoded on an optical disc. We note that this approach can potentially improve any storage technology, because it is independent of the technical parameters such as wavelength, numerical aperture, and the like.

4.6 Measuring a Signal in an Optimal Way

In this last section, we would like to complete the optical read-out analysis by determining the optimum detection system for a given measurement. So far we have restricted ourselves to an array detector that only measures local intensities in an optical image. In some cases, this may not be the best detection strategy. We should then identify where the information comes from and optimize the detection system accordingly. Let us consider again the small displacement problem. Assume that a beam of light, with Gaussian transverse shape $E(\boldsymbol{r}) = \alpha u_0(\boldsymbol{r})$, is displaced by a distance d along the x-direction. The shape of the displaced beam can be written, to the first order in d, as

$$\begin{aligned} E_d(\boldsymbol{r}) &= E(\boldsymbol{r}) + d\frac{\partial E(\boldsymbol{r})}{\partial x} \\ &= \alpha\left(u_0(\boldsymbol{r}) + \frac{d}{w_0}u_1(\boldsymbol{r})\right), \end{aligned} \tag{4.7}$$

where u_1 is the TEM_{01} mode. This equation is valid because the TEM_{01} mode is proportional to the first derivative of the TEM_{00} mode. Consequently, one sees that on a Gaussian beam the size of the displacement is characterized by the amount of TEM_{01} mode in the displaced image. In order to optimally measure a displacement one must therefore measure the projection of the image onto this mode. Using a homodyne detection whose local oscillator is a TEM_{01} mode, we measure the fluctuations of the TEM_{01} mode of the signal beam. One can show that this system is 100% efficient for detecting displacement, whereas the quadrant detector system is only 80% efficient [10]. In Fig. 4.7, we present the results of such an experiment. As in the previous experiment, a Gaussian beam of light is mixed with negligible losses with a TEM_{01} squeezed vacuum detection mode by means of a modified Mach–Zehnder interferometer. A modulation is induced on the beam by means of a mirror mounted on a piezo-electric actuator. The beam is then detected using the previously described spatial homodyne detector. The figure plots the recorded signal versus the phase of the local oscillator. The blue curve

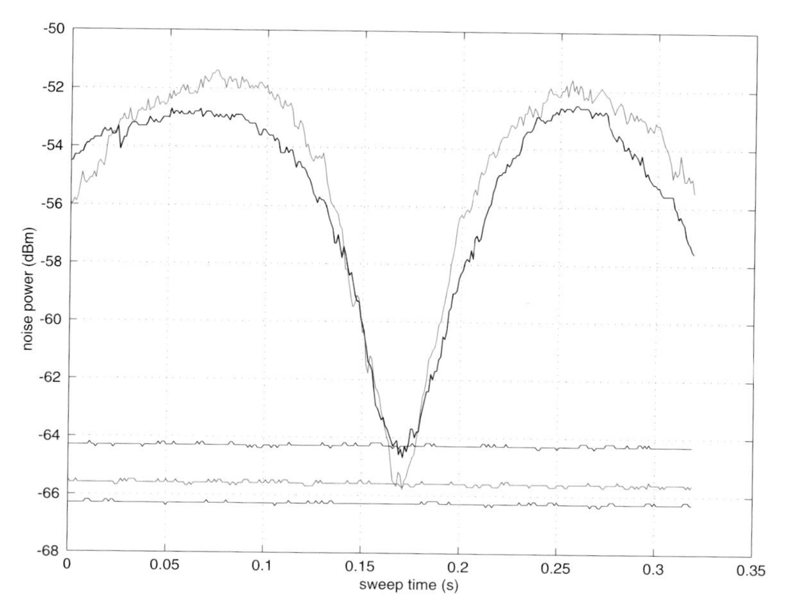

Fig. 4.7. Spatial homodyne to measure a small displacement with squeezed TEM_{01} light.

is obtained with a coherent detection state. The displacement signal appears when the relative phase at the homodyne wave plate is zero, corresponding to the dip of the blue curve where there is no observable signal. The maximum of the blue curve, at $\pi/2$ phase shift, can be shown to correspond to the amplitude of the tilt induced onto the beam [11]. Indeed, one can show that the tilt and the displacement are a pair of the conjugated variables.

For the displacement measurement, one can study the red curve that corresponds to a beam mixed with a squeezed vacuum detection mode. The displacement signal dip is lower than in the previous case. However, this dip does not reach the quantum noise floor suggesting that there is a displacement signal that could not be measured with a coherent beam. Hence this graph has demonstrated the method and its efficiency at detecting very week displacement modulation amplitudes.

4.7 Conclusion

In this chapter we have shown that multimode quantum light can be produced by the synthesis of many single-mode nonclassical beams. The modal synthesis of images can be applied to problems where considerable a priori information is available, thus allowing the improvement of the detection sensitivity below the quantum noise limit.

We have shown that techniques in quantum imaging can be applied to improve the optical read-out, either using array detectors or detection schemes, such as the homodyne detector, fitted to the parameters to be measured. Using this last method we can also access the conjugate variable of the detected parameter, which then leads to the possibility of spatial entanglement.

Finally, let us mention that this work, developed within the frame of the quantum imaging formalism, can also be applied to many other systems. As soon as several modes of some physical parameters—spatial, temporal, frequency, polarization, or others—are present, quantum modal synthesis is applicable. Hence, any measurement that is multimode and is limited by quantum noise can have its sensitivity improved using the techniques derived from this chapter.

References

1. M. I. Kolobov, Rev. Mod. Phys. **71**, 1539 (1999).
2. L. A. Lugiato and P. Grangier, J. Opt. Soc. Am. B **14**, 225 (1997).
3. M. Martinelli, N. Treps, S. Ducci, A. Maître, and C. Fabre, Phys. Rev. A **67**, 023808 (2003).
4. N. Treps, V. Delaubert, A. Maître, J.-M. Courty, and C. Fabre, Phys. Rev. A **71**, 013820 (2005).
5. N. Treps, N. Grosse, W. Bowen, C. Fabre, H. A. Bachor, and P. K. Lam, Science **301**, 940 (2003).
6. N. Treps, N. Grosse, W. P. Bowen, M. T. L. Hsu, A. Maître, C. Fabre, H. A. Bachor, and P. K. Lam, J. Opt. B **6**, 664 (2004).
7. M. I. Kolobov and C. Fabre, Phys. Rev. Lett. **85**, 3789 (2000).
8. M. Bertero and E. R. Pike, Opt. Acta, **29**, 727 (1982).
9. B. Richards and E. Wolf, Proc. R. Soc. A **253**, 359 (1959).
10. M. T. L. Hsu, V. Delaubert, P. K. Lam, and W. P. Bowen, J. Opt. B: Q. Semiclass. Opt. **6**, 495 (2004).
11. M. T. L. Hsu, W. P. Bowen, N. Treps, and P. K. Lam, quant-phys/0501144 (2005).

5 Ghost Imaging

Alessandra Gatti[1], Enrico Brambilla[1], Morten Bache[2], and Luigi A. Lugiato[1]

[1] INFM, Dipartimento di Fisica e Matematica, Universitá dell'Insubria, Via Valleggio 11, 22100, Como, Italy
luigi.lugiato@uninsubria.it
[2] Optical Fiber Group, Research Center COM, Technical University of Denmark, DK-2800 Lyngby, Denmark

5.1 Introduction

The topic of *ghost imaging* (GI) has attracted noteworthy attention in recent years [1–26]. Invented by Klyshko many years ago [1] with the idea of exploiting the quantum entanglement in photon pairs generated by Parametric Down-Conversion (PDC), this technique was also called entangled (two-photon) imaging until recently [1–11]. It is by now clear that appropriate classically correlated beams also can be used to implement such a technique [12–25]; the interesting relation between the two kinds of approaches will be discussed in the last two sections of this chapter.

In a standard imaging configuration one has a source that illuminates the object, an imaging system, and a detection system. In GI, instead, one exploits the correlation between two beams to retrieve information about an unknown object. Let us describe this technique in the case of entangled photon pairs as originally conceived by Klyshko [1] and later systematized in [5–7]. The photons of a pair are spatially separated and each propagates through a distinct imaging system, usually called the test and the reference arms (Fig. 5.1a). Information is not obtained by direct measurement of photon 1, because, for example, detector D_1 is pointlike and is held fixed (i.e., it is not scanned in the transverse plane), or D_1 is a "bucket" detector that measures total intensity of beam 1 and is therefore unable to reveal the transverse position of photon 1. Information is retrieved, instead, from the coincidences of signal–idler photon pairs as a function of the transverse position of the photon 2, because detector 2 is pointlike and is scanned in the transverse plane. The name "ghost imaging" just originates from the fact that the result is obtained by scanning the position of the photon which never passed through the object.

By changing the optical elements in the two arms, one can obtain a different kind of information about the object, for example., the intensity distribution of the object (ghost image) or the modulus square of the Fourier transform of the object (ghost diffraction).

The GI technique can find its applications in such situations where it is not easy to act on the test arm and/or to locate in that arm an array

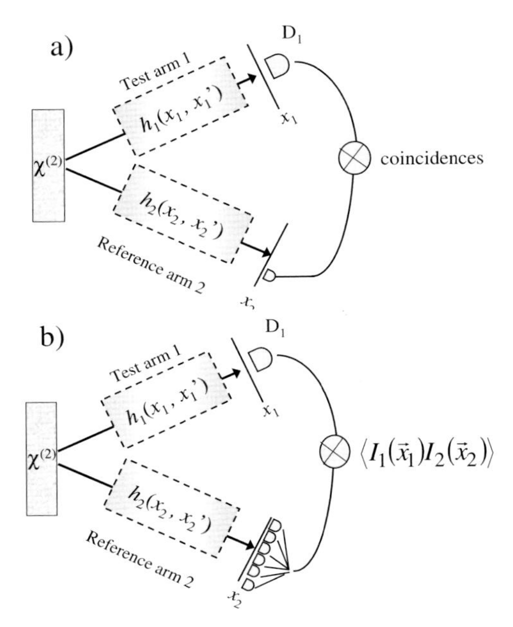

Fig. 5.1. Ghost imaging with entangled photon pairs. The object, about which information must be retrieved, is located in test arm 1. The functions h_1 and h_2 are the impulse response functions that take into account the propagation of light, the presence of optical elements (e.g., lenses, apertures), and the object. (a) The case of extremely low gain, in which one detects coincidences between the signal and the idler photons. The information is obtained by scanning the pointlike detector D_2 in the transverse plane $\boldsymbol{x}_2$ of the reference arm, whereas D_1 is fixed. (b) The case of high gain. The information is obtained from the intensity correlation function $\langle \hat{I}_1(\boldsymbol{x}_1)\hat{I}_2(\boldsymbol{x}_2)\rangle$ as a function of $\boldsymbol{x}_2$ by keeping D_1 fixed and using an array of pointlike detectors in the reference arm.

of pointlike detectors (e.g., when we wish to keep the detection hidden), hence we wish to keep the configuration of the optical elements and of the detection system in the test arm as simple as possible and fixed once forever, whereas we have full ease of acting on the reference arm (e.g., of scanning detector D_2 or, equivalently, of locating an array of pointlike detectors in arm 2, or of modifying arm 2 to pass from a ghost image to a ghost diffraction configuration).

Such a two-arm configuration provides more flexibility in comparison with standard imaging procedures. For example, there is a possibility of illuminating the object at a given light frequency in the test arm and of performing a spatially resolved detection in the other arm with a different light frequency, or of processing the information from the object by only operating on the imaging system of the reference arm [11]. In addition, it opens a possibility for performing coherent imaging by using, in a sense, spatially incoherent light, because each of the two down-converted beams taken separately is described by a thermallike mixture and only the two-beam state is pure (see Section 2.2.2 of Chapter 2).

In the high-gain regime of a large number of photon pairs, which has been already considered in Chapter 2, this technique is generalized to the measurement of the signal–idler correlation function of the intensity fluctuations [8,9,16,17] as a function of position $\boldsymbol{x}_2$ in the reference arm (Fig. 5.1b). The general theory, spanning from the extremely low-gain regime in which one detects coincidences to the high-gain regime, will be described in Section 5.2. The other sections, in which we discuss GI with entangled beams, are Section 5.3 (wave-particle aspects associated with ghost imaging), Section 5.4 (spatial averaging technique that enhances the speed of retrieval and increases the spatial bandwidth of the system in ghost diffraction), and Section 5.5 (ghost imaging with homodyne detection instead of direct intensity detection).

The "quantum" and the "classical" parts of this chapter are divided in Section 5.6, which illustrates the debate on whether quantum entanglement is necessary in GI. In the last two sections, we discuss the case of GI obtained by using splitt thermallike beams, first from a theoretical viewpoint (Section 5.7) and, next from an experimental standpoint (Section 5.8).

The use of thermallike light in GI schemes also inspired a topic that became of a certain interest, known as "quantum lithography with classical beams," or "sub-wavelength interference with classical beams." The quantum version of this started with a paper by Boto et al. [27] claiming that N-photon entangled states could be used for improving the resolution of lithography by a factor of N. A proof-of-principle experiment using $N = 2$ in the PDC case was provided by [28] where a halving of the period of the interference fringes was observed in a "ghost diffraction" pattern. In [8] we observed that the same effect may be observed when thermallike beams are used, and that in both the entangled and the thermal case the subwavelength interference relies on a simple geometrical artifact. We therefore questioned whether the experiment [28] really proves the entangled protocol of [27]. Subwavelength interference using thermal beams was then theoretically discussed in [29], and experimentally demonstrated in [30,31].

5.2 General Theory of Ghost Imaging with Entangled Beams

In the analytical treatment of this section we consider for simplicity only spatial variables and ignore the time argument, which corresponds to using a narrow frequency filter. A theory that includes time and frequency can be found in [9]. In addition, we assume translational invariance in the transverse plane, which amounts to requiring that the cross-section of the source is much larger than the object and all the optical elements.

In the entangled case, the signal and the idler fields are generated in a type-II $\chi^{(2)}$ crystal by a PDC process. Our starting point is the input-output relations of the crystal, which in the plane-wave pump approximation reads (see [8, 9] and references quoted therein)

$$\hat{a}_{i,\mathrm{out}}(\boldsymbol{q}) = U_i(\boldsymbol{q})\hat{a}_{i,\mathrm{in}}(\boldsymbol{q}) + V_i(\boldsymbol{q})\hat{a}^{\dagger}_{j,\mathrm{in}}(-\boldsymbol{q}), \qquad i \neq j = 1,2\,, \tag{5.1}$$

which coincides with Eqs. (2.13) in Chapter 2 provided $U_i(\boldsymbol{q})$ and $V_i(\boldsymbol{q})$ are identified with $U_i(\boldsymbol{q}, \Omega = 0)$ and $V_i(\boldsymbol{q}, \Omega = 0)$ in Eqs. (2.14–2.17) of Chapter 2. As usual in type-II media, the signal field $\hat{a}_1$ and the idler field $\hat{a}_2$ are distinguished by their orthogonal polarization.

Each of the two outgoing beams travels through a distinct imaging system, described by its impulse response functions $h_1(\boldsymbol{x}_1, \boldsymbol{x}'_1)$ and $h_2(\boldsymbol{x}_2, \boldsymbol{x}'_2)$, respectively (see Fig. 5.1). The fields at the detection planes are given by

$$\hat{c}_i(\boldsymbol{x}_i) = \int \mathrm{d}\boldsymbol{x}'_i h_i(\boldsymbol{x}_i, \boldsymbol{x}_i{}')\hat{a}_{i,\mathrm{out}}(\boldsymbol{x}_i{}') + \hat{L}_i(\boldsymbol{x}_i), \qquad i = 1,2\,, \tag{5.2}$$

where $\hat{L}_1, \hat{L}_2$ account for possible losses in the imaging systems, and depend on vacuum field operators uncorrelated from $\hat{a}_{i,\mathrm{out}}$. Information about the object is extracted by measuring the spatial correlation function of the intensities detected by D_1 and D_2, as a function of the position $\boldsymbol{x}_2$ of the pixel of D_2:

$$\langle \hat{I}_1(\boldsymbol{x}_1)\hat{I}_2(\boldsymbol{x}_2)\rangle = \langle \hat{c}^{\dagger}_1(\boldsymbol{x}_1)\hat{c}_1(\boldsymbol{x}_1)\hat{c}^{\dagger}_2(\boldsymbol{x}_2)\hat{c}_2(\boldsymbol{x}_2)\rangle\,. \tag{5.3}$$

All the object information is concentrated in the correlation function of intensity fluctuations:

$$G(\boldsymbol{x}_1, \boldsymbol{x}_2) = \langle \hat{I}_1(\boldsymbol{x}_1)\hat{I}_2(\boldsymbol{x}_2)\rangle - \langle \hat{I}_1(\boldsymbol{x}_1)\rangle\langle \hat{I}_2(\boldsymbol{x}_2)\rangle\,, \tag{5.4}$$

where $\langle \hat{I}_i(\boldsymbol{x}_i)\rangle = \langle \hat{c}^{\dagger}_i(\boldsymbol{x}_i)\hat{c}_i(\boldsymbol{x}_i)\rangle$ is the mean intensity of the ith beam. When using a bucket detector in arm 1, the measured quantity corresponds to the integral over $\boldsymbol{x}_1$ of both sides of Eq. (5.4). Because $\hat{c}_1$ and $\hat{c}^{\dagger}_2$ commute, all the terms in Eqs. (5.3), (5.4) are normally ordered and $\hat{L}_1, \hat{L}_2$ can be neglected, thus obtaining

$$\begin{aligned} G(\boldsymbol{x}_1, \boldsymbol{x}_2) = & \int \mathrm{d}\boldsymbol{x}'_1 \int \mathrm{d}\boldsymbol{x}''_1 \int \mathrm{d}\boldsymbol{x}'_2 \int \mathrm{d}\boldsymbol{x}''_2\; h^*_1(\boldsymbol{x}_1, \boldsymbol{x}''_1)h_1(\boldsymbol{x}_1, \boldsymbol{x}'_1) \\ \times\; & h^*_2(\boldsymbol{x}_2, \boldsymbol{x}''_2)h_2(\boldsymbol{x}_2, \boldsymbol{x}'_2)\Big[\langle \hat{a}^{\dagger}_{1\mathrm{out}}(\boldsymbol{x}''_1)\hat{a}_{1\mathrm{out}}(\boldsymbol{x}'_1)\hat{a}^{\dagger}_{2\mathrm{out}}(\boldsymbol{x}''_2)\hat{a}_{2\mathrm{out}}(\boldsymbol{x}'_2)\rangle \\ -\; & \langle \hat{a}^{\dagger}_{1\mathrm{out}}(\boldsymbol{x}''_1)\hat{a}_{1\mathrm{out}}(\boldsymbol{x}'_1)\rangle\, \langle \hat{a}^{\dagger}_{2\mathrm{out}}(\boldsymbol{x}''_2)\hat{a}_{2\mathrm{out}}(\boldsymbol{x}'_2)\rangle\Big]\,. \end{aligned} \tag{5.5}$$

The four-point correlation function in Eq. (5.5) has special factorization properties. As can be obtained from Eq. (5.1) (see also [1] in Chapter 2),

$$\begin{aligned} & \langle \hat{a}^{\dagger}_{1\mathrm{out}}(\boldsymbol{x}''_1)\hat{a}_{1\mathrm{out}}(\boldsymbol{x}'_1)\hat{a}^{\dagger}_{2\mathrm{out}}(\boldsymbol{x}''_2)\hat{a}_{2\mathrm{out}}(\boldsymbol{x}'_2)\rangle \\ =\; & \langle \hat{a}^{\dagger}_{1\mathrm{out}}(\boldsymbol{x}''_1)\hat{a}_{1\mathrm{out}}(\boldsymbol{x}'_1)\rangle\, \langle \hat{a}^{\dagger}_{2\mathrm{out}}(\boldsymbol{x}''_2)\hat{a}_{2\mathrm{out}}(\boldsymbol{x}'_2)\rangle \\ +\; & \langle \hat{a}^{\dagger}_{1\mathrm{out}}(\boldsymbol{x}''_1)\hat{a}^{\dagger}_{2\mathrm{out}}(\boldsymbol{x}''_2)\rangle\, \langle \hat{a}_{1\mathrm{out}}(\boldsymbol{x}'_1)\hat{a}_{2\mathrm{out}}(\boldsymbol{x}'_2)\rangle\,. \end{aligned} \tag{5.6}$$

By inserting this result in Eq. (5.5), one obtains

$$G_{PDC}(\boldsymbol{x}_1, \boldsymbol{x}_2) = \left| \int \mathrm{d}\boldsymbol{x}_1' \int \mathrm{d}\boldsymbol{x}_2' \, h_1(\boldsymbol{x}_1, \boldsymbol{x}_1') h_2(\boldsymbol{x}_2, \boldsymbol{x}_2') \langle \hat{a}_{1\mathrm{out}}(\boldsymbol{x}_1') \hat{a}_{2\mathrm{out}}(\boldsymbol{x}_2') \rangle \right|^2 , \tag{5.7}$$

where by using relations (5.1) and taking into account that the fields $\hat{a}_{i,\mathrm{in}}$ are in the vacuum state,

$$\langle \hat{a}_{1\mathrm{out}}(\boldsymbol{x}_1') \hat{a}_{2\mathrm{out}}(\boldsymbol{x}_2') \rangle = \int \frac{\mathrm{d}\boldsymbol{q}}{(2\pi)^2} \mathrm{e}^{\mathrm{i}\boldsymbol{q}\cdot(\boldsymbol{x}_1' - \boldsymbol{x}_2')} U_1(\boldsymbol{q}) V_2(-\boldsymbol{q}) \, . \tag{5.8}$$

The correlation length, or transverse coherence length x_{coh}, is determined by the inverse of the bandwidth of the function $U_1(\boldsymbol{q})V_2(-\boldsymbol{q})$ (see Eq. (1.23) in Chapter 1). An essential feature is that in Eq. (5.7) the modulus is outside the integral, which ensures the possibility of coherent imaging via correlation measurement.

5.2.1 Specific Imaging Schemes

There is an infinity of choices for the configuration of the optical elements in the test and reference arms, that is, for the functions h_1 and h_2.

Let us now analyze two paradigmatic examples of imaging systems sketched in Fig. 5.2. In both examples the set-up of arm 1 is fixed, and con-

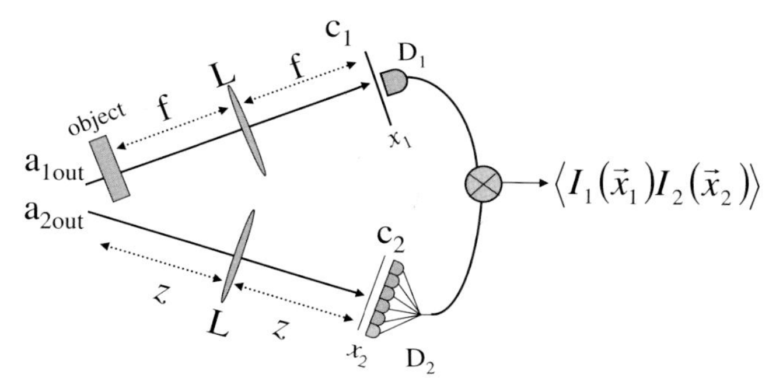

Fig. 5.2. Imaging scheme: L denotes two identical lenses of focal length f; D_1 is a pointlike detector. The distance z is either $z = f$ or $z = 2f$.

sists of an object, described by a complex transmission function $T(\boldsymbol{x})$, and a lens located at the focal distance f from the object and from the detection plane. Hence,

$$h_1(\boldsymbol{x}_1, \boldsymbol{x}_1') = -\frac{\mathrm{i}}{\lambda f} \exp\!\left(-\frac{2\pi\mathrm{i}}{\lambda f} \boldsymbol{x}_1 \cdot \boldsymbol{x}_1' \right) T(\boldsymbol{x}_1') \, , \tag{5.9}$$

with λ being the wavelength. In arm 2 there is a single lens placed at a distance z both from the source and from detection plane 2; for simplicity we take the two lenses to be identical.

Example 1: Ghost Diffraction Scheme

In the first example we assume $z = f$ so that

$$h_2(\boldsymbol{x}_2, \boldsymbol{x}'_2) = -\frac{i}{\lambda f} \exp\!\left(-\frac{2\pi i}{\lambda f}\boldsymbol{x}_2 \cdot \boldsymbol{x}'_2\right).$$

By inserting these propagators into Eq. (5.7) and taking into account Eq. (5.9), we obtain

$$G_{PDC}(\boldsymbol{x}_1, \boldsymbol{x}_2) \propto \left| U_1(-\boldsymbol{x}_2 \frac{2\pi}{\lambda f}) V_2(\boldsymbol{x}_2 \frac{2\pi}{\lambda f})\, \tilde{T}\left((\boldsymbol{x}_2 + \boldsymbol{x}_1)\frac{2\pi}{\lambda f}\right)\right|^2 , \tag{5.10}$$

where $\tilde{T}(\boldsymbol{q}) = \int \mathrm{d}\boldsymbol{x}/2\pi \mathrm{e}^{-\mathrm{i}\boldsymbol{q}\cdot\boldsymbol{x}} T(\boldsymbol{x})$ is the amplitude of the diffraction pattern from the object. Note that a shift of position $\boldsymbol{x}_1$ of the detector D_1 produces a translation of the pattern. The whole diffraction pattern from the object can be reconstructed via the correlation function provided that the spatial bandwidth q_0 is larger than the maximal transverse wave-number q in the diffraction pattern, or equivalently, provided that $x_{coh} < l_o$, where l_o is the smallest scale of variation of the object spatial distribution. In contrast, when $x_{coh} < l_o$ no information about the diffraction pattern of the object can be obtained without the correlations, that is, if we detect the light intensity distribution in arm 1 with an array of pixels. In fact, one can easily obtain that

$$\langle \hat{I}_1(\boldsymbol{x}_1)\rangle \propto \int \frac{\mathrm{d}\boldsymbol{q}}{(\lambda f)^2} \left| \tilde{T}(\boldsymbol{x}_1 \frac{2\pi}{\lambda f} - \boldsymbol{q})\right|^2 U_1(\boldsymbol{q}) V_2(-\boldsymbol{q}) \; . \tag{5.11}$$

For $x_{coh} < l_o$, $U_1(\boldsymbol{q})V_2(-\boldsymbol{q})$ can be taken out of the integral, and the resulting expression does not depend on $\boldsymbol{x}_1$ any more.

Example 2: Ghost Image Scheme

In the second example, we set $z = 2f$, so that

$$h_2(\boldsymbol{x}_2, \boldsymbol{x}'_2) = \delta(\boldsymbol{x}_2 + \boldsymbol{x}'_2) \exp\!\left(-\mathrm{i}|\boldsymbol{x}_2|^2 \frac{\pi}{\lambda f}\right).$$

Inserting this in Eq. (5.7) and taking into account Eq. (5.9), we get

$$G_{PDC}(\boldsymbol{x}_1, \boldsymbol{x}_2) \propto \left| \int \mathrm{d}\boldsymbol{x}'_1 \langle \hat{a}_1(\boldsymbol{x}'_1) \hat{a}(-\boldsymbol{x}_2)\rangle T^*(\boldsymbol{x}'_1)\, \mathrm{e}^{\mathrm{i}2\pi/\lambda f \boldsymbol{x}'_1 \cdot \boldsymbol{x}_1} \right|^2 \tag{5.12}$$

$$\propto \left| U_1\left(\frac{2\pi \boldsymbol{x}_1}{\lambda f}\right) V_2\left(-\frac{2\pi \boldsymbol{x}_1}{\lambda f}\right)\right|^2 |T(-\boldsymbol{x}_2)|^2 , \tag{5.13}$$

where in the second line $x_{coh} < l_o$ was assumed. Because the correlation function $\langle \hat{a}_1(\boldsymbol{x}'_1) \hat{a}(-\boldsymbol{x}_2)\rangle$, which depends on $\boldsymbol{x}'_1 + \boldsymbol{x}_2$ (see Eq. (5.8)), is nonzero in a region of size x_{coh} around $\boldsymbol{x}'_1 = -\boldsymbol{x}_2$, this condition ensures that $T(\boldsymbol{x}'_1)$ is roughly constant in this region and it can be taken out from the integral in Eq. (5.12), thus obtaining Eq. (5.13). In this example the intensity correlation function provides information about the image of the object. In the general case (5.12), the image reconstructed via the correlation function is a convolution of the object image with the second-order correlation function (5.8); therefore, the thermal coherence length x_{coh} fixes the resolution of the imaging scheme.

Hence, we have shown that the cross-correlation of the two beams allows us to reconstruct both the image and the diffraction pattern of an object and we can pass from one to the other by only operating on the optical set-up in the reference arm.

For the ghost image scheme if, instead of a pointlike detector, in arm 1 one uses a "bucket" detector that collects all the radiation in arm 1, one measures the quantity $\int \mathrm{d}\boldsymbol{x}_1 G_{PDC}(\boldsymbol{x}_1, \boldsymbol{x}_2)$ and one obtains, for $x_{coh} < l_o$,

$$\int \mathrm{d}\boldsymbol{x}_1 G_{PDC}(\boldsymbol{x}_1, \boldsymbol{x}_2) \propto |T(-\boldsymbol{x}_2)|^2, \tag{5.14}$$

which, again, provides the image of the object. An advantage of the bucket detector case is that the lens in the test arm can be avoided and the relative position of the detection plane and the object in the test arm become immaterial (provided the detection plane is beyond the object). In this case, one can vary at will the position of the object in the test arm and the position of the lens in the reference arm and achieve the reconstruction of the image of the object provided that the distance p_1 between the object and the lens in the reference arm (calculated as the sum of the distance between the object and the $\chi^{(2)}$ slab along the test arm and the distance between the slab and the lens along the reference arm) and the distance p_2 between the lens and the detection plane in the reference arm obey the thin lens law [1, 4]:

$$\frac{1}{p_1} + \frac{1}{p_2} = \frac{1}{f}. \tag{5.15}$$

We observe finally that, if we start from the general expression (5.12) and integrate over x_1, we obtain

$$\int \mathrm{d}\boldsymbol{x}_1 G_{PDC}(\boldsymbol{x}_1, \boldsymbol{x}_2) = \int \mathrm{d}\boldsymbol{x}'_1 |\langle \hat{a}_{1\mathrm{out}}(\boldsymbol{x}'_1) \hat{a}_{2\mathrm{out}}(-\boldsymbol{x}_2) \rangle|^2 |T(\boldsymbol{x}'_1)|^2, \tag{5.16}$$

which shows that using a bucket detector the imaging becomes incoherent (the modulus square is inside the integral).

5.3 Wave-Particle Aspect

In this section we discuss some fundamental aspects. Together with the scheme in Fig. 5.2, we can consider the alternative scheme in Fig. 5.3, which is used in [31] in the coincidence regime, using a double slit as an object. In this case, contrary to Fig. 5.2, in the test arm there is an array of pointlike detectors, and in the reference arm a pointlike detector. The information about the object is still contained in $G(\boldsymbol{x}_1, \boldsymbol{x}_2)$, but in this case one keeps $\boldsymbol{x}_2$ fixed and varies $\boldsymbol{x}_1$. Let us first consider the case $z = f$ in Fig. 5.3. If

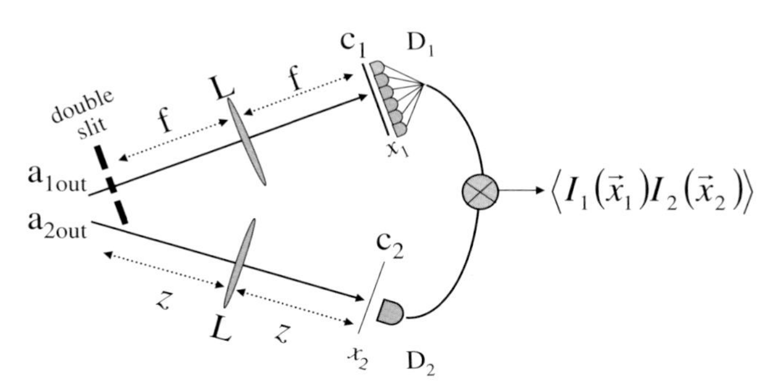

Fig. 5.3. Imaging scheme: L denotes two identical lenses of focal length f; D_1 is an array of pointlike detectors. The distance z is either $z = f$ or $z = 2f$.

the I reference field is not detected, there is no possibility of observing the interference fringes by direct measurement of the reference arm as shown by Eq. (5.11). In the coincidence regime it was argued [32] that in principle one could detect the I photon, and obtain "which-path" information on the S photon, and this is enough to cancel the fringes. We argue more generally that because the S beam alone is in an incoherent thermal mixture (see Eq. (2.29) in Chapter 2), the interference fringes are not visible due to the lack of coherence. However, in order to make fringes visible, it is enough to condition the S beam measurement to a measurement of the I beam by a single pointlike detector.

As a matter of fact, the result is given by Eq. (5.10) and, as we have shown in Section 5.2.1, one detects the whole interference pattern in the case of the scheme of Fig. 5.2, provided $x_{coh} < l_o$. For the scheme of Fig. 5.3, $\boldsymbol{x}_2$ is kept constant and the function $\left|\tilde{T}(\boldsymbol{x}_2 + \boldsymbol{x}_1)2\pi/\lambda f\right|^2$ is symmetrical with respect to $\boldsymbol{x}_1$ and $\boldsymbol{x}_2$, hence the interference pattern is visible in the same way (by scanning $\boldsymbol{x}_1$ instead of $\boldsymbol{x}_2$), even independently of the condition $x_{coh} < l_o$.

In the coincidence regime the fringes become visible, as explained in [1], because detection of the I photon in the far field determines the S photon momentum before the double slit, due to momentum entanglement, providing a quantum erasure [33] of any which-path information. In the macroscopic high-gain regime the fringes remain visible in the same way.

Consider now the scheme of Fig. 5.3 in the $z = 2f$ case, in which the D_2 detector is in the image plane with respect to the object, and the measurement exploits the S/I spatial correlation in the near field. In the coincidence regime fringes are not visible because the detection of the I photon in the near field, due to position entanglement, provides perfect which-path information about the S photon [32]. Our general result is given by Eq. (5.13). If we keep $\boldsymbol{x}_2$ fixed and vary $\boldsymbol{x}_1$, as is the case in the scheme of Fig. 5.3, we do not obtain any information about the object.

As a general conclusion from Sections 5.2.1 and 5.3 we can say that the results for imaging and the wave-particle duality features, which have been demonstrated in the microscopic case (coincidence regime) persist in the macroscopic case (high-gain regime of PDC). A more complete discussion on this point can be found in Ref. [8].

5.4 Spatial Average in Ghost Diffraction: Increase of Spatial Bandwidth and of Speed in Retrieval

This section is based on Ref. [11]. From Eq. (5.10) we see that the correlation provides information about the diffraction pattern of the object if we fix $\boldsymbol{x}_1$ and scan $\boldsymbol{x}_2$, but because the gain also depends on $\boldsymbol{x}_2$ there is a limit to the information we can extract. Precisely, the gain

$$\gamma(\boldsymbol{x}_2) = U_1\left(-\boldsymbol{x}_2\frac{2\pi}{\lambda f}\right) V_2\left(\boldsymbol{x}_2\frac{2\pi}{\lambda f}\right), \tag{5.17}$$

introduces a cutoff of the reproduced spatial Fourier frequencies at the imaging bandwidth of the PDC source q_0.

We showed in [11] how to circumvent this source-related limitation on the imaging bandwidth. We may in a suitable way average over the position of the test-arm detector $\boldsymbol{x}_1$: first, a change of variables is introduced as $\boldsymbol{x} \equiv \boldsymbol{x}_1 + \boldsymbol{x}_2$, and then an average over $\boldsymbol{x}_1$ is performed. We then obtain

$$\begin{aligned} G^{SA}_{PDC}(\boldsymbol{x}) &\equiv \int \mathrm{d}\boldsymbol{x}_1 G_{PDC}(\boldsymbol{x}_1, \boldsymbol{x}-\boldsymbol{x}_1) \propto \int \mathrm{d}\boldsymbol{x}_1 \left|\gamma(\boldsymbol{x}_1-\boldsymbol{x})\, T_{obj}\left(\boldsymbol{x}\frac{2\pi}{\lambda f}\right)\right|^2 \\ &= |\tilde{T}_{obj}\left(\boldsymbol{x}\frac{2\pi}{\lambda f}\right)|^2 \int \mathrm{d}\boldsymbol{x}_1\, |\gamma(\boldsymbol{x}_1-\boldsymbol{x})|^2 \qquad (5.18) \\ &\simeq \text{const} \times |\tilde{T}_{obj}(\boldsymbol{x}\left(\frac{2\pi}{\lambda f}\right))|^2, \qquad (5.19) \end{aligned}$$

where the superscript SA indicates that a spatial average has been carried out. The final approximation in Eq. (5.19) is that $|\gamma(\boldsymbol{x})|^2$ is a bound function inside the detection range of $\boldsymbol{x}_1$ implying that the integral evaluates to a constant. Thus, there is now no gain cutoff of the diffraction pattern, so the imaging bandwidth is substantially increased. Note that this average over $\boldsymbol{x}_1$ does not correspond to D_1 being a bucket detector. Instead, the change in variables $\boldsymbol{x} \equiv \boldsymbol{x}_1 + \boldsymbol{x}_2$ implies that the resulting correlation $G^{SA}_{PDC}(\boldsymbol{x})$ is a convolution of the signal and the idler intensity fluctuations, which in the numerics is easily calculated using the fast Fourier transforming technique.

In [11] one can find the interpretation of this average in terms of Klyshko's time-reversed picture [1] and several examples that illustrate the benefit of this bandwidth increase.

The spatial average technique works particularly well in the high-gain regime where many photon pairs are generated in each pump pulse. Thus the test arm has many photons per pulse transmitted by the object (in contrast, in the low-gain coincidence counting regime only one photon at a time is impinging on the object, and either it is transmitted or it is not), and therefore at the measurement plane they are scattered over the entire transverse plane. Because the spatial average technique employs an average over the position of the test detector, within a single shot we can get information about the image from all the test detector positions in the transverse plane containing photons. This implies that in the high-gain regime a much faster convergence rate is obtained using the spatial average technique, which represents another benefit in addition to the hugely improved bandwidth.

In reconstructing the image of the object a technique corresponding to the spatial average done for reconstructing the diffraction pattern would result in a largely increased image resolution. Unfortunately, it is not possible to carry out such a spatial average to achieve this.

In [11] it is shown that by using a bucket detector instead of a pointlike detector in the test arm one increases the speed of retrieval of the image, but no increase of image resolution is achieved. In addition, it is necessary to use a narrowband interference filter, otherwise significant degrading in resolution is observed.

The spatial averaging technique works in basically the same way in the case of ghost imaging with thermallike beams discussed later in this chapter.

5.5 Ghost Imaging with Homodyne Detection

In Ref. [10] we analyzed the scheme of Fig. 5.2 with the important difference that, in both the test and the reference arm, instead of performing a direct intensity detection, one performs a homodyne detection by introducing 50/50 beamsplitters and appropriate local oscillators, which select the phases of the observed quadrature components. In the case of ghost image detection, in the reference arm one must use a telescopic two-lens configuration instead of the single lens $z = 2f$ configuration shown in Fig. 5.2, whereas in the case of ghost diffraction the configuration remains as in Fig. 5.2 with $z = f$. An initial motivation for using a homodyne scheme for ghost imaging came from the need to circumvent the problems related to information visibility in the macroscopic regime. Specifically, when intensity detection is performed, the measured quantity $\langle \hat{I}_1(\boldsymbol{x}_1)\hat{I}_2(\boldsymbol{x}_2)\rangle$ includes the homogeneous background term $\langle \hat{I}_1(\boldsymbol{x}_1)\rangle\langle \hat{I}_2(\boldsymbol{x}_2)\rangle$. This term, which can be rather large, does not contain any information about the object and lowers the image visibility. Instead, by using homodyne detection the signal–idler correlation becomes second order instead of fourth order in the fields, and hence this background term is absent. Another advantage of homodyne detection is that arbitrary quadrature components of the test and the reference beams can be measured, which means that the homodyne detection scheme allows for both amplitude and phase measurements of the object. Figure 5.4 shows the result for the case

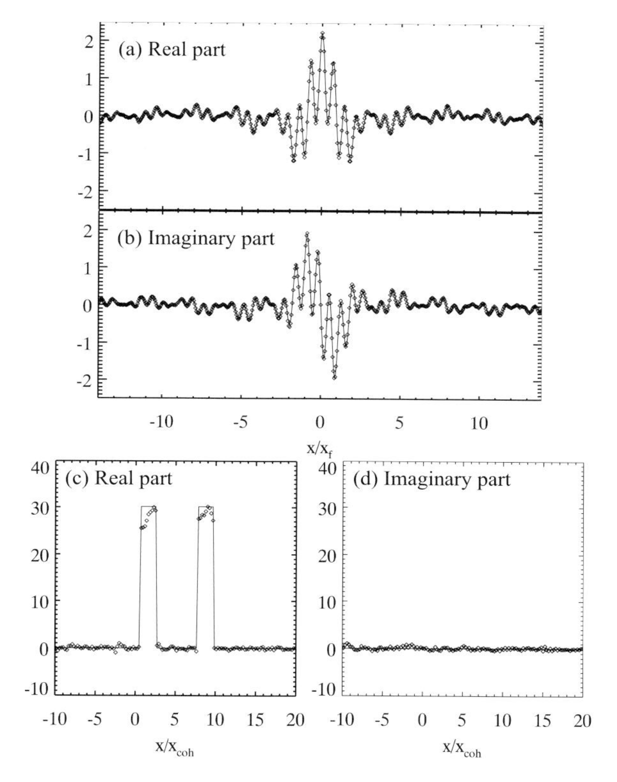

Fig. 5.4. Frames (a) and (b) show the real and the imaginary part of the Fourier transform of the double slit obtained using homodyne detection, spatial averages, and average over 200 pump pulses. The thin lines display the analytical Fourier transform. Frames (c) and (d) are obtained from (a) and (b) by taking the inverse Fourier transform.

in which the object is a double slit. In the frames (a) and (b), by adjusting the phases of the local oscillators, one detects the real and imaginary parts of the Fourier transform of the object. Owing to the use of the spatial average illustrated in Section 5.4, the convergence rate of the retrieval increases by a factor 10 and, in addition, because of the large bandwidth available one is capable of reproducing the Fourier pattern very precisely even far away from the central part of the pattern.

Another very interesting point about homodyne detection is that by measuring both quadratures in the the far-field distribution, we may reconstruct the complete near-field object distribution from this information by using the inverse Fourier transform. We have done this for the data in Fig. 5.4a,b and the result is shown in Fig. 5.4c,d. The real part (c) follows the object profile very precisely. This is because we now have access to high-frequency components. Consequently, as the far-field imaging bandwidth is large, the near-field resolution in the ghost image obtained using the inverse Fourier transform turns out to be much better than when one observes the ghost image directly in the near field using homodyne detection with a telescopic configuration in the reference arm (or using direct detection as in Fig. 5.2 with $z = 2f$).

In two dimensions these features become even more impressive. In Fig. 5.5 we used two different objects: (a) an amplitude transmission mask with the letters INFM, and (b) a more complicated amplitude transmission mask showing a picture of a wolf. We show the ghost images obtained by inverse Fourier transforms for different numbers of shots, and evidently the simple mask (a) converges faster than the more complicated mask (b). Nevertheless, in both cases a good sharp image is obtained after very few shot repetitions, implying that the corresponding far-field diffraction patterns converge very fast and with a very large bandwidth. After additional averaging over shots (using here 500 shots, as shown in the last frames) the irregularities gradually disappear.

Many more details and results can be found in [10].

5.6 Debate: Is Quantum Entanglement Really Necessary for Ghost Imaging?

The question was addressed rather early [4]. A more recent theoretical analysis [5] gave arguments that the ghost imaging scheme truly requires entanglement. The topic became hot after the ghost image experiment of Ref. [4] was successfully reproduced using classically correlated beams [12]. In this experiment a classical source produced pairs of single-mode angularly correlated pulses that served as classical analogues of momentum-correlated pairs of photons produced by PDC. In the accompanying theoretical discussion, the authors presented arguments that although the results of any single experiment in quantum imaging could be reproduced by classical sources with proper statistical correlation, a given classical source cannot emulate the behavior of a quantum entangled source for any arbitrary test and reference systems.

In [8] we addressed this question starting from the consideration that a key feature of the entangled state produced by PDC is the simultaneous

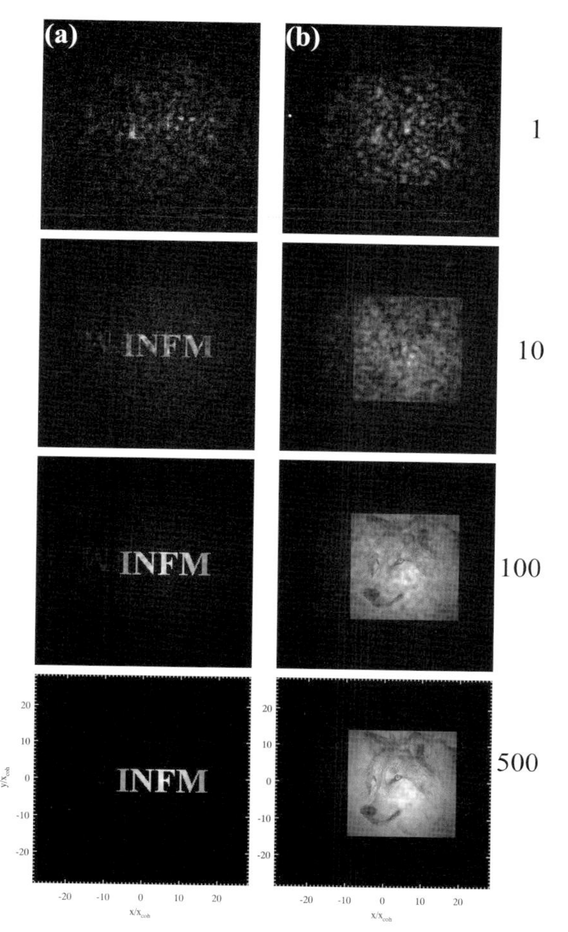

Fig. 5.5. Using the spatial average technique in an f–f setup with two different objects (a) and (b) to obtain the ghost image via the inverse Fourier transform. The correlations are calculated from a full $3+1$D simulation, and averaging additionally over the number of repeated pump shots shown on the right.

presence of (ideally) perfect spatial correlation in the near field and the far field of the signal–idler beams (see, e.g., [1] in Chapter 2). Let us see what happens if we replace the pure entangled state (see Eqs. (2.28) in Chapter 2) by a classical mixture. It is natural to focus on the two mixtures:

$$W = \prod_{\boldsymbol{q}} \left\{ \sum_{n=0}^{\infty} |c_n(\boldsymbol{q})|^2 |n, \boldsymbol{q}\rangle_1 |n, -\boldsymbol{q}\rangle_{22} \langle n, \boldsymbol{q}|_1 \langle n, -\boldsymbol{q}| \right\} , \qquad (5.20)$$

$$W' = \prod_{\boldsymbol{x}} \left\{ \sum_{n=0}^{\infty} |c_n(\boldsymbol{q}=0)|^2 |n, \boldsymbol{x}\rangle_1 |n, \boldsymbol{x}\rangle_{22} \langle n, \boldsymbol{x}|_1 \langle n, \boldsymbol{x}| \right\} . \qquad (5.21)$$

The mixture (5.20) preserves the local S/I spatial intensity correlations in the far field, and the intensity correlation function is completely delocalized in the near field. As shown in Ref. [8], by using the mixture (5.20) instead of the pure state (2.28) of Chapter 2, one obtains the same result (5.10) for the diffraction pattern in the $z = f$ configuration of Fig. 5.2, whereas in the $z = 2f$ ghost image configuration one obtains no information at all about the object (contrary to the result (5.13) of the pure state, which provides the image of the object). Conversely, the mixture (5.21) preserves the S/I local intensity correlations only in the near field. Not surprisingly, in this case the $z = 2f$ scheme provides the image of the object, as with the pure state, but in the $z = f$ case the diffraction pattern is not visible. The key point is that only the pure EPR state displays perfect S/I spatial correlation both in the near and in the far field. This analysis agrees with the basic conclusion of Ref. [12], that the result of each single experiment in entangled photon imaging can be reproduced by a classically correlated source. On the basis of these results, in Ref. [8] we argued that only in the presence of quantum entanglement is it possible to produce both the image and the diffraction pattern of an object by using a single source and by solely operating on the reference arm. We also pointed out the importance of performing in combination the two experiments with $z = f$ and $z = 2f$ in Fig. 5.2. This interpretation was received rather well in the quantum imaging community and was generally viewed as a possibility for discriminating between the presence of quantum entanglement and classical correlations in the source. In particular, in Refs. [13] and [14] combined experiments for ghost diffraction and ghost image with entangled beams were carried out successfully.

However, in later works [15–17] we found a basic counterexample, which partially contradicts the picture emerging from Refs. [8, 12]. Namely, we predicted that by exploiting the classical correlations between the two beams obtained by dividing a thermallike beam with the help of a beamsplitter, it is possible to perform in combination both experiments with $z = f$ and $z = 2f$. This is illustrated in the following section in which we will show, first of all, that there is a profound analogy in ghost imaging between the case of entangled beams from PDC and of the split thermallike beam. In the latter case, the correlation between the two beams is not perfect in the near nor in the far field, but it is enough to perform ghost imaging very well.

We have also demonstrated thermallike ghost imaging experimentally [23, 24] (see also [22]), and this concludes the debate in the sense that quantum entanglement is not necessary for ghost imaging, even if it bears an important advantage in special situations as discussed in the next sections.

5.7 Ghost Imaging by Split Thermallike Beams: Theory [15–17]

As we mentioned already, ghost imaging with PDC beams offers a possibility of performing coherent imaging using, in a sense, incoherent beams, because both the signal and the idler beams, taken separately, are incoherent. In this

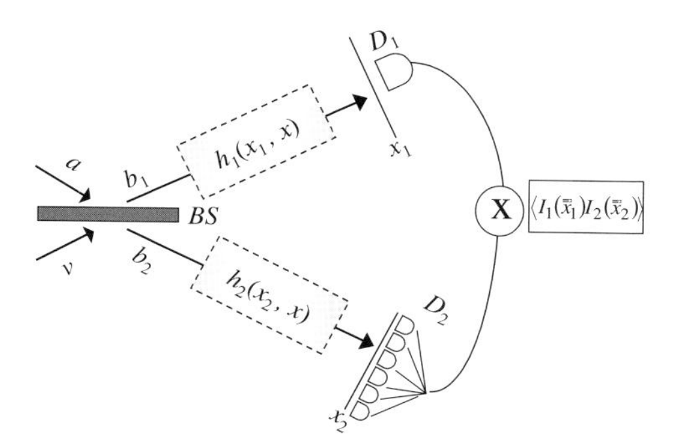

Fig. 5.6. Ghost imaging with incoherent thermal light. The thermal beam a at the beamsplitter BS is divided into two beams, b_1 and b_2, which travel through, respectively, a test and a reference system. The rest of the figure is identical to Fig. 5.1b.

section we show that ghost imaging can be implemented using truly incoherent light, as the radiation produced by a thermal (or thermallike) source. Here we consider a scheme (Fig. 5.6) appropriate for correlated imaging, in which a thermal beam is divided by a beamsplitter (BS) and the two outgoing beams are handled in the same way as the PDC beams in entangled imaging.

5.7.1 Analogy Between Thermal and Entangled Beams in Ghost Imaging [15–17]

In the thermal case, we start from the input-output relations of a beamsplitter

$$\hat{b}_1(\boldsymbol{x}) = r\hat{a}(\boldsymbol{x}) + t\hat{v}(\boldsymbol{x}) , \qquad \hat{b}_2(\boldsymbol{x}) = t\hat{a}(\boldsymbol{x}) + r\hat{v}(\boldsymbol{x}) , \tag{5.22}$$

where t and r are the complex transmission and reflection coefficients of the mirror, $\hat{a}$ is a thermal field, and $\hat{v}$ is a vacuum field uncorrelated with $\hat{a}$. Equation (5.22) coincides with Eq. (1.22) of Chapter 1, except that here we indicate $\hat{a}$ and $\hat{v}$ instead of $\hat{a}_1$ and $\hat{a}_2$, and the beamsplitter in general is not 50/50. We assume that the thermal state $\hat{a}(\boldsymbol{x})$ is characterized by a Gaussian field statistics, in which any correlation function of arbitrary order is expressed via the second-order correlation function [34]:

$$\begin{aligned} \Gamma(\boldsymbol{x}, \boldsymbol{x}') &= \langle \hat{a}^\dagger(\boldsymbol{x})\hat{a}(\boldsymbol{x}')\rangle \\ &= \int \frac{\mathrm{d}\boldsymbol{q}}{(2\pi)^2} \mathrm{e}^{-\mathrm{i}\boldsymbol{q}\cdot(\boldsymbol{x}-\boldsymbol{x}')} \langle n(\boldsymbol{q})\rangle_{\mathrm{th}} . \end{aligned} \tag{5.23}$$

Here $\langle \hat{n}(\boldsymbol{q})\rangle_{\mathrm{th}}$ denotes the expectation value of the photon number in mode $\boldsymbol{q}$ in the thermal state. In writing the second line of this equation, we implicitly used the hypothesis of translational invariance of the source, under which

$\Gamma(\boldsymbol{x},\boldsymbol{x}') = \Gamma(\boldsymbol{x}-\boldsymbol{x}')$. In particular, the following factorization property holds [34].

$$\langle : \hat{a}^\dagger(\boldsymbol{x})\hat{a}(\boldsymbol{x}')\hat{a}^\dagger(\boldsymbol{x}'')\hat{a}(\boldsymbol{x}''') : \rangle = \\ \langle \hat{a}^\dagger(\boldsymbol{x})\hat{a}(\boldsymbol{x}')\rangle\langle \hat{a}^\dagger(\boldsymbol{x}'')\hat{a}(\boldsymbol{x}''')\rangle + \langle \hat{a}^\dagger(\boldsymbol{x})\hat{a}(\boldsymbol{x}''')\rangle\langle \hat{a}^\dagger(\boldsymbol{x}'')\hat{a}(\boldsymbol{x}')\rangle . \quad (5.24)$$

By using Eq. (5.22) with $\hat{a}_{i,\mathrm{out}}$ replaced by $\hat{b}_i$, (i=1,2), Eqs. (5.3) and (5.4), we arrive again at Eq. (5.7), with $\hat{a}_{i,\mathrm{out}}$ replaced by $\hat{b}_i$. Hence, by taking into account the transformation (5.22) and that $\hat{v}$ is in the vacuum state, $\hat{b}_1$ and $\hat{b}_2$ in Eq. (5.5) can be simply replaced by $r\hat{a}$ and $t\hat{a}$, respectively. Hence, by using Eq. (5.24), we arrive at the final result

$$G_{\mathrm{th}}(\boldsymbol{x}_1,\boldsymbol{x}_2) = |tr|^2 \left| \int \mathrm{d}\boldsymbol{x}_1' \int \mathrm{d}\boldsymbol{x}_2' h_1^*(\boldsymbol{x}_1,\boldsymbol{x}_1') h_2(\boldsymbol{x}_2,\boldsymbol{x}_2') \langle \hat{a}^\dagger(\boldsymbol{x}_1')\hat{a}(\boldsymbol{x}_2')\rangle \right|^2 , \quad (5.25)$$

where $\langle \hat{a}^\dagger(\boldsymbol{x}_1')\hat{a}(\boldsymbol{x}_2')\rangle$ is given by Eq. (5.23). At this point the analogy between the results in the two cases of entangled beams and split thermal beams clearly emerges. Apart from the numerical factor $|tr|^2$ and the presence of h_1^* instead of h_1, the thermal second-order correlation function $\langle \hat{a}^\dagger(\boldsymbol{x})\hat{a}(\boldsymbol{x}')\rangle$ in Eq. (5.25) plays the same role as the PDC signal–idler correlation function $\langle \hat{a}_{1\mathrm{out}}(\boldsymbol{x})\hat{a}_{2\mathrm{out}}(\boldsymbol{x}')\rangle$ in Eq. (5.7). Consequently, from Eqs. (5.23) and (5.8), the thermal mean photon number $\langle \hat{n}(\boldsymbol{q})\rangle_{\mathrm{th}}$ plays the same role as $U_1(\boldsymbol{q})V_2(-\boldsymbol{q})$ in the PDC case. The correlation function $\langle \hat{a}^\dagger(\boldsymbol{x})\hat{a}(\boldsymbol{x}')\rangle$ governs the properties of spatial coherence of the thermal source [10, 34, 35]. The correlation length, or transverse coherence length x_{coh}, is determined by the inverse of the bandwidth q_0 of the function $\langle \hat{n}(\boldsymbol{q})\rangle_{\mathrm{th}}$. The same comments hold for the correlation function $\langle \hat{a}_{1\mathrm{out}}(\boldsymbol{x})\hat{a}_{2\mathrm{out}}(\boldsymbol{x}')\rangle$ and the function $U_1(\boldsymbol{q})V_2(-\boldsymbol{x})$ in the entangled case. On the basis of this precise analogy, referring to the imaging scheme of Fig. 5.2, we can expect that all the results for the detection of the diffraction pattern of the object (see Eqs. (5.10) and (5.11)), as well as the result for the ghost image case (see Eq. (5.13)) still hold provided we replace $U_1V_2$with $\langle \hat{n}\rangle_{\mathrm{th}}$. This is true apart from the important feature that in the diffraction pattern result (5.10) the argument of $\tilde{T}$ is $(\boldsymbol{x}_1-\boldsymbol{x}_2)2\pi/\lambda f$ instead of $(\boldsymbol{x}_1+\boldsymbol{x}_2)2\pi/\lambda f$ as a consequence of the fact that in Eq. (5.24) there is $\langle \hat{a}^\dagger(\boldsymbol{x}_1')\hat{a}(\boldsymbol{x}_2')\rangle$, whereas in Eq. (5.7) one has $\langle \hat{a}_{1\mathrm{out}}(\boldsymbol{x}_1')\hat{a}_{2\mathrm{out}}(\boldsymbol{x}_2')\rangle$. Hence, instead of Eq. (5.10) we have

$$G_{th}(\boldsymbol{x}_1,\boldsymbol{x}_2) \propto \left| \langle \hat{n}\left(-\boldsymbol{x}_2\frac{2\pi}{\lambda f}\right)\rangle_{th} \tilde{T}\left((\boldsymbol{x}_1-\boldsymbol{x}_2)\frac{2\pi}{\lambda f}\right)\right|^2 . \quad (5.26)$$

In Eq. (5.12) $\langle \hat{a}_{1\mathrm{out}}(\boldsymbol{x}_1')\hat{a}_{2\mathrm{out}}(-\boldsymbol{x}_2')\rangle$ is obviously replaced by $\Gamma(\boldsymbol{x}_1', -\boldsymbol{x}_2')$ given by Eq. (5.24). In both cases of PDC beams and thermallike beams, the resolution in the ghost image retrieval is determined by the transverse coherence length x_{coh}. In addition to the difference between Eq. (5.10) and Eq. (5.26), the replacement of $\hat{a}_{1\mathrm{out}}$ in Eq. (5.7) by $\hat{a}^\dagger$ in Eq. (5.25) brings another difference: in the thin lens law (5.15), in the thermal case, p_1 is calculated as the difference, instead of the sum, of the distance between the beamsplitter and the lens along the reference arm and the distance of the object and the beamsplitter along the reference arm [19, 22, 23].

In conclusion, the classical correlation of the thermal beams offers imaging capabilities similar to those of the entangled PDC beams; both the image and the diffraction pattern of an object can be reconstructed and we can pass from one to the other by only operating on the optical set-up in the reference arm.

As a special example of thermal source, in [15–17] we considered the signal field generated by PDC. In [15,17] one finds a detailed numerical comparison between the results obtained by exploiting entanglement of the two PDC beams, and the classical correlation of the two beams obtained by splitting the signal beam symmetrically on the other hand. The object is a double slit, and the simulation takes into account the finite transverse size of the pump beam and its pulsed character, and it includes the temporal variable. The parameters in the two cases are the same, apart from the fact that it is ensured that the mean photon number of the two correlated beams are approximately identical in the two simulations. The results turn out to be close to each other for both ghost diffraction and ghost image, when the statistics for the correlation function G is obtained from the same number of pump pulses.

5.7.2 Resolution Aspects

As we noted in the previous section, the resolution in the near field (ghost image retrieval) is determined by the transverse coherence length x_{coh} (i.e., the speckle size), exactly as in the case of PDC beams. Hence, the more incoherent is the thermallike beam, the better is the resolution. An example of the "thermal" light whose coherence properties can be engineered is offered by, for example, chaotic radiation obtained by scattering laser light through random media [36].

On the other hand, in the far field (ghost diffraction retrieval) the resolution is determined by the transverse coherence length (or speckle size) $x'_{coh} \propto \lambda f / w_s$ in the far field, where w_s is the transverse size of the thermallike source [23]. In the PDC case it is the same with w_s given by the transverse size of the pump. Note that in the idealized case of translational invariance considered in our analytical formulas one has $w_S = \infty$ and $x'_{coh} = 0$. In any realistic case the transverse coherence length is, of course, finite.

5.7.3 Relations with the Classic Hanburry–Brown and Twiss Correlation Technique [37]

The approach of [15–17] is reminiscent of the Hanburry–Brown and Twiss (HBT) interferometric method for determining the stellar diameter [34, 37].

However, a basic difference is that the measurement of the stellar diameter relies on the coherence acquired by thermal radiation emitted by the star during the propagation to the earth. On the contrary in our "thermal" ghost imaging approach we exploit just the incoherence of the thermal radiation, and the light that illuminates the object must have a transverse coherence length x_{coh} smaller than the smallest scale of variation l_o of the object spatial distribution.

In such a way

— In the ghost image experiment, one can detect the image of the object with a good resolution.
— For the observation of the diffraction pattern of the object, a direct detection scheme does not work because the illumination is spatially incoherent with respect to the object (see the equivalent of Eq. (5.11) in the thermal case), whereas the ghost diffraction scheme works perfectly (see Eq. (5.26)).

Another fundamental difference is the following. In a standard HBT scheme the object is placed in the thermal beam before the beamsplitter, and not in the test arm as in our case and as suggested by the ghost imaging approach. This feature introduces a basic difference: in the HBT configuration one would retrieve the Fourier transform of the modulus square of the object transmission function instead of the Fourier transform of the object, thus losing any phase information. In our scheme instead, where the object is located in only one arm of the two, phase information about the object can be extracted and, for example, the diffraction pattern from a pure phase object can be reconstructed.

In [16] we show the numerical simulation for the case of a pure phase object.

5.7.4 Correlation Aspects

The imaging schemes described in Fig. 5.2 and Fig. 5.7 have a peculiar feature. In the $z = f$ scheme the diffraction pattern reconstruction is made possible by the presence of spatial correlations in the far field of the correlated beams (momentum correlations of the photons). In the $z = 2f$ scheme, it is the presence of spatial correlations in the near field (position correlations of the photons) that ensures the possibility of reconstructing the image. Our results for the thermal case may hence appear surprising if one has in mind the case of a coherent beam impinging on a beamsplitter, where the two outgoing fields are uncorrelated (i.e., $G(\boldsymbol{x}_1, \boldsymbol{x}_2) = 0$). However, when the input field is an intense thermal beam, that is, the photon number per mode is not too small, the two outgoing beams are well correlated in space both in the near-field and in the far-field planes.

To prove this point, let us consider the number of photons detected in two small identical portions S ("pixels") of the thermal beams in the near field immediately after the beamsplitter, $\hat{N}_i = \int_S \mathrm{d}\boldsymbol{x}\, \hat{b}_i^\dagger(\boldsymbol{x})\hat{b}_i(\boldsymbol{x})$, $i = 1, 2$, and the difference $\hat{N}_- = \hat{N}_1 - \hat{N}_2$. Using Eq. (5.22), for $|r|^2 = |t|^2 = 1/2$ it can be proven that the variance $\langle(\delta\hat{N}_-)^2\rangle = \langle\hat{N}_-^2\rangle - \langle\hat{N}_-\rangle^2$ is given by

$$\langle(\delta\hat{N}_-)^2\rangle = \langle\hat{N}_1\rangle + \langle\hat{N}_2\rangle \,, \tag{5.27}$$

which corresponds exactly to the shot-noise level. Remarkably, Eq. (5.27) holds regardless of statistical properties of the input beam $\hat{a}$ provided that in the other input port there is the vacuum. On the other hand, by using the identity $\langle(\delta\hat{N}_-)^2\rangle = \langle(\delta\hat{N}_1)^2\rangle + \langle(\delta\hat{N}_2)^2\rangle - 2\langle\delta\hat{N}_1\delta\hat{N}_2\rangle$, and taking into account that $\langle(\delta\hat{N}_1)^2\rangle = \langle(\delta\hat{N}_2)^2\rangle$ for $|r|^2 = |t|^2$, the degree of spatial correlations is described by

$$C \stackrel{\text{def}}{=} \frac{\langle\delta\hat{N}_1\delta\hat{N}_2\rangle}{\sqrt{\langle(\delta\hat{N}_1)^2\rangle}\sqrt{\langle(\delta\hat{N}_2)^2\rangle}} = 1 - \frac{\langle\hat{N}_1\rangle}{\langle(\delta\hat{N}_1)^2\rangle} \,. \tag{5.28}$$

For any state $0 \leq |C| \leq 1$, where the upper bound is imposed by the Cauchy–Schwarz inequality. The lower bound corresponds to the coherent state, for which $\langle(\delta\hat{N}_1)^2\rangle = \langle\hat{N}_1\rangle$. For the thermal state, there is always some excess noise with respect to the coherent state $\langle(\delta\hat{N}_1)^2\rangle > \langle\hat{N}_1\rangle$, so that the correlation (5.28) never vanishes. Remarkably, a high degree of spatial correlation between the beams $\hat{b}_1$ and $\hat{b}_2$ is ensured by the presence of a high level of excess noise in the input beam. As shown in detail in the Appendix of [17], for thermal systems with a large number of photons, provided that the pixel size is on the order of x_{coh} or larger, $\langle\hat{N}_1\rangle/\langle(\delta\hat{N}_1)^2\rangle \ll 1$, and C can be made close to its maximum value (see Fig. 5.7).

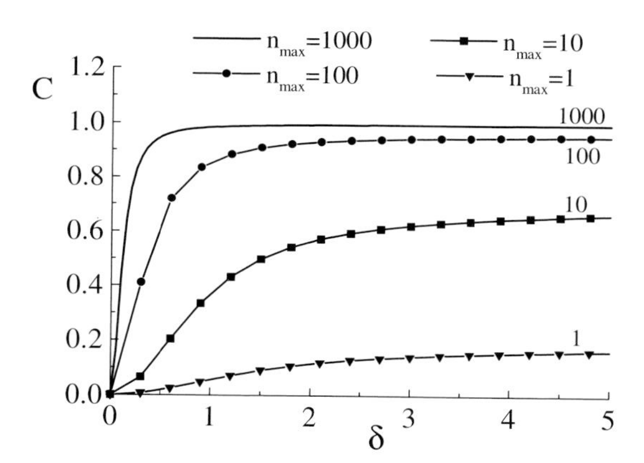

Fig. 5.7. Degree of spatial correlations C between two identical detection regions of the beams obtained by splitting thermal light, as a function of the ratio δ between the pixel size and the coherence length; $n_{\max}$ is the mean photon number in the most intense mode; $C = 1$ represents the maximum degree of correlation.

Even more important, in the absence of losses it is not difficult to show that Eqs. (5.27) and (5.28) hold in any plane linked to the near-field plane by a Fresnel transformation. Let us assume that the propagation of beams $\hat{b}_1, \hat{b}_2$ is described by a linear and unitary kernel H, $\hat{b}_{H,i}(\boldsymbol{x}) = \int \mathrm{d}\boldsymbol{x}' H(\boldsymbol{x}, \boldsymbol{x}')\hat{b}_i(\boldsymbol{x}')$, $i = 1, 2$. Then the form of the beamsplitter transformation (5.22) is preserved during the propagation, provided that the thermal field $\hat{a}$ is substituted by the propagated field $\hat{a}_H(\boldsymbol{x}) = \int \mathrm{d}\boldsymbol{x}' H(\boldsymbol{x}, \boldsymbol{x}')\, \hat{a}\,(\boldsymbol{x}')$. Hence, Eqs. (5.27) and (5.28) also hold for $\hat{b}_{H,1}$ and $\hat{b}_{H,2}$, because these equations are just a consequence of the beamsplitter transformation (5.22) with $|r|^2 = |t|^2 = 1/2$. Moreover, the field $\hat{a}_H$ after propagation is still described by a thermal statistics, because the Gaussian statistics and the factorization property (5.24) of the fourth-order correlation function are preserved by a linear unitary transformation. Therefore also in the far field the correlation can be very good, provided that the size of the detection region is on the order of x'_{coh} or larger and the thermal beam is intense enough.

We remark that despite the fact that C can be made close to 1 by increasing the mean number of photons, it never reaches the quantum level, as shown by Eq. (5.27).

5.7.5 Visibility Aspects

An important issue is the visibility of the information in the PDC and the thermal regimes. The information about the object is retrieved by subtracting the background term $\langle \hat{I}_1(\boldsymbol{x}_1)\rangle\langle \hat{I}_2(\boldsymbol{x}_2)\rangle$ from the measured correlation function (5.3), as indicated in Eq. (5.4). A measure of this visibility is given by evaluating the following quantity in relevant positions,

$$\mathcal{V} = \frac{G(\boldsymbol{x}_1, \boldsymbol{x}_2)}{\langle \hat{I}_1(\boldsymbol{x}_1)\hat{I}_2(\boldsymbol{x}_2)\rangle} = \frac{G(\boldsymbol{x}_1, \boldsymbol{x}_2)}{\langle \hat{I}_1(\boldsymbol{x}_1)\rangle\langle \hat{I}_2(\boldsymbol{x}_2)\rangle + G(\boldsymbol{x}_1, \boldsymbol{x}_2)}\,, \tag{5.29}$$

with $0 \le \mathcal{V} \le 1$.

A first remark concerns the presence of $\langle \hat{n}(\boldsymbol{q})\rangle_{\mathrm{th}}$ in Eq. (5.25) in place of $U_1(\boldsymbol{q})V_2(-q)$ in Eq. (5.7). As a consequence, in the thermal case $G_{\mathrm{th}}(\boldsymbol{x}_1, \boldsymbol{x}_2)$ scales as $\langle \hat{n}(\boldsymbol{q})\rangle^2_{\mathrm{th}}$. In the entangled case, $G_{\mathrm{PDC}}(\boldsymbol{x}_1, \boldsymbol{x}_2)$ scales as $|U_1(\boldsymbol{q})V_2(-q)|^2 = \langle \hat{n}(\boldsymbol{q})\rangle_{\mathrm{PDC}} + \langle \hat{n}(\boldsymbol{q})\rangle^2_{\mathrm{PDC}}$, where $\langle \hat{n}(\boldsymbol{q})\rangle_{\mathrm{PDC}} = |V_2(-\boldsymbol{q})|^2 = |V_1(\boldsymbol{q})|^2$ is the mean number of photons per mode in the PDC beams, and $|U_1(\boldsymbol{q})|^2 = 1 + |V_1(\boldsymbol{q})|^2$ (see, e.g., Ref. [1] of Chapter 2). The difference between the two cases is immaterial when the mean photon number is large, whereas it emerges clearly in the small photon number regime, $\langle \hat{n}(\boldsymbol{q})\rangle \ll 1$. Actually, in the thermal case the visibility does not exceed the value 1/2, whatever the value of $\langle \hat{n}(\boldsymbol{q})\rangle_{\mathrm{th}}$, because $G_{\mathrm{th}}(\boldsymbol{x}_1, \boldsymbol{x}_2)$ scales in the same way as the background term. On the contrary, in the PDC case the visibility can approach the value 1 in the small photon number regime, because in this case the leading scale of $G_{\mathrm{PDC}}(\boldsymbol{x}_1, \boldsymbol{x}_2)$ is $\langle \hat{n}(\boldsymbol{q})\rangle_{\mathrm{PDC}}$ and this term becomes dominant with respect to the background $\langle \hat{I}_1(\boldsymbol{x}_1)\rangle\langle \hat{I}_2(\boldsymbol{x}_2)\rangle \propto \langle \hat{n}(\boldsymbol{q})\rangle^2_{\mathrm{PDC}}$. Hence, in the regime of single photon-pair detection the entangled case presents much better visibility of the information with respect to classically correlated thermal beams (see also [38]).

In addition, very important for the visibility are the duration time and the size of the detection pixel. Standard calculations [34] show that the visibility scales as the ratio of the coherence time of the source $\tau_{\rm coh}$ to the detection time (see also [5, 9]). This implies that conventional thermal sources with very small coherence times are not suitable for the schemes studied here. A suitable source should present a relatively long coherence time, as, for example, a sodium lamp, for which $\tau_{\rm coh} \approx 10^{-10}$ s [38], or the chaotic light produced by scattering a laser beam through a random medium [36].

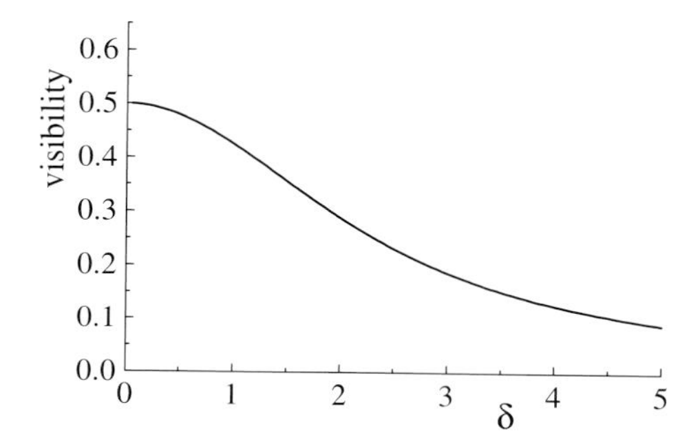

Fig. 5.8. Visibility $\mathcal{V}_S$ of the spatial correlation between two identical detection regions of the beams obtained by splitting thermal light, as a function of the ratio δ of the pixel size to the coherence length.

Similarly, the visibility scales as the ratio of the transverse coherence length to the detected pixel size [5, 24] (see also Fig. 5.8), and this feature points to the same conclusions as inferred from the temporal scaling.

We note that the comparison of Fig. 5.7 and Fig. 5.8 shows the presence of a trade-off between the correlation and the visibility concerning the pixel size. In the same way, by decreasing the coherence length the resolution improves, but the visibility deteriorates, hence there is a trade-off also between the resolution and the visibility [17,20,24], and this feature arises in the same way in the near field (ghost image retrieval) and in the far field (ghost diffraction retrieval).

Pseudo-thermal sources [36] appear as the best candidates to overcome major visibility problems, because of the possibility of engineering the speckle size independently in the near and in the far field.

5.7.6 Some Historical Considerations

An important precursor of our works [15–17] is represented by Ref. [5], which identifies some fundamental relations between the case of PDC beams and that of a thermal beam. It is important to remark, in this connection, that [5] did not introduce an essential ingredient for the analogy between ghost imaging with PDC beams and with split thermal beams, that is, the beamsplitter in Fig. 5.6. Without the beamsplitter, the configuration becomes equivalent to that of HBT (see Section 5.7.3), in which the object is located in the thermal beam before the beamsplitter. In this case, as shown in Section 5.7.3 one cannot detect, for example, a pure phase object.

The analysis of [5] focuses on the existence of a duality, rather than analogy, between the PDC and the thermal case. In addition, the authors of [5] pointed out the visibility problem of the thermal case associated with the detection time and the detection size (see Section 5.7.5). In conclusion, their analysis does not suggest that the thermal case can be an alternative to the PDC case.

After the appearance of the e-print [15] (which led to both publications [16, 17]), a lively interest arose on the topic of "thermal" ghost imaging, with a number of important theoretical [18–20] and experimental [21–25] contributions.

5.7.7 Rule-of-Thumb Comparison Between Entangled and "Thermal" Ghost Imaging

In Section 5.7.5 we illustrated extensively the visibility problems in the "thermal" configuration and pointed out, as already done in [5], that such problems do not affect the PDC case, in which the visibility can approach the level of unity in the coincidence regime. The fact that the quantum configuration turns out to be superior is certainly not a surprise. In this section we try, on the basis of qualitative arguments, to identify the situations in which such a superiority becomes significant. At the same time, we point out that in other situations the "thermal" approach can be more convenient.

The visibility problems related to the detection time and the pixel detection area can be completely circumvented by using pseudo-thermal light [36], due to the fact that one can easily engineer the coherence time as well as the transverse coherence length in both the near and the far field. However, as observed in Section 5.7.5, in the thermal case the visibility is always smaller than 0.5 and, in almost all cases, is much smaller than unity. In this problem, the signal-to-noise ratio basically coincides with the visibility. It is important to note, on the other hand, that despite this small visibility it is perfectly possible to retrieve the image or the diffraction pattern of the object provided that a large number of measurements are done in order to obtain the correlation function $G(\boldsymbol{x}_1, \boldsymbol{x}_2)$. In addition, in some cases it is possible to reduce the number of measurements by employing the spatial averaging technique described in Section 5.4.

In the case of ghost imaging with entangled PDC beams in the coincidence regime, the main problem is represented by the long time necessary to retrieve the information, because coincidences are rare and a huge number of measurements are necessary.

As we said before, the visibility problem in the thermal case also eventually becomes a problem for the time necessary to retrieve the information. Therefore, the comparison between the PDC and the thermal approach is best done in terms of time.

When the (amplitude or phase) modulation of the object is macroscopic, we expect that the procedure can be faster when one uses a pseudo-thermal source. Additional advantages are that one can easily achieve a high resolution and that the experimental apparatus is much cheaper.

On the other hand, when the modulation is greatly reduced, the signal-to-noise ratio issue becomes dominant and one can expect a crossover between the retrieval times in the two approaches. In the limit where high precision measurements are necessary, the "thermal" approach becomes impractical and the superiority of the quantum configuration becomes manifest. Other cases of superiority arise when the ghost imaging technique should be applied to quantum information schemes, for example, when the information needs to be hidden from a third party.

5.8 Ghost Imaging with Split Thermal Beams: Experiment

In this section we illustrate some of the experimental results reported in [23, 24].

The experimental set-up is sketched in Fig. 5.9. The source of pseudo-thermal light is provided by a scattering medium illuminated by a He–Ne or ND–Yag laser beam. The medium is a slowly rotating ground glass placed in front of a scattering cell containing a turbid solution of $3\,\mu$m latex spheres. When this is illuminated with a large collimated He–Ne laser beam ($\lambda = 0.6328\,\mu$m, diameter $D_0 \approx 10\,$mm), the stochastic interference of the waves emerging from the source produces at large distance ($z \approx 600\,$mm) a time-dependent speckle pattern, characterized by a chaotic statistics and by a correlation time τ_{coh} on the order of 0.5 s (for an introduction to laser speckle statistics, see, e.g., [39]). Notice that the ground glass can be used alone to produce chaotic speckles, whose correlation time depends on the speed of rotation of the ground-glass disk and on the laser diameter, as in classical experiments with pseudo-thermal light [36, 40]. Indeed, in some part of the experiments described in the following it will be used alone. This, however, presents a problem that the generated speckle patterns reproduce themselves after a whole tour of the disk, which can be partially avoided by shifting the disk laterally at each tour. The turbid solution provides an easy way to generate a truly random statistics of light, because of the random motion of particles in the solution, allowing a huge number of independent patterns to be generated and used for statistics. Notice that the turbid medium cannot be used alone, because a portion of the laser light would not be scattered, thus leaving a residual coherent contribution. At a distance $z_0 = 400\,\mathrm{mm}$ from

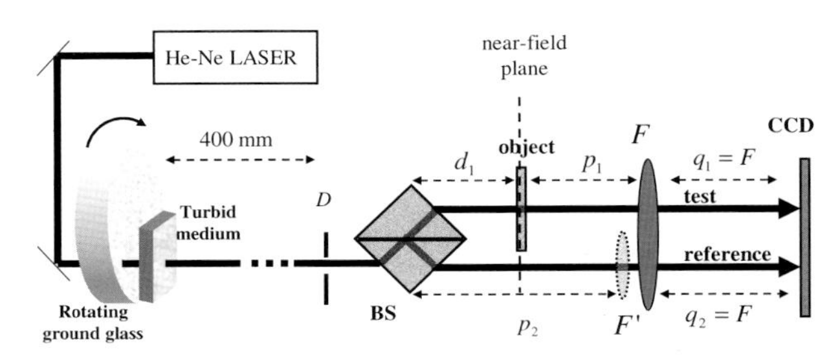

Fig. 5.9. Scheme of the experimental set-up (see the text for details).

the thermal source, a diaphragm of diameter $D = 3\,\mathrm{mm}$ selects an angular portion of the speckle pattern, allowing the formation of an almost collimated speckle beam characterized by a huge number (on the order of 10^4) of speckles of the size $\Delta x \approx \lambda z_0/D_0 \approx 25\,\mu\mathrm{m}$ [39]. The speckle beam is separated by the beamsplitter into two "twin" speckle beams that exhibit a high (although classical) level of spatial correlations. The two beams emerging from the BS have slightly noncollinear propagation directions, and illuminate two different nonoverlapping portions of the CCD camera. The data are acquired with an exposure time (1–3 ms) much shorter than τ_{coh}, allowing the recording of high-contrast speckle patterns. The frames are taken at a rate of 1 Hz, so that each data acquisition corresponds to uncorrelated speckle patterns.

5.8.1 High-Resolution Ghost Imaging [23]

In [23] we performed a reconstruction of both the image (Fig. 5.10) and the diffraction pattern (Fig. 5.11) of the object by operating only on the optical set-up of the reference arm and by using a single classical source. The optical set-up of object arm 1 is fixed. An object, consisting of a thin needle of 160 μm diameter inside a rectangular aperture 690 μm wide, is placed in this arm at a distance d_1 from the BS.

The object plane, located at a distance $\approx$200 mm from the BS, will be taken as the reference plane, and referred to as the *near-field* plane (this is not to be confused with the source near field, as the object plane is in the far zone with respect to the source). A single lens of the focal distance $F = 80$ mm is placed after the object, at a distance p_1 from the object and

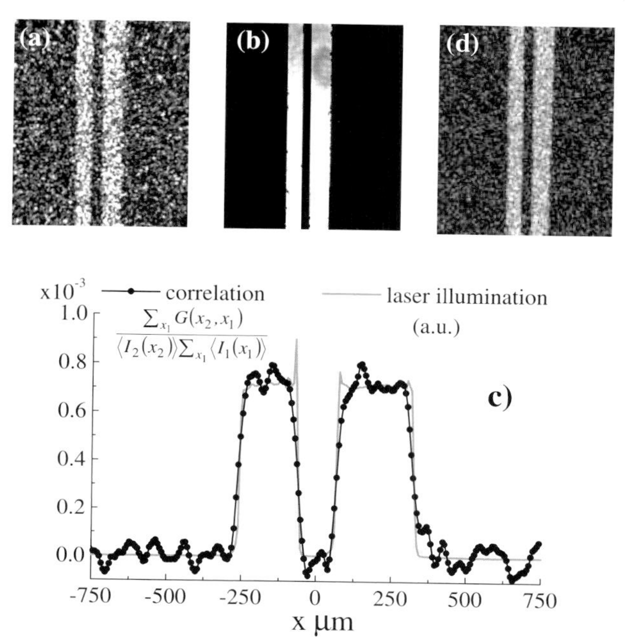

Fig. 5.10. Reconstruction of the object: (a) correlations of the intensity fluctuations; (b) image observed using laser light; (c) averages of 500 horizontal data sections from (a) and (b); (d) the same as (a) but with 30,000 data acquisitions instead of 5000.

$q_1 = F$ from the CCD. Hence, the CCD is in the far-field plane with respect to the object. However, because the light is incoherent, the diffraction pattern of the object is not visible on the CCD camera, as shown in Fig. 5.11a. We consider two different setups for reference arm 2. In the first one, an additional lens of the focal distance F' is inserted in arm 2 immediately before the lens F. The equivalent focal distance F_2 of the two-lens system is smaller than its distance from the CCD camera $q_2 = F$, being $1/F_2 = 1/F + 1/F'$. This allowed us to locate the position of the plane conjugate to the CCD plane, by temporarily inserting the object in arm 2 and determining the position that produced a well-focused image on the CCD camera with laser illumination (Fig. 5.10b). The object was then translated into the object arm. The distances in the reference arm approximately obey a thin-lens equation of the form $1/(p_2 - d_1) + 1/q_2 \approx 1/F_2$, providing a demagnification factor $m \approx 1.2$. The data of the intensity distribution of the reference arm are acquired, and each pixel is correlated with the total photon counts of arm 1, which corresponds to having a bucket detector there. Averages performed over 5000 data acquisitions show a well-resolved image of the needle (Fig. 5.10a) that can be compared with the image obtained with laser illumination (Fig. 5.10b). Figure 5.10c compares the corresponding horizontal sections, averaged over

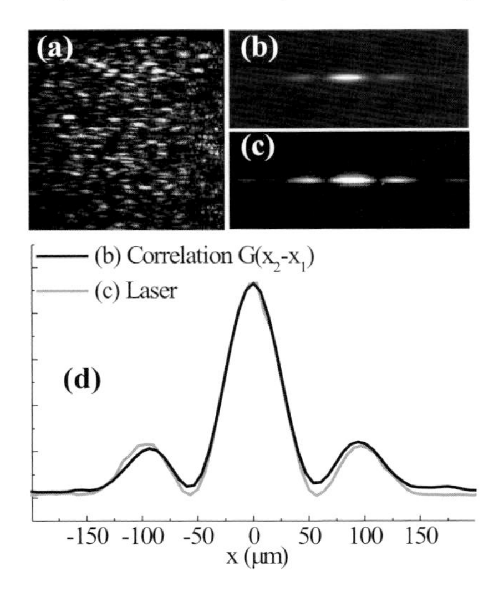

Fig. 5.11. Reconstruction of the diffraction pattern: (a) single-shot intensity distribution in the arm 1; (b) correlation function $G(\boldsymbol{x}_2 - \boldsymbol{x}_1)$; (c) object diffraction pattern observed using laser light; (d) horizontal cut of (b) and (c).

500 pixels in the vertical direction. The spatial resolution shown by correlated imaging with incoherent light is comparable with that obtained via coherent illumination. When the average is performed over 30,000 data acquisitions, the contrast increases significantly (Fig. 5.10d).

In the second set-up the lens F' is simply removed from the scheme of Fig. 5.9, so that the CCD camera is in the focal plane of the lens F also in arm 2. The spatial cross-correlation of the intensities is calculated as a function of the displacement $\boldsymbol{x}_2 - \boldsymbol{x}_1$ between the pixel positions in the two arms, by making an additional average over pixel positions at each fixed $\boldsymbol{x}_2 - \boldsymbol{x}_1$ [10,11]. Thus, averages over only 500 independent frames are enough to show a sharp reproduction of the diffraction pattern of the object (Fig. 5.11). This is comparable with the diffraction pattern obtained by laser illumination (Fig. 5.11c). Horizontal sections of the two patterns display a very good agreement (Fig. 5.11b,d).

As shown in Section 5.7.2, relevant to the resolution of the ghost image and ghost diffraction schemes are the speckle sizes in the near and far fields. These spatial coherence properties can be investigated by measuring the fourth-order correlation functions in the absence of the object. The autocorrelation function of the reference beam $\langle \hat{I}_2(\boldsymbol{x})\hat{I}_2(\boldsymbol{x}')\rangle$ was first measured in the set-up with the lens F' inserted, so that the reference beam recorded by the CCD is the (demagnified) image of the near field. This is plotted in Fig. 5.12 (squares) as a function of $|\boldsymbol{x} - \boldsymbol{x}'|$. Neglecting the shot-noise contribution at $\boldsymbol{x} = \boldsymbol{x}'$, and using the Siegert formula for Gaussian

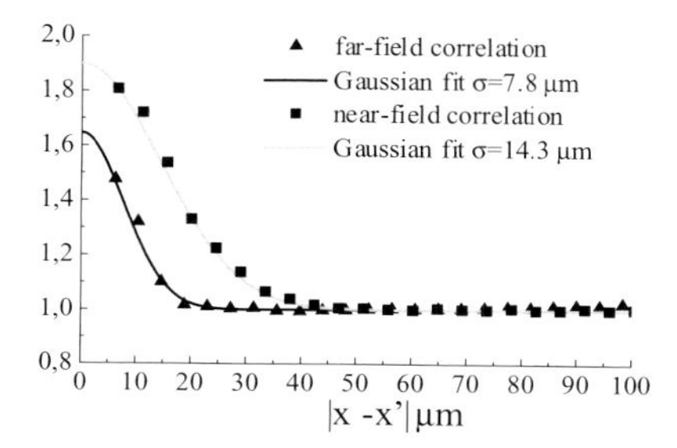

Fig. 5.12. Normalized fourth-order autocorrelation function in the near-field and far-field planes. The full lines are Gaussian fits of the correlation peaks.

statistics, we get

$$\langle \hat{I}_2(\boldsymbol{x})\hat{I}_2(\boldsymbol{x}')\rangle = \langle \hat{I}_2(\boldsymbol{x})\rangle\langle \hat{I}_2(\boldsymbol{x}')\rangle + |r|^4 \left|\Gamma_n(m\boldsymbol{x}, m\boldsymbol{x}')\right|^2 , \tag{5.30}$$

where Γ_n is the correlation function (5.23) in the near field. The baseline in Fig. 5.12 corresponds to the product of the mean intensities, and the narrow peak around $\boldsymbol{x} = \boldsymbol{x}'$ is the second term at the r.h.s. of Eq. (5.30). A Gaussian fit of this peak gave a variance $\sigma_n = (14.3 \pm 0.2)\,\mu\text{m}$, implying a coherence length in the near-field plane $\Delta x_n \approx 2m\sigma_n = (34.3 \pm 0.6)\,\mu\text{m}$. The triangles in Fig. 5.12 plot the intensity correlation function in the far-field plane, obtained by measuring the autocorrelation function of beam 1 in the focal plane of the lens F. A Gaussian fit gave $\sigma_f = (7.8 \pm 0.3)\,\mu\text{m}$, from which we infer a far-field coherence length $\Delta x_f \approx 2\sigma_f = (15.6 \pm 0.6)\,\mu\text{m}$. This in turn corresponds to a spread in transverse wave vectors $\Delta q = (2\pi/\lambda F)\Delta x_f = (1.94 \pm 0.07) \times 10^{-3}\,\mu\text{m}^{-1}$. Hence we find for our classical beams

$$\Delta x_n \Delta q = 0.066 \pm 0.003 , \tag{5.31}$$

whereas previously it was believed that the bound $\Delta x_n \Delta q > 1$ could be overcome only using entangled beams (see, e.g., [13, 14]). In addition, the result (5.31) is roughly four times smaller than the results reported in Refs. [13, 14], where entangled photons were used. Notice that Eq. (5.31) is not violating any EPR bound, contrary to the experiments carried out with single photon pairs [14, 41]. In fact, in any plane, the probability of detecting a photon at position $\boldsymbol{x}_2$ in beam 2 conditioned to the detection of a photon at $\boldsymbol{x}_1$ in the beam 1 is:

$$\begin{aligned} P(\boldsymbol{x}_2|\boldsymbol{x}_1) &\propto \langle \hat{I}_2(\boldsymbol{x}_2)\hat{I}_1(\boldsymbol{x}_1)\rangle / \langle \hat{I}_1(\boldsymbol{x}_1)\rangle \\ &= \langle \hat{I}_2(\boldsymbol{x}_2)\rangle + |rt|^2 \left|\Gamma(\boldsymbol{x}_1, \boldsymbol{x}_2)\right|^2 / \langle \hat{I}_1(\boldsymbol{x}_1)\rangle . \end{aligned} \tag{5.32}$$

The two terms in Eq. (5.32) have roughly the same height (see Fig. 5.12), but the first one, originating from the background, is much broader than the second, because the beam diameter is much larger than the coherence length. Hence in good approximation the conditional variance in the position is the beam spot size. The product of the variances in the near and the far zones satisfies a Fourier relation in accordance with the bound derived in [13]. The crucial point is that the conditional variance and the resolution of ghost imaging do not coincide in general, because the resolution is determined by the coherence length of the field correlation function $\Gamma(\boldsymbol{x}, \boldsymbol{x}')$. They do coincide only in the special case where the background is negligible, as, for example, in the coincidence-count regime of PDC considered by [13,14]. Only in this case, which in principle corresponds to 100% visibility, the bound of [13] holds also for resolutions.

In our experiment, where the visibility is limited to 50%, the product of the near- and the far-field resolutions is not bounded, because the coherence lengths (the speckle sizes) in the two planes are independent quantities. In the near field, the size of the speckle depends on the laser diameter D_0, and on the distance z from the source, $\Delta x_n \propto \lambda z / D_0$ [39]. As we checked, the diaphragm being close enough to the near field, its diameter D does not much affect Δx_n. It determines, instead, the speckle size in the far field, roughly given by $\Delta x_f \propto \lambda F/D$ [39]. Using the values of our set-up, we find $\Delta x_n \approx 30\,\mu$m and $\Delta x_f \approx 17\,\mu$m, in good agreement with the values estimated from the correlation (Fig. 5.12). Two aspects of our experiment are crucial: (i) the presence in the near field of a large number of small speckles inside a broad beam, and (ii) a measurement time $\ll \tau_{coh}$. This allows the formation by interference of a far-field speckle pattern, characterized by a small coherence length, because $\Delta x_f \propto 1/D$. In this respect our source differs from the classical one used in [12,13], where each shot consists of a single narrow pulse and the product of resolutions is bounded by the pulse diffraction.

We observe finally that in [26], where the experimental observation of a pure phase object by entangled beams is reported, in the introduction there is the erroneous interpretation that the illumination of the object in our experiment [23] is coherent due to the presence of the diaphragm which selects a reduced number of speckles. As specified above, the number of speckles in the beam transmitted by the diaphragm is, instead, huge ($\simeq 10^4$). In addition, in the object plane (near field) the speckle size is on the order of 30 μm, as shown above, much less than the spatial scale that characterizes the object (e.g., the diameter of the needle is 160 μm, as mentioned before). Hence, the illumination is incoherent, as is clear also from the absence of the interference pattern in Fig. 5.11a. In the next subsection we will show the changes that arise when the illumination becomes coherent.

We note that the illumination is incoherent also from a temporal viewpoint, even if the acquisition time is smaller than the coherence time. As a matter of fact, to retrieve the ghost diffraction pattern one averages over thousands of data acquisitions, that is, over a time interval much longer than the coherence time. Also this feature will appear clearly in the discussion of the following subsection.

5.8.2 The Ghost Diffraction Experiment: Complementarity Between Coherence and Correlation [24]

In Ref. [24], we used a ND–Yag laser instead of a He–Ne laser, but the geometrical configuration remains the same as in Fig. 5.9 and [23], and the same is true for the object. Let us focus on the ghost diffraction case (Fig. 5.9 without the lens F').

In a first set of measurements the source size is $D_0 = 10\,\mathrm{mm}$, and the object is illuminated by a large number of speckles whose size $\Delta x_n = 36$ μm is much smaller than the slit separation. The light is spatially incoherent as described in the previous section. The results are shown in the first row of Fig. 5.13. The frame (a) is the instantaneous intensity distribution of the object beam, showing a speckle pattern, with no interference fringes from the double slit, as expected for incoherent illumination. At closer inspection,

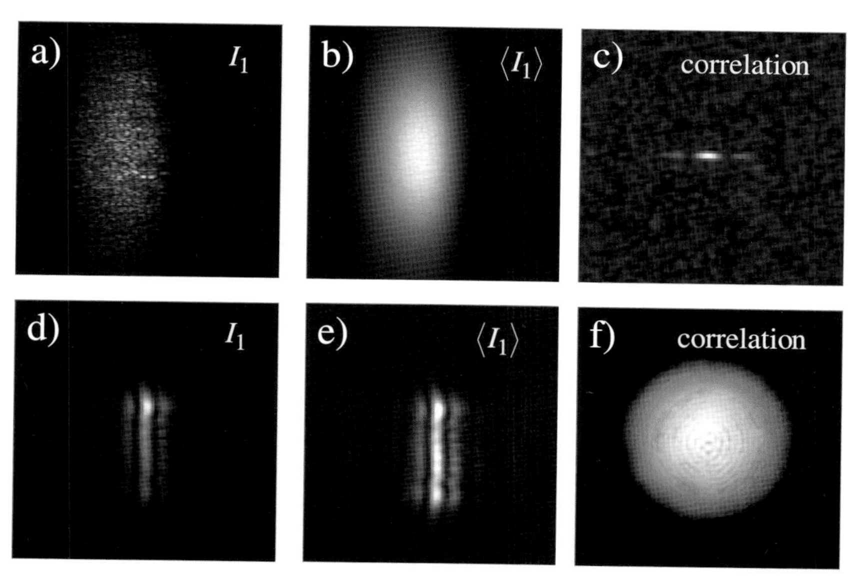

Fig. 5.13. Ghost diffraction set-up: transition from incoherent light to partially coherent light. In the three upper frames (a)–(c) the source size is $D_0 = 10\,\mathrm{mm}$, with near-field speckles $\Delta x_n = 36\,\mu\mathrm{m}$. In the three lower frames (d)–(f) the source size is $D_0 = 0.1\,\mathrm{mm}$, with $\Delta x_n = 3.2\,\mathrm{mm}$; (a) and (d): instantaneous intensity distribution I_1 of the object beam; (b) and (e): intensity distribution $\langle I_1 \rangle$, averaged over 350 shots; (c) and (f): correlation function $G(\boldsymbol{x}_1, \boldsymbol{x}_2)$ as a function of $\boldsymbol{x}_2$, for a fixed $\boldsymbol{x}_1$, averaged over 20,000 shots.

the shape of the speckles resembles the interference pattern of the double slit, but because these speckles move randomly in the transverse plane from shot to shot, an average over several shots displays a homogeneous broad spot (Fig. 5.13b). The frame (c) is a plot of $G(\boldsymbol{x}_1, \boldsymbol{x}_2)$ as a function of the reference pixel position $\boldsymbol{x}_2$, and shows the result of correlating the intensity distribution in the reference arm with the intensity collected from a single fixed pixel in the object arm. Notice that differently from [23], no spatial average [10, 11] is employed: this makes the convergence rate slower but the scheme is closer to the spirit of ghost diffraction in which the information is retrieved by only scanning the reference pixel position. The *ghost diffraction* pattern emerges after a few thousands of averages, and is well visible after 20,000 averages. This is confirmed by the data of Fig. 5.14a which compare the horizontal section of the diffraction pattern from a correlation measurement to that obtained with laser illumination. In a second set of measurements the

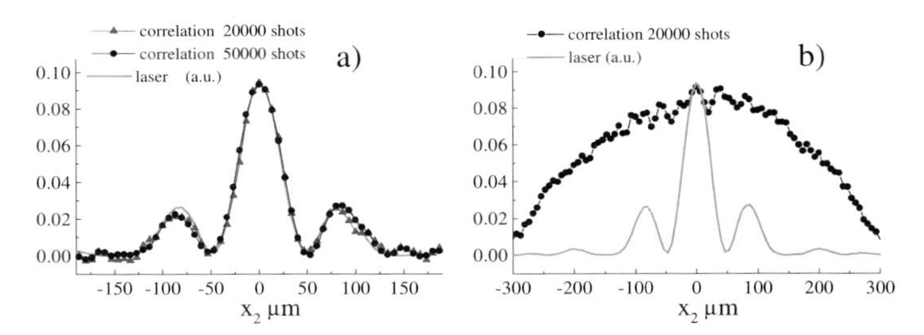

Fig. 5.14. Horizontal sections of the correlation function $G(\boldsymbol{x}_1, \boldsymbol{x}_2)$ as a function of x_2, for a fixed $\boldsymbol{x}_1$ (see Fig. 5.13c,f): (a) the case of incoherent light, $D_0 = 10\,\text{mm}$; the data are obtained with an average over 20,000 shots (triangles) and 50,000 shots (circles); (b) the case of partially coherent illumination, $D_0 = 0.1\,\text{mm}$ (20,000 shots). The light full line is for comparison with the diffraction pattern observed with a laser.

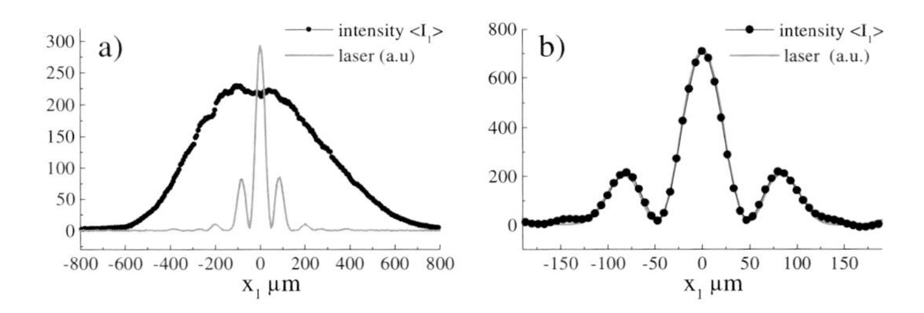

Fig. 5.15. Horizontal sections of the average intensity distribution $\langle \hat{I}_1(\boldsymbol{x}_1) \rangle$ in the object arm (see Fig. 5.13b,e): (a) is obtained for incoherent light with $D_0 = 10\,\text{mm}$ (350 shots), and (b) plots the case of partially coherent illumination, with $D_0 = 0.1\,\text{mm}$ (500 shots). The light full line is for comparison with the diffraction pattern observed with a laser.

source size is reduced to $D_0 = 0.1\,\mathrm{mm}$ by inserting a small pinhole after the ground glass. As a result, the spatial coherence of the light illuminating the object is increased. As the speckle size at the diaphragm D is now $\approx 3\,\mathrm{mm}$, on average the object is illuminated by a single speckle of much larger size than the slit separation. The results are reported in the second raw of Fig. 5.13. As expected, the interference fringes are now visible in the instantaneous intensity distribution of the object beam 1 (frame (d)), and become sharper after averaging over some hundreds of shots (frame (e)). Notice that the shape of the interference pattern is now elongated in the vertical direction, because the light emerging from the small source is not collimated. Horizontal sections of $\langle \hat{I}_1 \rangle$, plotted in Fig. 5.14b, show a very good agreement with the diffraction pattern from laser illumination. Instead, no interference fringes at all appear in the correlation function of the intensities in the two arms, when plotted as a function of $\boldsymbol{x}_2$ (frame (f)). Notice that in this set of measurements the turbid medium was removed in order to increase the power. This is feasible in this case, because the very small size of the source allows a large number of independent patterns to be generated in a single tour of the glass disk.

Figures 5.13–5.15 evidence a remarkable complementarity between the observation of interference fringes in the correlation function (ghost diffraction), and in the intensity distribution of the object beam (ordinary diffraction). They also show the fundamental role played by the spatial incoherence of the source in producing a ghost diffraction pattern: the more incoherent is the source, the more the two beams are spatially correlated and the more information about the object is available in the ghost diffraction pattern. The more coherent is the source, the flatter is the spatial correlation function of the two beams and the less information about the object is contained in the ghost diffraction. This is completely analogous to the complementarity between the one-photon and two-photon interference in Young's double-slit experiments with photons from a PDC source [42], which was explained as a complementarity between the coherence and the entanglement. In our case of thermal beams, the complementarity is rather between the coherence and the spatial correlations, showing that also in this respect the classical spatial correlation produced by splitting thermal light plays the same role as the entanglement of PDC photons.

These results can be easily understood by using the formalism developed in Section 5.7.1, and in particular by inspection of Eq. (5.25) for the correlation function of the intensity fluctuations $G(\boldsymbol{x}_1, \boldsymbol{x}_2)$. In the limit of spatially coherent light the field correlation function $\Gamma(\boldsymbol{x}_1, \boldsymbol{x}_2) = \langle \hat{a}^\dagger(\boldsymbol{x}_1)\hat{a}(\boldsymbol{x}_2)\rangle$ becomes constant in space in the region of interest, and the two integrals in Eq. (5.25) factor to the product of two ordinary imaging schemes, showing the diffraction pattern of the object only in object arm 1. As a result, by plotting the correlation as a function of $\boldsymbol{x}_2$, no object diffraction pattern can be observed; that is, no ghost diffraction occurs. The same observation can be made with respect to the PDC case, explaining thus the analogy between the role of the coherence in the PDC and in the thermal case.

We note that, as shown in Fig. 12 of [24], when the incoherence level of the radiation in the near field is increased, the visibility of the diffraction pattern increases, contrary to what happens for the ghost image of the object.

We observe finally that an experiment of ghost imaging with standard thermal light has been realized recently [25].

References

1. D. N. Klyshko, Zh. Eksp. Teor. Fiz. **94**, 82 (1988) [Sov. Phys. JETP **67**, 1131 (1988)]; A. V. Belinskii and D. N. Klyshko, Zh. Eksp. Teor. Fiz. **487**, 105 (1994) [JETP **78**, 259 (1994)].
2. P. H. S. Ribeiro, S. Padua, J. C. Machado da Silva, and G. A. Barbosa, Phys. Rev. A **49**, 4176 (1994).
3. D. V. Strekalov, A. V. Sergienko, D. N. Klyshko, and Y. H. Shih, Phys. Rev. Lett. **74**, 3600 (1995).
4. T. B. Pittman, Y. H. Shih, D. V. Strekalov, and A. V. Sergienko, Phys. Rev. A **52**, R3429 (1995).
5. B. E. A. Saleh, A. F. Abouraddy, A. V. Sergienko, and M. C. Teich, Phys. Rev. A **62**, 043816 (2000).
6. A. F. Abouraddy, B. E. A. Saleh, A. V. Sergienko, and M. C. Teich, Phys. Rev. Lett. **87**, 123602 (2001).
7. A. F. Abouraddy, B. E. A. Saleh, A. V. Sergienko, and M. C. Teich, J. Opt. Soc. Am. B **19**, 1174 (2002).
8. A. Gatti, E. Brambilla, and L. A. Lugiato, Phys. Rev. Lett. **90**, 133603-1 (2003).
9. A. Gatti, E. Brambilla, M. Bache, and L. A. Lugiato, Laser Physics **15**, 176 (2005), special issue in honour of the 70th birthday of Herbert Walther.
10. M. Bache, E. Brambilla, A. Gatti, and L. A. Lugiato, Phys. Rev. A **70**, 023823 (2004).
11. M. Bache, E. Brambilla, A. Gatti, and L. A. Lugiato, Opt. Expr. **12**, 6067 (2004).
12. R. S. Bennink, S. J. Bentley, and R. W. Boyd, Phys. Rev. Lett. **89**, 113601 (2002).
13. R. S. Bennink, S. J. Bentley, R. W. Boyd, and J. C. Howell, Phys. Rev. Lett. **92** 033601 (2004).
14. M. D'Angelo, Y.-H. Kim, S. P. Kulik, and Y. Shih, Phys. Rev. Lett. **92**, 233601 (2004).
15. A. Gatti, E. Brambilla, M. Bache, and L. A. Lugiato, quant-ph/0307187.
16. A. Gatti, E. Brambilla, M. Bache, and L. A. Lugiato, Phys. Rev. Lett. **93**, 093602 (2004).
17. A. Gatti, E. Brambilla, and L. A. Lugiato, Phys. Rev. A **70**, 013802 (2004).
18. J. Cheng and S. Han, Phys. Rev. Lett. **92**, 093903 (2004).
19. D.-Z. Cao, J. Xiong, and K. Wang, Phys. Rev. A **71**, 013801 (2005).
20. Y. Cai and S.-Y. Zhu, Opt. Lett. **29**, 2716 (2004).
21. G. Scarcelli, A. Valencia, and Y. Shih, Phys. Rev. A **70**, 051802(R) (2004).

22. A. Valencia, G. Scarcelli, M. D'Angelo, and Y. Shih, Phys. Rev. Lett. **94**, 063601 (2005), reports a ghost image experiment with thermal photons but does not consider the ghost diffraction and the resolution issues.
23. F. Ferri, D. Magatti, A. Gatti, M. Bache, E. Brambilla, and L. A. Lugiato, Phys. Rev. Lett. **94**, 183602 (2005).
24. A. Gatti, M. Bache, D. Magatti, E. Brambilla, F. Ferri, and L. A. Lugiato, J. Mod. Opt., **53**, 739 (2006).
25. D. Zhang, X.-H. Chen, Y.-H. Zhai, and L.-A. Wu, Phys. Rev. Lett. **94**, 173601 (2005).
26. A. F. Abouraddy, P. R. Stone, A. V. Sergienko, B. E. Saleh, and M. C. Teich, Phys. Rev. Lett. **93**, 213903 (2004).
27. A. N. Boto, P. Kok, D. S. Abrams, S. L. Braunstein, C. P. Williams, and J. P. Dowling, Phys. Rev. Lett. **85**, 2733 (2000).
28. M. D'Angelo, M. V. Chekhova, and Y. Shih, Phys. Rev. Lett. **87**, 013602 (2001).
29. K. Wang and D.-Z. Cao, Phys. Rev. A **70**, 041801 (2004); D.-Z. Cao, Z. Li, Y.-H. Zhai, and K. Wang, Eur. Phys. J. D **33**, 137 (2005).
30. J. Xiong, F. Huang, D. Z. Cao, H. G. Li, X. J. Sun, and K. Wang, Phys. Rev. Lett. **94**, 173601 (2005).
31. G. Scarcelli, A. Valencia, and Y. Shih, Europhys. Lett. **68**, 618 (2004).
32. A. Zeilinger, Rev. Mod. Phys. **71** S288 (1999).
33. M. O. Scully and K. Druehl, Phys. Rev. A **25**, 2208 (1982).
34. L. Mandel and E. Wolf, *Optical Coherence and Quantum Optics* (Cambridge University Press, Cambridge, 1995).
35. O. Svelto, *Principles of Lasers* (Plenum, New York, 1982).
36. W. Martiessen and E. Spiller, Am. J. Phys. **32**, 919 (1964).
37. R. Hanburry-Brown and R. Q. Twiss, Nature (London) **177**, 27 (1956).
38. A. V. Belinsky and D. N. Klyshko, Phys. Lett. A **166**, 303 (1992).
39. J. W. Goodman, *Laser speckle and related phenomena*, Topics in Applied Physics, Vol. 9, ed. by J. Dainty (Springer-Verlag, Berlin, 1975), p. 9.
40. F. T. Arecchi, Phys. Rev. Lett. **15**, 912 (1965).
41. J. C. Howell, R. S. Bennink, S. J. Bentley, and R. W. Boyd, Phys. Rev. Lett. **92** 210403 (2004).
42. A. F. Abouraddy, M. B. Nasr, B. E. Saleh, A. V. Sergienko, and M. C. Teich, Phys. Rev. A **63**, 063803 (2001).

6 Quantum Limits of Optical Super-Resolution

Mikhail I. Kolobov

Laboratoire PhLAM, Université de Lille 1, F-59655 Villeneuve d'Ascq Cedex, France
mikhail.kolobov@univ-lille1.fr

6.1 Super-Resolution in Classical Optics

The classical limit of resolution of an optical instrument was formulated in the well-known works by Abbe and Rayleigh at the end of the nineteenth century [1]. This classical limit states that the resolution of an optical system is limited by diffraction on the system pupil. Because of diffraction a point source at the input of the system creates a diffraction pattern of finite size on its output. When two point sources are placed closer and closer to each other, their diffraction patterns start to overlap and it becomes more and more difficult to discriminate these patterns. The smallest distance between two input point sources that allows for discrimination depends on many factors and is difficult to quantify. Several criteria have been proposed for such a discrimination, and the most famous one is the classical Rayleigh criterion. According to it, the diffraction patterns of two point sources are considered to be just resolved if the central maximum of the first one coincides with the first minimum of the second. For the case of the Airy pattern, corresponding to the Fraunhofer diffraction of a point source at a circular aperture, this gives the smallest distance between two input point sources equal to $R = 0.61\lambda/\alpha$, where λ is the wavelength of the light and α is the ratio of the radius of the system pupil to the distance between the pupil and the image plane. The distance R is known as the Rayleigh resolution limit.

As follows from this argument, the classical Rayleigh resolution limit is based on a simple visual observation and presumed resolving capabilities of a human eye. It is not a fundamental physical limit such as the speed of light or a Heisenberg uncertainty relation. Today it is recognized that modern CCD cameras allow us to achieve performance very much exceeding that of a visual observation. For example, experimental measurements of displacements in the nanometer range have been performed to detect deflection of glass fibers [2–4], microscopic phase objects [5], movement of biological subcellular vesicles [6], measurement of ultraweak absorption using the mirage effect [7], or in atomic force microscopy [8]. In all these measurements the resolution is superior to the classical Rayleigh limit and is determined not by diffraction, but by a different type of fluctuation in the experimental scheme. The possibility of improving the resolution beyond the diffraction limit is generally called "super-resolution" and often has different meanings. Below we shall give a rigorous definition of the term *super-resolution* in the sense used in this chapter.

In modern classical optics the resolution of an optical system is characterized not by the two-point Rayleigh resolution criterion, but in terms of its spatial transmission bandwidth. A typical optical system has a finite band of spatial frequencies that are transmitted through the system up to some cut-off frequency determined by the size of the system pupil. The optical system is then said to be bandlimited or diffraction-limited because diffraction effects on its pupil are responsible for finite resolution.

A coherent diffraction-limited imaging system in classical optics can be described by a linear equation relating the complex amplitude $a(s)$ of an input object with the complex amplitude $e(s)$ of the image [9],

$$e(s) = \int_{-\infty}^{\infty} h(s, s')a(s')ds', \tag{6.1}$$

The impulse response function $h(s, s')$ that appears in this integral equation represents the image at point s in the image plane from a point-source at point s' in the object plane. For translationally invariant or isoplanatic systems the impulse response depends only on the difference $s - s'$ and the integral in (6.1) becomes convolution,

$$e(s) = \int_{-\infty}^{\infty} h(s - s')a(s')ds'. \tag{6.2}$$

In optics, the impulse response $h(s - s')$ is usually called *the point-spread function* (PSF) of the system, and its Fourier transform *the transfer function* (TF). For bandlimited optical systems the transfer function is identically zero outside the transmission band of the system. *Super-resolution* is defined as technique of restoring the spatial frequencies of the object outside the transmission band, or in other words, enhancing the resolution beyond the diffraction limit [9]. It is important to underline that in the case where the object and the image fields are related by the convolution (6.2), super-resolution in this sense is impossible. To achieve super-resolution one needs some a priori information about the input object. In this chapter we shall use as the a priori information the assumption that the object has finite spatial size. In this case, the spatial Fourier spectrum of the object is an entire analytical function and therefore, it can be completely determined by the analytic continuation from the part of the spectrum transmitted by the system pupil [10, 11]. Such out-of-band extrapolation of the spatial spectrum of the object is equivalent to the resolution enhancement beyond the Rayleigh limit. This idea of super-resolution dates back to the 1960s and has been intensively discussed in the literature [10–16].

However, it was recognized that such an analytic continuation of the spatial spectrum of the object is extremely sensitive to the presence of noise in the system. In fact, the problem of out-of-band extrapolation of the spatial spectrum is a typical case of the so-called "ill-posed problem" [17]. This property seriously hampers the potential of super-resolution. In practice, to achieve super-resolution one has to detect the diffraction-limited image at the output of the optical system and then try to reconstruct the original object using specially designed numerical algorithms. In the general case, an attempt to obtain significant super-resolution beyond the Rayleigh limit leads to a drastic decrease of the signal-to-noise ratio in the reconstructed object as compared to that in the original one. The main conclusions that one can derive from the numerous papers on classical super-resolution are [9]:

(i) Significant super-resolution in the sense of out-of-band extrapolation is possible only in the case when the size of the original object is not too large compared with the Rayleigh resolution distance;
(ii) The amount of super-resolution increases logarithmically, that is, rather weakly, with the signal-to-noise ratio in the original object.

6.2 Quantum Theory of Super-Resolution

This and the following sections are based on Refs. [18,19] where the quantum theory of optical super-resolution was developed. We refer the reader to these references for further details.

6.2.1 Quantum Theory of Optical Imaging

We shall first review the classical theory of optical imaging and introduce the basis functions and physical parameters that will be used in its quantum counterpart.

The optical scheme of diffraction-limited coherent optical imaging is shown in Fig. 6.1. For simplicity we consider the one-dimensional case. The object of finite size X is placed in the object plane. The first lens L_1 performs the spatial Fourier transform of the object into the pupil plane with a pupil of finite size d. Diffraction on this pupil is a physical origin of the finite resolution in our scheme (we neglect diffraction on the imaging lenses). The second lens L_2 performs the inverse Fourier transform and creates a diffraction-limited image in the image plane.

As mentioned above, to achieve super-resolution one needs some a priori information about the object. In our case we know a priori that the object is confined within the area of size X and is identically zero outside. The spatial Fourier transform of such an object is an entire analytical function. Therefore, knowing the part of the Fourier spectrum within the area d of the pupil allows for an analytic continuation of the total spectrum and, therefore, for unlimited resolution.

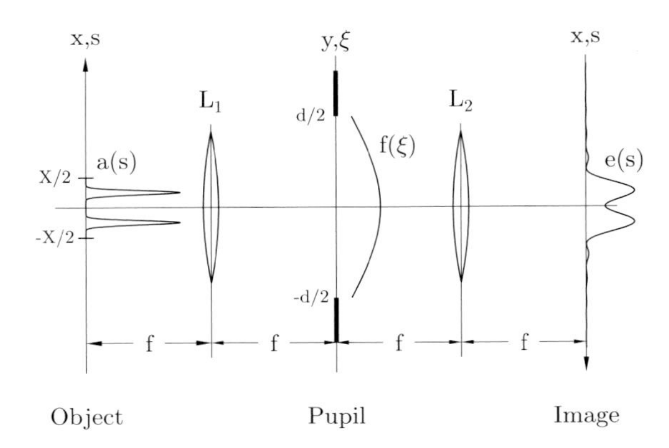

Fig. 6.1. Optical scheme of one-dimensional coherent diffraction-limited optical imaging.

Let us introduce the dimensionless spatial coordinates in the object and the image plane as $s = 2x/X$, and in the pupil plane as $\xi = 2y/d$ (see Fig. 6.1). In terms of dimensionless coordinates s the transformation L of the classical object amplitude $a(s)$ into the classical image amplitude $e(s)$ reads:

$$e(s) = (La)(s) = \int_{-1}^{1} \frac{\sin[c(s-s')]}{\pi(s-s')} a(s')\, ds', \qquad -\infty < s < \infty. \tag{6.3}$$

Here $c = \pi dX/2\lambda f$ is the space-bandwidth product.

The problem of reconstruction of the object $a(s)$ from a detected image $e(s)$ in the absence of noise is equivalent to inversion of the integral operator L. The operator $L^\star$ adjoint to L is given by [20]

$$(L^\star f)(s) = \int_{-\infty}^{\infty} \frac{\sin[c(s-s')]}{\pi(s-s')} f(s')\, ds', \qquad |s| \le 1. \tag{6.4}$$

The product $A = L^\star L$ is the self-adjoint operator,

$$(Af)(s) = \int_{-1}^{1} \frac{\sin[c(s-s')]}{\pi(s-s')} f(s')\, ds', \qquad |s| \le 1, \tag{6.5}$$

studied by Slepian and Pollak [21]. The orthonormal system of eigenfunctions of A is given by

$$\varphi_k(s) = \begin{cases} \dfrac{1}{\sqrt{\lambda_k}} \psi_k(s) & |s| \le 1, \\ 0 & |s| > 1, \end{cases} \tag{6.6}$$

where $\psi_k(s)$ are the linear prolate spheroidal functions [16, 21], and λ_k are the corresponding eigenvalues. The functions $\varphi_k(s)$ form a basis in $L^2(-1,1)$

and may be considered "elements of information" of the input object. The eigenvalues λ_k are an infinite set of real positive numbers obeying $1 \geq \lambda_0 > \lambda_1 > \cdots > 0$. For small k the λ_k fall off slowly with k until the index reaches the critical value $k = S$, called the Shannon number,

$$S = \frac{2c}{\pi} = \frac{dX}{\lambda f}, \tag{6.7}$$

beyond which the λ_k rapidly approach zero.

Using the properties of prolate spheroidal functions,

$$\int_{-1}^{1} \frac{\sin[c(s-s')]}{\pi(s-s')} \psi_k(s')\, ds' = \lambda_k \psi_k(s), \tag{6.8}$$

$$\int_{-\infty}^{\infty} \frac{\sin[c(s-s')]}{\pi(s-s')} \psi_k(s')\, ds' = \psi_k(s), \tag{6.9}$$

we obtain

$$L\varphi_k = \sqrt{\lambda_k}\psi_k, \qquad L^\star \psi_k = \sqrt{\lambda_k}\varphi_k. \tag{6.10}$$

Expanding the object amplitude over the functions $\varphi_k(s)$ and the image amplitude over $\psi_k(s)$, we can easily find the relation between the expansion coefficients of the object and the image. Indeed, because the functions $\varphi_k(s)$ form a complete orthonormal set in $[-1, 1]$ we can write the object amplitude as

$$a(s) = \sum_{k=0}^{\infty} a_k \varphi_k(s), \qquad |s| \leq 1, \tag{6.11}$$

with the coefficients a_k given by

$$a_k = \int_{-\infty}^{\infty} a(s)\varphi_k(s)\, ds. \tag{6.12}$$

A similar expansion can be written for the image amplitude in terms of functions $\psi_k(s)$,

$$e(s) = \sum_{k=0}^{\infty} e_k \psi_k(s), \qquad -\infty < s < \infty, \tag{6.13}$$

with the coefficients e_k given by

$$e_k = \int_{-\infty}^{\infty} e(s)\psi_k(s)\, ds. \tag{6.14}$$

Substituting these expansions into Eq. (6.3) and using the first of Eqs. (6.10) we obtain the following relation between a_k and e_k,

$$e_k = \sqrt{\lambda_k} a_k. \tag{6.15}$$

In quantum theory the classical object amplitude $a(s)$ becomes the dimensionless photon annihilation operator in the object plane $\hat{a}(s)$ and the classical image amplitude $e(s)$ the corresponding photon annihilation operator in the image plane $\hat{e}(s)$. These operators obey the standard commutation relations,

$$[\hat{a}(s), \hat{a}^\dagger(s')] = \delta(s - s'), \qquad [\hat{e}(s), \hat{e}^\dagger(s)] = \delta(s - s'), \tag{6.16}$$

and are normalized so that $\langle \hat{a}^\dagger(s)\hat{a}(s)\rangle$ gives the mean photon number per unit dimensionless length in the object plane and $\langle \hat{e}^\dagger(s)\hat{e}(s)\rangle$ in the image plane.

In the quantum theory we can use Eqs. (6.11),(6.13) now treating the expansion coefficients a_k and e_k as photon annihilation operators $\hat{a}_k$ and $\hat{e}_k$. Using the properties of the prolate spheroidal functions it can be shown that the operators $\hat{a}_k$ in the object plane obey the following commutation relations,

$$[\hat{a}_k, \hat{a}_l^\dagger] = \delta_{kl}, \qquad [\hat{a}_k, \hat{a}_l] = 0. \tag{6.17}$$

The same commutation relations must be satisfied by the photon annihilation operators $\hat{e}_k$ in the image plane. However, Eq. (6.15) does not preserve the commutation relations (6.17). The reason for this is that the classical imaging equation (6.3) takes into account only nonzero field amplitude in the region $|s| \leq 1$ of the object plane. The rest of this plane $|s| > 1$ is ignored because there the classical field amplitude is zero. In quantum theory this region must be taken into account to guarantee the conservation of the commutation relations.

To obtain the canonical transformation of photon annihilation and creation operators from the object into the image plane we shall split the coordinate s into two regions, the "core", $|s| \leq 1$, corresponding to the area of localization of the classical object, and the "wings", $|s| > 1$, outside this area. The orthonormal bases in these areas of the object plane are given by [18]

$$\varphi_k(s) = \begin{cases} \frac{1}{\sqrt{\lambda_k}}\psi_k(s) & |s| \leq 1, \\ 0 & |s| > 1, \end{cases} \qquad \chi_k(s) = \begin{cases} 0 & |s| \leq 1, \\ \frac{1}{\sqrt{1-\lambda_k}}\psi_k(s) & |s| > 1. \end{cases} \tag{6.18}$$

In terms of two sets $\{\varphi_k(s)\}$ and $\{\chi_k(s)\}$ we can write the annihilation operators in the object plane as

$$\hat{a}(s) = \sum_{k=0}^{\infty} \hat{a}_k \varphi_k(s) + \sum_{k=0}^{\infty} \hat{b}_k \chi_k(s). \tag{6.19}$$

Here $\hat{b}_k$ are the annihilation operators of the prolate modes χ_k in the wings region, expressed through the field operator $\hat{a}(s)$ by

$$\hat{b}_k = \int_{-\infty}^{\infty} \hat{a}(s)\chi_k(s)ds. \tag{6.20}$$

Substituting the expansion (6.19) into Eq. (6.3) we obtain the following relation between the photon annihilation operators in the object and the image plane,

$$\hat{e}_k = \sqrt{\lambda_k}\hat{a}_k + \sqrt{1-\lambda_k}\hat{b}_k. \tag{6.21}$$

It is easy to verify that this transformation preserves the commutation relations of the operators, $[\hat{a}_k, \hat{a}_l^\dagger] = [\hat{b}_k, \hat{b}_l^\dagger] = [\hat{e}_k, \hat{e}_l^\dagger] = \delta_{kl}$.

Equation (6.21) is completely equivalent to the transformation performed by a beamsplitter. Indeed, if we consider the operators $\hat{a}_k$ and $\hat{b}_k$ as the photon annihilation operators in the modes defined by prolate spheroidal waves incoming to the beamsplitter with the amplitude transmission coefficient $\sqrt{\lambda_k}$ and the reflection coefficient $\sqrt{1-\lambda_k}$, then e_k is the photon annihilation operator in the kth mode of the transmitted wave.

From Fig. 1 one may think that the vacuum fluctuations coming into the image plane from the region $|\xi| > 1$ of the Fourier plane outside the pupil should also be taken into account. Indeed, when treating the field in the Fourier plane as an operator we must include the contribution from this region into the resulting field in the image plane. However, the advantage of expansion (6.13) is that the field from this region does not contribute to the expansion coefficients $\hat{e}_k$ of the image because it is orthogonal to the prolate spheroidal wave functions. This property was pointed out by Bertero and Pike in [20] for out-of-band classical noise and remains valid in quantum theory.

6.2.2 Quantum Theory of Optical Fourier Microscopy

One can decompose the input object and the output image over these eigenfunctions and obtain the relation between the decomposition coefficients. Then detecting the output image with, for example, a sensitive CCD camera, one can evaluate the decomposition coefficients of the image. Using the relation between the decomposition coefficients of the image and the object one can reconstruct the latter with resolution better than the classical diffraction limit.

Our numerical simulations in Ref. [23] have shown, however, that for evaluation of the decomposition coefficients one has to detect the output image over an unrealistically large area in the image plane due to the oscillating behavior of the prolate functions. That is why in Ref. [22] we have proposed a modified version of the scheme where the CCD camera is placed in the pupil plane instead of the image plane, and one detects the spatial Fourier spectrum. We have called this modified scheme super-resolving Fourier microscopy. The advantage of this set-up is that now the spatial Fourier spectrum is measured over the finite region within the pupil. To understand the role of the quantum fluctuations on the resolution of Fourier microscopy we need to formulate the quantum theory of this modified scheme.

The dimensionless photon annihilation and creation operators in the pupil plane will be denoted as $\hat{f}(\xi)$ and $\hat{f}^\dagger(\xi)$. These operators obey the standard commutation relations,

$$[\hat{f}(\xi), \hat{f}^\dagger(\xi')] = \delta(\xi - \xi'), \tag{6.22}$$

and are normalized so that $\langle \hat{f}^\dagger(\xi)\hat{f}(\xi)\rangle$ gives the mean photon number per unit dimensionless length in the pupil plane. The spatial Fourier transform $(T\hat{a})(\xi)$ performed by lens L_1, in terms of these dimensionless variables reads as follows,

$$\hat{f}(\xi) = (T\hat{a})(\xi) = \sqrt{\frac{c}{2\pi}} \int_{-\infty}^{\infty} \hat{a}(s) e^{-ics\xi} ds. \tag{6.23}$$

It is important to note that the limits of integration in this equation are over the whole object plane because it concerns operators and not the classical c-numbers.

In terms of two sets $\{\varphi_k(s)\}$ and $\{\chi_k(s)\}$ we can write the annihilation operators in the pupil plane as

$$\hat{f}(\xi) = \sum_{k=0}^{\infty} \hat{f}_k \varphi_k(\xi) + \sum_{k=0}^{\infty} \hat{g}_k \chi_k(\xi). \tag{6.24}$$

Here $\hat{f}_k$ are the annihilation operators of the prolate modes φ_k in the core region of the pupil plane, and $\hat{g}_k$ are the annihilation operators of the prolate modes χ_k in the wings region.

In our analysis we shall use the following properties of prolate spheroidal functions [16],

$$\int_{-1}^{1} \varphi_k(s) e^{-ics\xi} ds = (-i)^k \sqrt{\frac{2\pi}{c}} \psi_k(\xi), \tag{6.25}$$

$$\int_{-\infty}^{\infty} \psi_k(s) e^{-ics\xi} ds = (-i)^k \sqrt{\frac{2\pi}{c}} \varphi_k(\xi). \tag{6.26}$$

Using (6.18), the field transform (6.23) between the object and the pupil plane and these properties, we find the following propagation relations for the core and wings of the light wave,

$$(T\varphi_k)(\xi) = (-i)^k \left[\sqrt{\lambda_k}\varphi_k(s) + \sqrt{1-\lambda_k}\chi_k(s)\right], \tag{6.27}$$

$$(T\chi_k)(\xi) = (-i)^k \left[\sqrt{1-\lambda_k}\varphi_k(s) - \sqrt{\lambda_k}\chi_k(s)\right]. \tag{6.28}$$

Substituting Eqs. (6.19) and (6.24) into the field transform (6.23) and using Eqs. (6.27) and (6.28), we arrive at the following relations between the photon annihilation operators of the prolate modes in the object and the pupil planes,

$$\hat{f}_k = (-i)^k (\sqrt{\lambda_k}\hat{a}_k + \sqrt{1-\lambda_k}\hat{b}_k), \tag{6.29}$$

$$\hat{g}_k = (-i)^k(\sqrt{1-\lambda_k}\hat{a}_k - \sqrt{\lambda_k}\hat{b}_k). \tag{6.30}$$

These relations are similar to the transformation performed by a beamsplitter with the amplitude transmission coefficients $(-i)^k\sqrt{\lambda_k}$ and the reflection coefficients $(-i)^k\sqrt{1-\lambda_k}$, and preserve the commutation relation of the annihilation and creation operators in the pupil plane.

Let us assume that we can detect the spatial Fourier amplitudes $\hat{f}(\xi)$ in the pupil plane within the transmission area of the pupil using a sensitive CCD camera. This transmitted part of the spatial Fourier spectrum is given by the first sum in Eq. (6.24); the term given by the second sum is absorbed by the opaque area of the pupil. It should be emphasized that, because we need the complex field amplitudes and not the intensities, one should use the homodyne detection scheme with a local oscillator. Using Eqs. (6.24), (6.20), and(6.29) we can calculate the operator-valued coefficients $\hat{a}_k^{(r)}$ of the reconstructed object as

$$\hat{a}_k^{(r)} = \frac{\hat{f}_k}{(-i)^k\sqrt{\lambda_k}} = \hat{a}_k + \sqrt{\frac{1-\lambda_k}{\lambda_k}}\hat{b}_k, \tag{6.31}$$

where the superscript (r) stands for "reconstructed." As follows from Eq. (6.31), the reconstruction of the input object is not exact because of the second term in Eq. (6.31). This term contains the annihilation operators $\hat{b}_k$ responsible for the vacuum fluctuations of the electromagnetic field in the area outside the object. It is important to notice that these vacuum fluctuations prevent reconstruction of the higher and higher coefficients $\hat{a}_k$ in the object because of the multiplicative factor $\sqrt{(1-\lambda_k)/\lambda_k}$. Indeed, the eigenvalues λ_k rapidly become very small after the index k has attained some critical value. This leads to rapid "amplification" of the vacuum fluctuations in the reconstructed object that limits the number of the reconstructed coefficients $\hat{a}_k$.

6.3 Quantum Limits in Reconstruction of Optical Objects

6.3.1 Reconstruction of Classical Noise-Free Objects

In this section we shall illustrate numerically the role of quantum fluctuations on the reconstruction of simple objects with super-resolution beyond the classical diffraction limit. However, before taking into account quantum fluctuations of light in the input object, we first demonstrate the potential of the super-resolution technique with prolate spheroidal functions for reconstruction of noise-free classical objects, that is, when the quantum fluctuations are neglected. This case corresponds to the classical limit of quantum theory developed in the previous section and can be simply obtained by taking mean values of the operators.

In what follows we shall denote the classical complex amplitudes corresponding to the quantum-mechanical operators by the same letters without carets; for example, $a(s) = \langle \hat{a}(s) \rangle$, $a_k = \langle \hat{a}_k \rangle$, and so on. Because the classical complex amplitude of the object is zero outside the area $|s| \leq 1$, we have $\langle \hat{b}_k \rangle = 0$. Using Eq. (6.19) we can write this classical amplitude as

$$a(s) = \sum_{k=0}^{\infty} a_k \varphi_k(s). \tag{6.32}$$

The classical complex amplitude $f(\xi)$ of the field in the pupil plane is obtained from Eq. (6.23),

$$f(\xi) = \sqrt{\frac{c}{2\pi}} \int_{-1}^{1} a(s) e^{-ics\xi} ds, \tag{6.33}$$

with the integration limits over the object area, $|s| \leq 1$. Taking into account the property of the prolate functions given by Eq. (6.25), we can write the spatial Fourier spectrum $f(\xi)$ as the following decomposition,

$$f(\xi) = \sum_{k=0}^{\infty} (-i)^k a_k \psi_k(s), \qquad -\infty < \xi < \infty. \tag{6.34}$$

This spatial Fourier spectrum of the object spreads outside the transmission area of the pupil $|\xi| \leq 1$. The spatial Fourier components in the opaque area are absorbed and cannot be detected by the CCD camera placed in the pupil plane. Super-resolution attempts to reconstruct these absorbed Fourier components. From Eq. (6.31) we obtain the classical reconstructed coefficients,

$$a_k^{(r)} = a_k. \tag{6.35}$$

Because we have neglected the quantum fluctuations, the reconstructed coefficients are identical to those of the input object. The classical amplitude of the reconstructed object $a^{(r)}(s)$ can be written as the following decomposition over the prolate functions,

$$a^{(r)}(s) = \sum_{k=0}^{L-1} a_k \varphi_k(s). \tag{6.36}$$

Because in practice one can never have infinitely many coefficients a_k, we have restricted the summation in this equation to L first prolate functions. When $L \to \infty$, the reconstructed object approaches the exact one, $a^{(r)}(s) \to a(s)$. In practice the super-resolution over the Rayleigh limit is determined by the number L of terms used in the decomposition (6.36).

Alternatively to reconstruction of the object itself one can try to reconstruct its spatial Fourier spectrum as

$$f^{(r)}(\xi) = \sum_{k=0}^{L-1} (-i)^k a_k \psi_k(s), \qquad -\infty < \xi < \infty. \tag{6.37}$$

Similarly to the reconstruction of the object, when $L \to \infty$, the reconstructed spectrum approaches the exact one, $f^{(r)}(\xi) \to f(\xi)$.

For numerical simulations we have taken a simple object of two narrow Gaussian peaks,

$$a(s) = A\left[\exp\left(-\frac{(s-s_0)^2}{2\sigma^2}\right) + \exp\left(-\frac{(s+s_0)^2}{2\sigma^2}\right)\right], \qquad |s| \le 1, \tag{6.38}$$

of width σ separated by distance $2s_0$. We choose $2s_0 = 1$ and $\sigma = 0.1$, so that two peaks are well separated in the input object. The normalization constant A is chosen so that the integral of the object intensity over the area of the object is equal to the total mean number of photons $\langle \hat{N} \rangle$ in the object,

$$\int_{-1}^{1} a^2(s) ds = \langle \hat{N} \rangle. \tag{6.39}$$

The Rayleigh resolution distance $R = \pi X/(2c)$ in dimensionless coordinates s is equal to π/c, where c is the space-bandwidth product. In our simulations we work with $c = 1$. In this situation for $2s_0 < \pi$ we are beyond the Rayleigh limit.

In Fig. 6.2a we have plotted the normalized input object $a(s)/\sqrt{\langle \hat{N} \rangle}$, in Fig. 6.2b, its spatial Fourier spectrum $f(\xi)$ in the pupil plane, and in Fig. 6.2c, the output image $e(s)$ in the image plane. Comparing the input object with its image one can clearly see that it is impossible to resolve two Gaussian peaks in the image plane according to the Rayleigh criterion. In Fig. 6.2b we have shown by the grey color the opaque area of the pupil. The part of the spatial Fourier spectrum of the object in this area is absorbed and therefore cannot be detected by the CCD camera placed in the pupil plane. Below we shall illustrate the reconstruction of these absorbed Fourier components by the prolate functions technique.

For numerical simulations we had to evaluate two sets of prolate spheroidal functions, $\varphi_k(s)$, defined on the interval $|s| \le 1$, and $\psi_k(s)$ defined for all s, $-\infty < s < \infty$. The first set is necessary for decomposition of the input object $a(s)$, and the second one is needed for reconstruction of the spatial Fourier spectrum $f^{(r)}(\xi)$. For numerical calculations of $\varphi_k(s)$ we have used the algorithm from Ref. [24]. In this algorithm the prolate functions $\varphi_k(s)$ are evaluated as the series with the Legendre polynomials $P_k(s)$,

$$\varphi_n(s) = \sum_{k=0}^{\infty} \gamma_k^{(n)} \sqrt{k + \frac{1}{2}} P_k(s), \qquad |s| \le 1. \tag{6.40}$$

The coefficients $\gamma_k^{(n)}$ are found as the eigenvectors of the symmetric matrix A with the following nonzero elements,

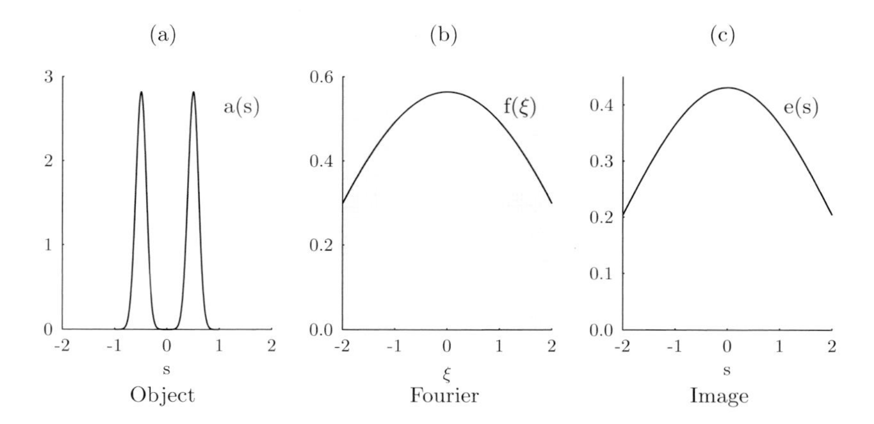

Fig. 6.2. Double-peak object $a(s)$ used in numerical simulations (a), its spatial Fourier spectrum $f(\xi)$ in the pupil plane (b), and the image $e(s)$ created in the image plane (c). Grey area shows the part of the spatial spectrum absorbed by the opaque area of the pupil.

$$A_{k,k} = k(k+1) + c^2 \frac{2k(k+1)-1}{(2k+3)(2k-1)}, \tag{6.41}$$

$$A_{k,k+2} = A_{k+2,k} = c^2 \frac{(k+2)(k+1)}{(2k+3)\sqrt{(2k+1)(2k+5)}}, \tag{6.42}$$

for all $k = 0, 1, 2 \ldots$.

We have written a numerical program in Mathematica that implements this algorithm. The advantage of this method is that it does not require the direct solution of the eigenproblem for $\varphi_k(s)$ and λ_k which is unstable due to the rapid decrease of the eigenvalues. This algorithm allows us to calculate for $c = 1$ at least 17 first prolate functions in spite of the fact that the eigenvalues of the higher-order functions become extremely small (e.g., $\lambda_{17} = 4.183 \times 10^{-50}$). In Fig. 6.3 we show the first 17 prolate functions $\varphi_k(s)$ evaluated by our numerical program.

For numerical calculation of the second set of the prolate spheroidal functions, $\psi_k(s)$, we have used the following property of the Legendre polynomials (see Eq. (10.1.14) in Ref. [25]),

$$\int_{-1}^{1} P_n(s) e^{-ics\xi} ds = 2i^n j_n(\xi), \tag{6.43}$$

where $j_n(s)$ is the spherical Bessel function of the first order [25]. Using this equation we can easily obtain the following representation of $\psi_n(\xi)$,

$$\psi_n(\xi) = \sqrt{\frac{2c}{\pi}} i^n \sum_{k=0}^{\infty} (-i)^k \gamma_k^{(n)} \sqrt{k + \frac{1}{2}} j_k(c\xi), \qquad -\infty < \xi < \infty. \tag{6.44}$$

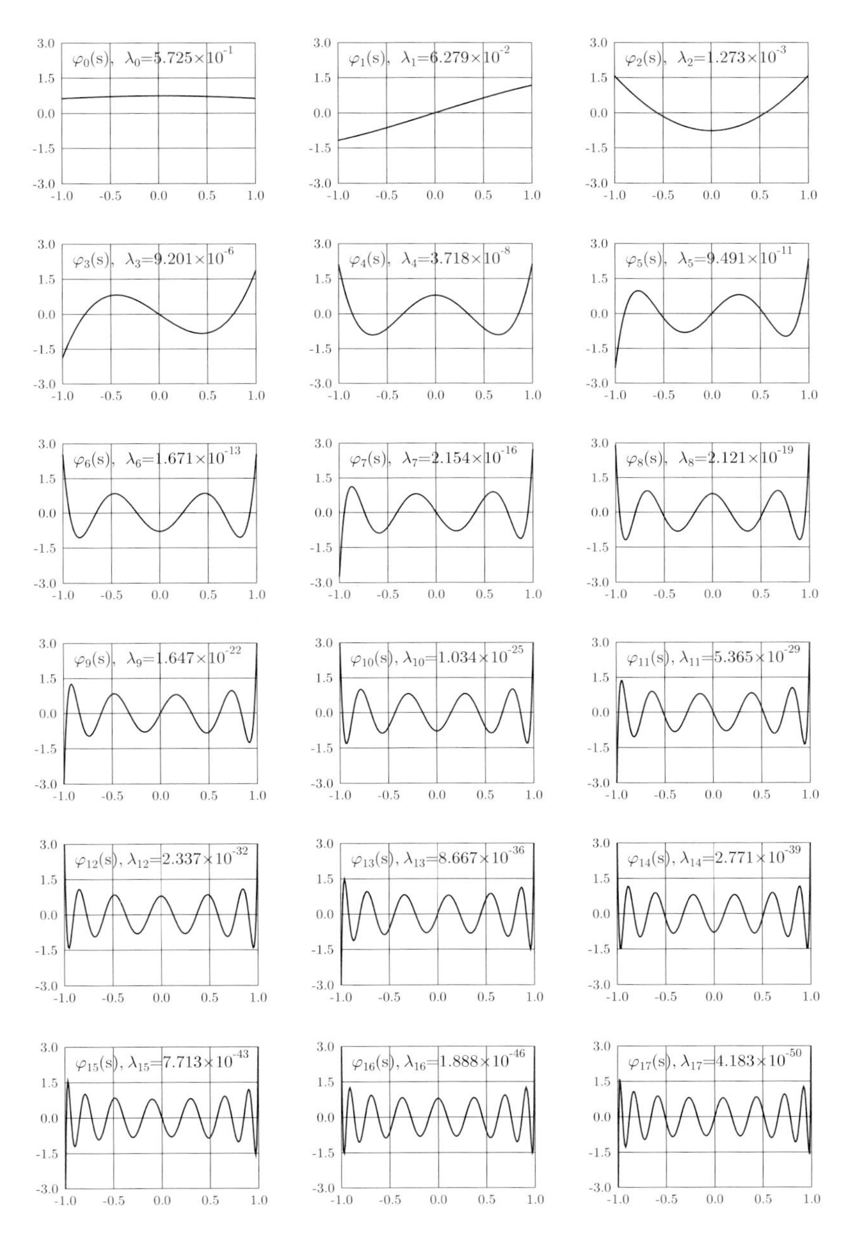

Fig. 6.3. Examples of the prolate spheroidal functions $\varphi_k(s)$ and the corresponding eigenvalues λ_k calculated using our numerical program.

We illustrate the result of reconstruction of the spatial Fourier spectrum of the input object in Fig. 6.4. In this figure we show the exact spatial Fourier spectrum of the input object, drawn as a solid line, as a function of dimensionless coordinate ξ in the pupil plane. Only part of this spectrum within the transmission area of the pupil, $|\xi| \leq 1$, is transmitted to the image plane. The spatial Fourier harmonics in the opaque area of the pupils, shown by the grey color, are absorbed. This is a reason for the of very large diffraction spread in the image plane shown in Fig. 6.2c. The dashed lines in Fig. 6.4 correspond to the spatial Fourier spectrum of the reconstructed object with $L = 5, 7$, and 11 prolate functions. One can see that the reconstructed spectrum approaches the exact one for ever higher spatial frequencies $|\xi|$ as the number of prolate functions increases.

With 7 prolate functions two spectra are very close to each other for spatial frequencies $|\xi| \leq 8$. This corresponds to a super-resolution factor of 8 over the Rayleigh limit.

6.3.2 Reconstruction of Objects with Quantum Fluctuations

For numerical simulations of quantum fluctuations we have chosen a c-number representation of the quantum mechanical operators $\hat{a}_k$ and $\hat{b}_k$ in Eq. (6.31) corresponding to the antinormal ordering of the creation and annihilation operators. In this representation the operators $\hat{a}_k$ and $\hat{b}_k$ become the c-number Gaussian stochastic variables α_k and β_k, respectively, which we shall write as

$$\alpha_k = a_k + \delta\alpha_k, \qquad \beta_k = \delta\beta_k. \tag{6.45}$$

Here $a_k = \langle \hat{a}_k \rangle$ is the mean value of the field coefficients in the object area, and $\delta\alpha_k$ and $\delta\beta_k$ are the stochastic Gaussian fluctuations. Note that the mean values $\langle \hat{b}_k \rangle$ are zero because the classical field components outside the object vanish. We have chosen the antinormally ordered representation because it remains valid even in the case of the multimode squeezed state of the light field at the input of the scheme.

We introduce the quadrature components of the fluctuations $\delta\alpha_k$ and $\delta\beta_k$ as follows,

$$\delta\alpha_k = \delta X_k^{\alpha} + i\delta Y_k^{\alpha}, \qquad \delta\beta_k = \delta X_k^{\beta} + i\delta Y_k^{\beta}. \tag{6.46}$$

When the input light is in the coherent state in the object area and in the vacuum state outside, the correlation functions of the quadrature fluctuations are equal to

$$\langle \delta X_k^{\mu} \delta X_{k'}^{\mu} \rangle = \langle \delta Y_k^{\mu} \delta Y_{k'}^{\mu} \rangle = \frac{1}{4}\delta_{kk'}, \tag{6.47}$$

with $\mu = \alpha, \beta$.

If instead of coherent light we use the multimode squeezed light for illumination of the object and multimode squeezed vacuum in the area outside with subsequent homodyne detection at the pupil plane, these correlation functions become

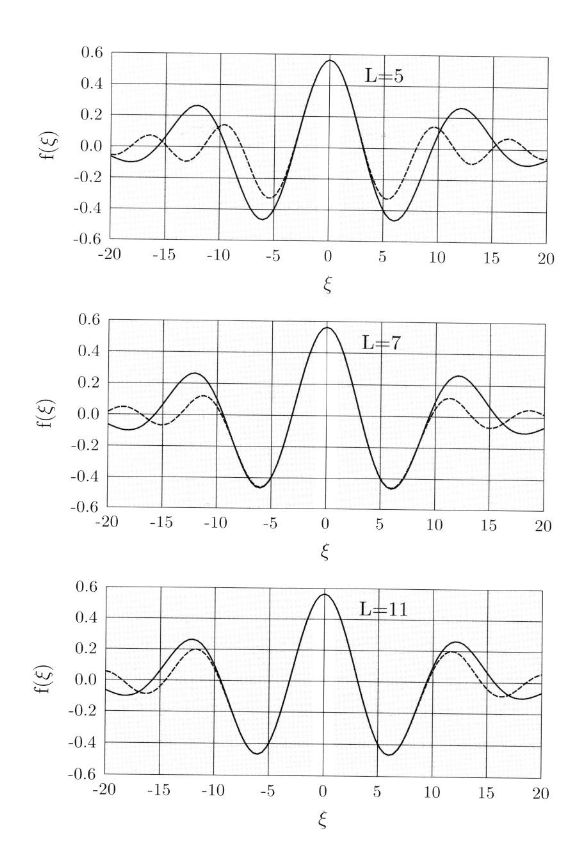

Fig. 6.4. Exact spatial Fourier spectrum of the object from Fig. 6.3a (solid line), and the spectra reconstructed with $L = 5, 7$, and 11 prolate functions (dashed lines). Here and in Fig. 6.5 the grey area indicates the absorbed part of the spatial spectrum.

$$\langle \delta X_k^\mu \delta X_{k'}^\mu \rangle = \frac{1}{4} e^{-2r} \delta_{kk'} \qquad \langle \delta Y_k^\mu \delta Y_{k'}^\mu \rangle = \frac{1}{4} e^{2r} \delta_{kk'}, \tag{6.48}$$

where r is the squeezing parameter. In these formulas we have assumed that the input light is amplitude squeezed and for simplicity have chosen the same squeezing parameter for all the essential modes that are used in the decomposition of the reconstructed object.

The relative value of quantum fluctuations depends on the signal-to-noise ratio in the input object which for the light in a coherent state is determined by the total mean number of photons passed through the object area during the observation time. For example, for a laser beam with $\lambda = 1064$ nm and optical power of 1 mW, and observation time of 1 ms we obtain the mean photon number of $\langle \hat{N} \rangle = 5.3 \cdot 10^{12}$.

In Fig. 6.5a we have shown the results of reconstruction of the spatial Fourier spectrum of the object from Fig. 6.2a when quantum fluctuations of a coherent state are taken into account. The solid line gives the exact spatial Fourier spectrum of the object. As in Fig. 6.4 the grey area shows the absorbing part of the pupil. We use 7 prolate functions and the mean photon number in the input object is taken $\langle \hat{N} \rangle = 10^{12}$. The five thin lines correspond to the five random Gaussian realizations of the quantum fluctuations in the coherent state of $\hat{a}_k$ and the vacuum fluctuations of $\hat{b}_k$. The dashed line corresponds to the reconstructed spectrum with 7 prolate functions without noise. One can observe that the role of quantum fluctuations becomes more and more important as one goes to the higher and higher spatial frequencies where the random realizations of the Fourier spectra deviate more and more from the mean value given by the dashed line.

In Fig. 6.5b we have increased the total mean value of photons to $\langle \hat{N} \rangle = 10^{13}$. This corresponds to an increased signal-to-noise ratio in the input object and should allow for better super-resolution. This is illustrated in Fig. 6.5b by reduced deviation of the random realizations from the mean value of the spectrum as compared to Fig. 6.5a.

The same result can be achieved by using multimode squeezed light instead of increasing the power of the source illuminating the object. This is illustrated in Fig. 6.5c where we have used $\langle \hat{N} \rangle = 10^{12}$ as in Fig. 6.5a, but have considered the light in a multimode squeezed state with the squeezing parameter $e^r = 10$ instead of the coherent state. As the result the fluctuations in the higher spatial frequencies are reduced, giving better super-resolution.

In the next section we shall give a quantitative characteristic of super-resolution as a function of the signal-to-noise ratio.

6.3.3 Point-Spread Function for Super-Resolving Reconstruction of Objects

For the reconstruction process we can write a similar relation between the reconstructed field operator $\hat{a}^{(r)}(s)$ and the object field operator $\hat{a}(s)$. Using an operator-valued equivalent of Eq. (6.36) together with Eq. (6.31) we arrive at the following result,

$$\hat{a}^{(r)}(s) = \int_{-1}^{1} h^{(r)}(s,s')\hat{a}(s')ds' + \sum_{k=0}^{L-1} \sqrt{\frac{1-\lambda_k}{\lambda_k}}\hat{b}_k\varphi_k(s). \tag{6.49}$$

Here the reconstruction point-spread function $h^{(r)}(s,s')$ is given by

$$h^{(r)}(s,s') = \sum_{k=0}^{L-1} \varphi_k(s)\varphi_k(s'). \tag{6.50}$$

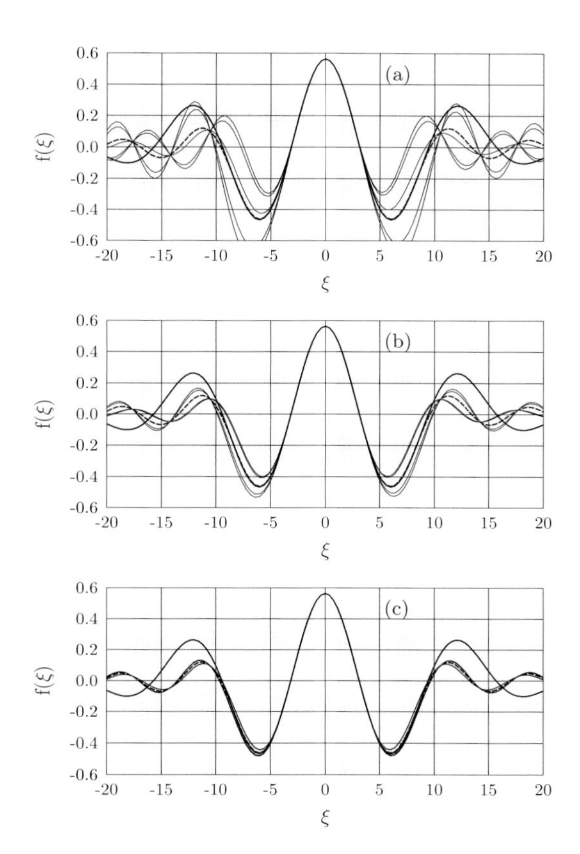

Fig. 6.5. Exact spatial Fourier spectrum of the object from Fig. 6.2a (solid line), the reconstructed spectrum with $L = 7$ prolate functions (dashed lines), and five random Gaussian realizations of the reconstructed spectrum with $L = 7$ prolate functions (thin lines): (a) coherent light with mean total photon number $\langle \hat{N} \rangle = 10^{12}$; (b) coherent light with $\langle \hat{N} \rangle = 10^{13}$; and (c) squeezed light with $\langle \hat{N} \rangle = 10^{12}$ and $\exp(r) = 10$.

As seen from this equation, the form of the reconstruction PSF and, in particular, its width depend on the number of terms L in the sum. When this number grows infinitely, $L \to \infty$, the reconstruction PSF tends to the δ-function,

$$\lim_{L \to \infty} h^{(r)}(s, s') = \sum_{k=0}^{\infty} \varphi_k(s)\varphi_k(s') = \delta(s - s'), \tag{6.51}$$

and we have unlimited super-resolution. However, this ideal situation is never realized practically due to the second term in Eq. (6.49) which grows infinitely when $L \to \infty$. Thus, Eq. (6.49) is a good illustration of the statement that the ultimate limit of super-resolution in the reconstructed object is given not by diffraction but by the quantum fluctuations of light represented by the second term.

The number L of terms in the sum (6.50) that determines the width of the reconstruction PSF, depends on the signal-to-noise ratio in the input object. To obtain the maximum L we shall compare the signal-to-noise ratio in the input object to that in the reconstructed object. As follows from Eq. (6.49) with increasing L the signal-to-noise ratio in the reconstructed object deteriorates. We shall assume that reconstruction of the object is possible until the limit when the signal-to-noise ratio in the reconstructed object becomes unity.

Let us define the signal-to-noise ratio in the input object as [26]

$$R = \frac{\langle \hat{N} \rangle^2}{\langle (\Delta \hat{N})^2 \rangle}, \tag{6.52}$$

where

$$\langle \hat{N} \rangle = \int_{-1}^{1} \langle \hat{a}^\dagger(s) \hat{a}(s) \rangle ds, \tag{6.53}$$

is the total mean number of photons in the input object, and $\langle (\Delta \hat{N})^2 \rangle$ its variance. Similarly we define the signal-to-noise ratio $R^{(r)}$ in the reconstructed object as

$$R^{(r)} = \frac{\langle \hat{N}^{(r)} \rangle^2}{\langle (\Delta \hat{N}^{(r)})^2 \rangle}, \tag{6.54}$$

where the mean number of photons in the reconstructed object is given by

$$\langle \hat{N}^{(r)} \rangle = \int_{-1}^{1} \langle \hat{a}^{(r)\dagger}(s) \hat{a}^{(r)}(s) \rangle ds. \tag{6.55}$$

The deterioration of the signal-to-noise ratio in the reconstructed object can be described by the noise figure F,

$$F = \frac{R}{R^{(r)}}, \tag{6.56}$$

that is commonly used in the literature about amplifiers. Because the signal-to-noise ratio $R^{(r)}$ in the reconstructed object is always smaller than that in the input object, the noise figure is always larger than unity. If we assume that the minimum value of $R^{(r)}$ that allows for reconstruction of the object is unity, this gives us the maximum noise figure $F_{\max} = R$ corresponding to the maximum super-resolution.

Let us consider an input object in a coherent state, so that $\langle \hat{a}(s) \rangle = a(s)$, $\langle \hat{a}^\dagger(s) \hat{a}(s) \rangle = |a(s)|^2$, $\langle \hat{a}^\dagger(s) \hat{a}^\dagger(s') \hat{a}(s') \hat{a}(s) \rangle = |a(s)|^2 |a(s')|^2$. It is easy to show that in this case the input signal-to-noise ratio R is equal to the mean total photon number in the input object,

$$R = \langle \hat{N} \rangle = \int_{-1}^{1} |a(s)|^2 ds. \tag{6.57}$$

On the other hand, for the signal-to-noise ratio $R^{(r)}$ in the reconstructed object in this case we obtain the following result,

$$R^{(r)} = \Big(\sum_{k=0}^{L-1} |a_k|^2 \Big)^2 / \Big(\sum_{k=0}^{L-1} \frac{|a_k|^2}{\lambda_k} \Big), \tag{6.58}$$

where a_k are the coefficients of decomposition of $a^{(r)}(s)$ over the prolate functions $\varphi_k(s)$ in Eq. (6.36).

As follows from Eq. (6.58), the signal-to-noise ratio $R^{(r)}$ and, therefore, the noise figure F depend on the shape of the input object. For numerical evaluation of the super-resolution factor as a function of the total mean number of photons in the input object we have taken a narrow rectangular object placed at the origin $s = 0$,

$$a_\epsilon(s) = \begin{cases} \sqrt{\dfrac{\langle \hat{N} \rangle}{\epsilon}} & |s| \leq \epsilon/2, \\ 0 & |s| > \epsilon/2, \end{cases} \tag{6.59}$$

Taking the width ϵ of this object ever smaller we arrive at a pointlike source, while keeping the total number of photons constant and equal to $\langle \hat{N} \rangle$. Such a pointlike object gives us the reconstruction PSF $h^{(r)}(0, s)$ at the output.

The degree of super-resolution in the reconstructed object can be characterized by the ratio of the width of the diffraction-limited imaging PSF to the width of the reconstruction PSF. In Fig. 6.6 we have shown the imaging PSF $h(s)$ and the reconstruction PSF $h^{(r)}(0, s)$ for $L = 7$ normalized to unity at their maxima. To define the super-resolution factor we shall introduce the half-widths W and W_L of these two PSF measured at their half-maxima. Then we define the super-resolution factor S as the ratio of W to W_L,

$$S = \frac{W}{W_L}. \tag{6.60}$$

For the example given in Fig. 6.6 these half-widths are equal to $W = 1.895$, $W_L = 0.252$, and $S = 7.5$.

In Fig. 6.7 we have plotted the super-resolution factor S as a function of the total mean number of photons $\langle \hat{N} \rangle$ in the input object for the case of coherent light and multimode squeezed light. As seen from this figure, for the same mean number of photons multimode squeezed light provides higher super-resolution than the coherent light.

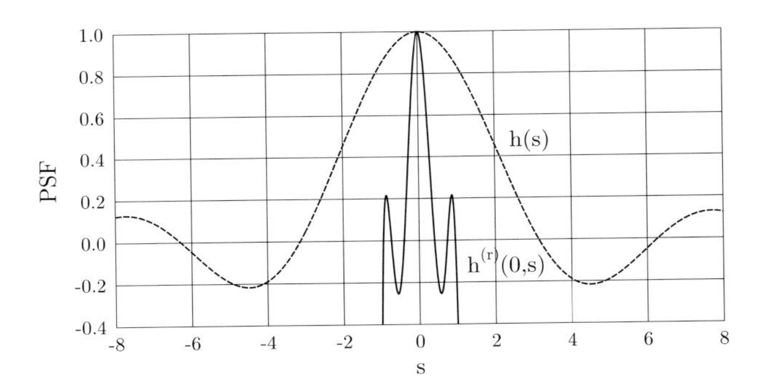

Fig. 6.6. Diffraction-limited imaging point-spread function $h(s)$ and the reconstruction point-spread function $h^{(r)}(0, s)$ using 7 prolate functions.

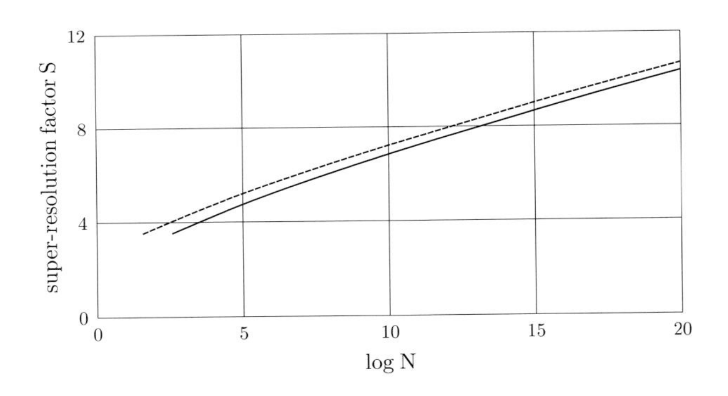

Fig. 6.7. Super-resolution factor S as a function of the total mean number of photons $\langle \hat{N} \rangle$ for coherent light (solid line), and multimode squeezed line with $\exp(r) = 10$ (dashed line).

6.4 Squeezed-Light Source for Microscopy with Super-Resolution

In previous sections we have presented a theoretical scheme that allows us to improve the super-resolution beyond the standard quantum limit in reconstruction of an optical object using multimode squeezed light. The theory was formulated in terms of prolate spheroidal functions that are the eigenwaves of the optical imaging scheme. To achieve super-resolution with multimode squeezed light one has to prepare these prolate waves in a squeezed state. The question remains on how to produce such squeezed prolate waves.

In this section we provide the answer to this question. Precisely we demonstrate that an optical parametric amplifier (OPA) with a properly chosen diaphragm on its output and a Fourier lens produces squeezed prolate spheroidal waves used in super-resolving microscopy. We investigate the quantum statistics of the squeezed prolate spheroidal waves in our scheme in dependence on physical parameters of the OPA and the optical configuration. We formulate simple estimates on the number of "object elements" to be reconstructed in connection with the number of degrees of freedom in a nonclassical illuminating light.

The scheme of one-dimensional optical imaging with multimode squeezed light is shown in Fig. 6.8.

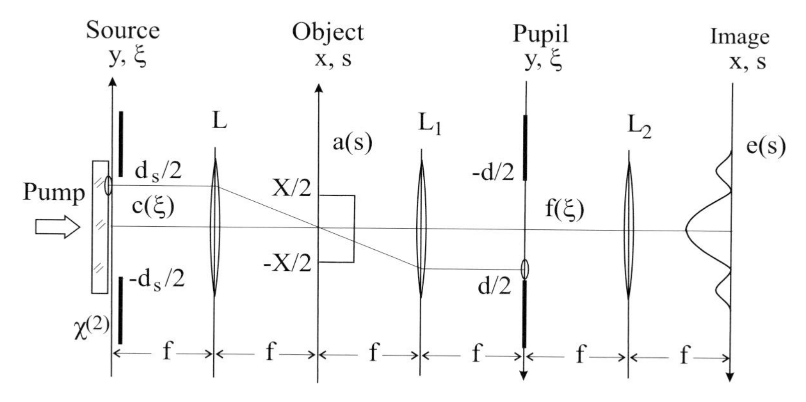

Fig. 6.8. Schematic of optical imaging with squeezed light.

The part of the scheme to the right of the object plane performs diffraction-limited imaging of an object of finite size X located in the object plane, investigated above. The part to the left of the object plane is an illumination scheme. It consists of a traveling-wave OPA placed in the source plane and a Fourier lens L. It is wellknown from the literature that a traveling-wave OPA with plane-wave pump and nonlinear crystal with large transverse area creates a multimode squeezed vacuum on its output [29]. A new feature of our scheme is a diaphragm of size d_s on the output of the OPA which serves for selection of the transverse modes in squeezed state. As we demonstrate below, when the size of this diaphragm matches the size of the pupil, $d_s \geq d$, this set-up squeezes exactly the prolate spheroidal waves that are the eigenmodes of the imaging scheme. This result can be easily understood qualitatively. Indeed, when all three lenses in the scheme have the same focal distance f, as in Fig. 6.1, the lenses L and L_1 create a geometrical image of the diaphragm d_s in the pupil plane. Therefore, it is intuitively clear that one has to match the diaphragm size d_s and the pupil size d to select the modes of the source that will pass through the imaging scheme.

Let us introduce the dimensionless coordinates in the source plane as $\xi = 2y/d$ (see Fig. 6.8). The dimensionless photon annihilation operators in the source plane, denoted as $\hat{c}(\xi)$, obey the standard commutation relations,

$$[\hat{c}(\xi), \hat{c}^\dagger(\xi')] = \delta(\xi - \xi'), \tag{6.61}$$

These operators are normalized so that $\langle \hat{c}^\dagger(\xi)\hat{c}(\xi)\rangle$, gives the mean photon number per unit dimensionless length in the source plane. The Fourier transform $(T\hat{c})(s)$ physically performed by the lens L in Fig. 6.8 reads as follows,

$$\hat{a}(s) = (T\hat{c})(s) = \sqrt{\frac{c}{2\pi}} \int_{-\infty}^{\infty} d\xi e^{-ics\xi} \hat{c}(\xi). \tag{6.62}$$

The photon annihilation operator $\hat{a}(s)$ in the object plane can be written as the decomposition over $\psi_k(s)$,

$$\hat{a}(s) = \sum_{k=0}^{\infty} \hat{A}_k \psi_k(s) + \hat{F}(s), \tag{6.63}$$

where the operator-valued coefficients $\hat{A}_k$ are evaluated as

$$\hat{A}_k = \int_{-\infty}^{\infty} ds\, \hat{a}(s)\psi_k(s). \tag{6.64}$$

The operators $\hat{A}_k$ and $\hat{A}_k^\dagger$ satisfy the standard commutation relations of the photon annihilation and creation operators for discrete modes,

$$[\hat{A}_k, \hat{A}_{k'}^\dagger] = \delta_{k,k'}, \qquad [\hat{A}_k, \hat{A}_{k'}] = 0. \tag{6.65}$$

The Fourier transform of the prolate spheroidal functions $\psi_k(s)$, performed by the lens L_1, is zero outside the interval $|\xi| \leq 1$. Consequently, the set of functions $\{\psi_k(s)\}$ is not complete in the Hilbert space $L^2(-\infty, \infty)$, and to satisfy the commutation relations for operators $\hat{a}(s)$ and $\hat{a}^\dagger(s)$ one has to add an additional term $\hat{F}(s)$. This term has a zero Fourier spectrum in the interval $|\xi| \leq 1$ and does not contribute to the coefficients $\hat{A}_k$,

$$\int_{-\infty}^{\infty} ds\, \psi_k(s)\hat{F}(s) = 0. \tag{6.66}$$

The field $\hat{F}(s)$ can be decomposed over the complementary set of prolate spheroidal functions $\{\theta_k(s)\}$, orthogonal to $\{\psi_k(s)\}$.

Physically speaking, the first term in Eq. (6.63) is the object field component that propagates in our scheme through the pupil to the image plane. The second object field component $\hat{F}(s)$ is absorbed outside the pupil and is not observed. Therefore, in what follows we shall omit $\hat{F}(s)$ in the object field (6.63).

It follows from (6.18) that the functions $\psi_k(s)$ and $\theta_k(s)$ can be written in the form

$$\psi_k(s) = \sqrt{\lambda_k}\varphi_k(s) + \sqrt{1-\lambda_k}\chi_k(s),$$
$$\theta_k(s) = \sqrt{1-\lambda_k}\varphi_k(s) - \sqrt{\lambda_k}\chi_k(s). \tag{6.67}$$

In terms of two sets $\{\varphi_k(s)\}$ and $\{\chi_k(s)\}$ we can write the annihilation operators in the source plane as

$$\hat{c}(\xi) = \sum_{k=0}^{\infty} \hat{c}_k \varphi_k(\xi) + \sum_{k=0}^{\infty} \hat{d}_k \chi_k(\xi), \tag{6.68}$$

Here $\hat{c}_k$ are the annihilation operators of the prolate waves φ_k in the core region, and $\hat{d}_k$ are the annihilation operators of the prolate modes χ_k in the wings region. The operators $\hat{c}_k$ are expressed through the field operator $\hat{c}(\xi)$ by

$$\hat{c}_k = \int_{-\infty}^{\infty} d\xi\, \hat{c}(\xi)\varphi_k(\xi), \qquad \hat{d}_k = \int_{-\infty}^{\infty} d\xi\, \hat{c}(\xi)\chi_k(\xi). \tag{6.69}$$

Taking into account (6.68) and (6.19) we obtain

$$\hat{a}(s) = (T\hat{c})(s) = \sum_{k=0}^{\infty} (-i)^k \left[\hat{c}_k \psi_k(s) + \hat{d}_k \theta_k(s)\right], \tag{6.70}$$

and

$$\hat{A}_k = (-i)^k \hat{c}_k. \tag{6.71}$$

As expected, there is no contribution into $\hat{A}_k$ from the second sum in Eq. (6.68) containing operators $\hat{d}_k$ and describing the illumination coming from the wings area of the source. If the source diaphragm is larger or matches the size of the pupil, $d_s \geq d$, it has no effect on the operator amplitudes $\hat{c}_k$ and $\hat{A}_k$ (see (6.20)), and we shall neglect the diaphragm in the calculation of $\hat{A}_k$.

To obtain explicitly the coefficients $\hat{A}_k$ in the case of illumination of the scheme by a traveling-wave OPA we shall use a simplified description of an OPA with a plane-wave undepleted pump (see Ref. [29]). In this approximation one can find analytically the spatial Fourier amplitudes $\hat{C}(q)$,

$$\hat{C}(q) = \int_{-\infty}^{\infty} d\xi e^{-iq\xi} \hat{c}(\xi), \tag{6.72}$$

of the field at the output of the crystal as a linear transformation of the corresponding input Fourier amplitudes $\hat{C}_{\text{in}}(q)$ and $\hat{C}^\dagger_{\text{in}}(q)$,

$$\hat{C}(q) = U(q)\hat{C}_{\text{in}}(q) + V(q)\hat{C}^\dagger_{\text{in}}(-q). \tag{6.73}$$

Here the operators $\hat{C}_{\rm in}(q)$ at the input of the OPA are in the vacuum state. The complex coefficients $U(q)$ and $V(q)$ depend on the nonlinear susceptibility of the crystal, its length, and the matching conditions in the OPA. These coefficients have the property

$$|U(q)|^2 - |V(q)|^2 = 1, \tag{6.74}$$

that guarantees the preservation of the commutation relations (6.61). Using the standard parameters of multimode squeezing (see Ref. [29]) we obtain:

$$U(q) \pm V^*(-q) = e^{-i\varphi(q)} \left[e^{\pm r(q)} \cos\theta(q) + i e^{\mp r(q)} \sin\theta(q) \right]. \tag{6.75}$$

The phase of the amplified (stretched) quadrature amplitude of the OPA output field $\tilde{c}(q)$ is $\theta(q)$. The phase factor $\exp(-i\varphi(q))$ specifies the phase of the quadrature amplitudes of the input field $\hat{C}_{\rm in}(q)$ that are squeezed or stretched. For the vacuum input field of the OPA the last phase is irrelevant.

The operator amplitudes $\hat{A}_k$ are found explicitly with the use of Eqs. (6.71) and (6.69). The quantities $\hat{c}_{\rm in}(\xi)$ and $\varphi_k(\xi)$ are expressed through their Fourier transforms (6.73) and (6.27). After some calculation we obtain,

$$\hat{A}_k = \frac{1}{\sqrt{2\pi c}} \int_{-\infty}^{\infty} dq\, \psi_k(q/c) \left[U(q)\hat{C}_{\rm in}(q) + V(q)\hat{C}^\dagger_{\rm in}(-q) \right]. \tag{6.76}$$

We shall introduce the real quadrature components of the field amplitudes $\hat{A}_k$ in the object plane as

$$\hat{A}_k = \hat{A}_{1k} + i\hat{A}_{2k}. \tag{6.77}$$

For the variances of these quadrature components we obtain

$$\langle \left(\Delta\hat{A}_{1k} \right)^2 \rangle = \frac{1}{4c} \int_{-\infty}^{\infty} dq\, \psi_k^2(q/c) \left[e^{\pm 2r(q)} \cos^2\theta(q) + e^{\mp 2r(q)} \sin^2\theta(q) \right], \tag{6.78}$$

$$\langle \left(\Delta\hat{A}_{2k} \right)^2 \rangle = \frac{1}{4c} \int_{-\infty}^{\infty} dq\, \psi_k^2(q/c) \left[e^{\mp 2r(q)} \cos^2\theta(q) + e^{\pm 2r(q)} \sin^2\theta(q) \right]. \tag{6.79}$$

Here the upper and the lower sign correspond, respectively, to the even and the odd prolate spheroidal functions ψ_k.

It follows from this result that the prolate spheroidal waves in the object illumination can be prepared in squeezed state. By proper choice of the squeezing phase $\theta(q)$ at low spatial frequencies q one can minimize quantum fluctuations in one of the quadrature amplitudes $A_{\sigma k}$, $\sigma = 1, 2$ (namely, in the one detected in the image plane of our scheme). Taking the degree and the phase of squeezing as constant, $r(q) = r$, $\theta(q) = \theta$, we can estimate the variance of the squeezed quadrature amplitude:

$$\langle (\Delta A_{\sigma k})^2 \rangle \sim e^{-2r}/4. \tag{6.80}$$

In reality the spatial-frequency band of multimode squeezing is limited by the phase-matching condition in the OPA. One has to take into account the spatial-frequency dispersion of squeezing, that is, the frequency dependence of the squeezing phase $\theta(q)$ due to diffraction in free space and inside the OPA. Both phenomena degrade squeezing of prolate spheroidal waves.

As shown in [28, 29], the effect of diffraction on squeezing can be almost perfectly compensated by means of adjustment of the lens array. If this is done, it follows from (6.78), (6.79), that the degree of squeezing of prolate spheroidal waves depends on the overlap in the object plane of two areas: (i) the area illuminated by the squeezed plane waves $\hat{C}(q)$, which are focused by lens L to the points $s = q/c$, and (ii) the area of the energy localization $\sim \psi_k^2(s)$ of the prolate spheroidal waves.

In analogy to other phenomena with spatially multimode squeezed light, in the even and odd components of the object field the different quadrature amplitudes are squeezed; for, say, $\theta = 0$, the squeezed quadratures are A_{1k} and A_{2k} for the odd and the even prolate waves, respectively.

Finally, we can formulate the conditions on multimode squeezing in object illumination in terms of the number N of independent degrees of freedom in the light field, propagating through the diaphragm of the OPA (we assume here the optimum size $d_s = d$ of the diaphragm). The properties of the OPA emission can be characterized by the coherence length l_c of the output field $\hat{c}(\xi)$. The spatial-frequency range q_c of effective squeezing is related to the coherence length by the estimate

$$|q_c| \leq \pi(d/2)/l_c. \tag{6.81}$$

The minimum requirement on the OPA is that the effectively squeezed waves $\hat{C}(q_c)$ illuminate, after passing the lens L, the object region $|s| \leq 1$. That is, the waves

$$|q_c|/c \geq 1, \tag{6.82}$$

should be squeezed. This gives the following estimate for the coherence length,

$$l_c \leq \pi d/2c, \tag{6.83}$$

and for the number N of independent degrees of freedom in the illuminating light, emitted from the region d:

$$N \sim d/l_c \geq \frac{dX}{\lambda f} = S. \tag{6.84}$$

Here S is the Shannon number of our optical scheme.

As seen from (6.67), the wave profiles $\psi_k(s)$ can be expanded in terms of two bases, $\{\varphi_k(s)\}$ and $\{\chi_k(s)\}$, representing the core and the wings of illuminating field in the object plane, where $\lambda_k \leq 1$ are the eigenvalues of the imaging transformation. As known from the theory of prolate spheroidal functions [16,21], these eigenvalues are close to 1 only for $k \leq S$, where S is the

Shannon number. For higher values of the index k the field energy in prolate spheroidal waves $\psi_k(s)$ is concentrated in the wings, that is, outside the object area $|s| \leq 1$. Hence, the condition $N \sim S$ provides an effective squeezing only of the prolate spheroidal waves with $k \leq S$. In order to minimize quantum noise of the higher prolate spheroidal waves (with $k > S$), it is necessary to use OPA with a large number of effectively squeezed spatial modes of radiation,

$$N \gg S, \tag{6.85}$$

and to illuminate by nonclassical light a spot in the object plane with the size much larger than the object itself.

References

1. J. Rayleigh, in *Collected Optics Papers of Lord Rayleigh*, part A (Opt. Soc. Am., Washington,1994), p. 117.
2. A. Flock and D. Strelioff, Nature, **310**, 397 (1984).
3. J. J. Art, A. C. Craftford, and R. Fettiplace, J. Physiol. (London)**371**, 18P (1986).
4. S. Kamimura, Appl. Opt. **26**, 3425 (1987).
5. W. Denk and W. W. Webb, Appl. Opt. **29**, 2382 (1990).
6. J. Jelles, B. J. Schnapp, and M. P. Scheetz, Nature, **331**, 450 (1988).
7. D. Fournier, A. Boccara, N. Amer, and R. Gerlach, Appl. Phys. Lett. **37**, 519 (1980).
8. C. Putman, B. De Grooth, N. Van Hulst, and J. Greve, J. Appl. Phys. **72**, 6 (1992).
9. M. Bertero, and C. De Mol, in *Progress in Optics* Vol. XXXVI, edited by E. Wolf (North-Holland, Amsterdam, 1996), p. 129.
10. H. Wolter, in *Progress in Optics* Vol. I, edited by E. Wolf (North-Holland, Amsterdam, 1996), ch. V.
11. J. L. Harris, J. Opt. Soc. Am. **54**, 931 (1964).
12. G. Toraldo di Francia, J. Opt. Soc. Am. **45**, 497 (1955).
13. C. W. McCutchen, J. Opt. Soc. Am. **57**, 1190 (1967).
14. C. K. Rushforth and R. W. Harris, J. Opt. Soc. Am. **58**, 539 (1968).
15. G. Toraldo di Francia, J. Opt. Soc. Am. **59**, 799 (1969).
16. B. R. Frieden, in *Progress in Optics*, Vol. IX, E. Wolf, ed. (North-Holland, Amsterdam, 1971), pp. 311-407.
17. A. N. Tikhonov and V. Y. Arsenin, *Solution of ill-posed problems* (Winston/Wiley, Washington, DC, 1977).
18. M. I. Kolobov and C. Fabre, Phys. Rev. Lett. **85**, 3789 (2000).
19. V. N. Beskrovnyy and M. I. Kolobov, Phys. Rev. A **71**, 043802 (2005).
20. M. Bertero and E. R. Pike, Opt. Acta **29**, 727 (1982).
21. D. Slepian and H. O. Pollak, Bell System Tech. J. **40**, 43 (1961).
22. P. Scotto, P. Colet, M. San Miguel, and M. I. Kolobov, in Proceedings of the European Quantum Electronics Conference EQEC 2003, Munich, (2003).
23. M. I. Kolobov, C. Fabre, P. Scotto, P. Colet, and M. San Miguel, in *Coherence and Quantum Optics VIII*, N. Bigelow, J. H. Eberly, C. R. Stroud, and I. A. Walmsley, eds. (Plenum, New York, 2003).

24. H. Xiao, V. Rokhlin, and N. Yarvin, Inverse Problems, **17**, 805 (2001).
25. M. Abramowitz and I. A. Stegun, *Handbook of Mathematical Functions*, 9th ed. (Dover, New York, 1970).
26. M. I. Kolobov and L. A. Lugiato, Phys. Rev. A **52**, 4930 (1995).
27. I. V. Sokolov and M. I. Kolobov, Optics Letters **29**, 703 (2004).
28. M. I. Kolobov and I. V. Sokolov, JETP **69**, 1097 (1989); M. I. Kolobov and I. V. Sokolov, Phys. Lett. A **140**, 101 (1989).
29. M. I. Kolobov, Rev. Mod. Phys. **71**, 1539 (1999).

7 Noiseless Amplification of Optical Images

Mikhail I. Kolobov[1] and Eric Lantz[2]

[1] Laboratoire PhLAM, Université de Lille 1, F-59655 Villeneuve d'Ascq Cedex, France
mikhail.kolobov@univ-lille1.fr
[2] Laboratoire d'Optique P. M. Duffieux, Université de Franche Comté, F-25030 Besançon Cedex, France
elantz@univ-fcomte.fr

7.1 Introduction

A possibility for noiseless amplification was demonstrated theoretically by Caves [1] using general quantum-mechanical considerations based on conservation of commutation relations for the Heisenberg operators in the amplification process. Concerning their noise characteristics all amplifiers can be divided into two groups: phase-insensitive amplifiers (PIA), whose gain is independent of the phase of the input signal, and phase-sensitive amplifiers (PSA), whose gain is phase-dependent. Caves has demonstrated that all phase-insensitive amplifiers introduce at least 3 dB of noise in the amplified signal, whereas phase-sensitive amplifiers can, under certain conditions, preserve the signal-to-noise ratio. This last quality is a reason why nowadays these amplifiers are called *noiseless*. Noiseless amplification of temporal optical signals was demonstrated experimentally in [2] for continuous-wave and in [3] for pulsed optical signals.

The first step towards noiseless amplification of images was made in [26], where it was theoretically demonstrated that such an amplifier can be realized by a ring-cavity optical parametric amplifier below threshold, operating as a phase-sensitive amplifier. The cavity-based geometry ensures a continuous-wave regime of amplification. Under certain conditions found in [4], such an amplifier preserves the signal-to-noise ratio of the amplified image. Recently, a cavity-based noiseless image amplification was studied theoretically for an optical parametric oscillator in a confocal cavity [5].

Traveling-wave noiseless amplification of images was investigated in Ref. [6]. In that paper the authors considered an optical parametric amplifier (OPA) without an external cavity. This geometry is more natural for possible practical realizations and has several advantages over the cavity-based scheme. The most important ones are greater frequency and spatial-frequency amplification bandwidths that are determined by the phase-matching conditions in the OPA. Greater frequency bandwidth allows for amplification of images in single-shot pulses, and greater spatial-frequency bandwidth provides better resolving power of noiseless amplification. In the noiseless image amplification scheme considered in Ref. [6], the OPA was placed in the spatial Fourier plane of the imaging scheme. An alternative configuration with a traveling-wave OPA where an input object is projected directly at the entrance of the OPA, was investigated in Ref. [7]. In this case under certain conditions one can also obtain noiseless image amplification.

Parametric image amplification has been studied by several experimental groups [8–14]. In the experiments performed by Lantz's group, parametric image amplification was achieved for a monochromatic near-infrared image with resolution 60×80 points in the amplified image with a mean gain of 15 dB [8]. Then the process of parametric amplification was applied for low-pass and bandpass spatial filtering of amplified images using the filtering properties of the transfer function of the amplifier [9], and for ultrahigh-speed imaging [10]. Parametric amplification has also been employed in time-gated image recovery [12, 13], and in biomedical imaging [14]. Quantum fluctuations in parametric image amplification were studied in [15–17]. In particular, in [16] noiseless image amplification was observed for the first time. However, in this experiment the authors did not address spatial distribution of quantum fluctuations in the image. Instead, they investigated the temporal behavior of quantum fluctuations recorded by a photodiode at the Fourier frequency of 27 MHz while the photodiode was scanned over the image. Spatially noiseless amplification of optical images was demonstrated for the first time in Ref. [17].

In this chapter we shall give a brief theoretical description of a traveling-wave noiseless image amplifier, following Ref. [6]. We shall also describe recent experimental results demonstrating noiseless amplification of images in single-shot pulses. In Section 7.2 we shall outline the optical scheme of the traveling-wave image amplifier. In Section 7.3 we give a physical explanation of the squeezing transformation that governs the field evolution in our scheme and show the possibility for optimization of the observation procedure. Section 7.4 is devoted to investigation of the gain and noise characteristics of the amplifier and establishing the conditions for its noiseless performance. In Sections 7.5 and 7.6 we shall describe experimental results of noiseless image amplification.

7.2 Traveling-Wave Scheme for Amplification of Images

We consider the optical scheme of a traveling-wave image amplifier, shown in Fig. 7.1. An input optical image of a finite area S_O is located in the object plane P_1. We shall assume that this image is imprinted into a faint spatial modulation of the wavefront of an optical wave with carrier frequency ω illuminating the plane P_1. The distance, separating the lens L_1 from the object plane and the plane $z = 0$ from the lens L_1 is equal to the focal length f of the lens. The position z_2 of the face plane P_2 of the nonlinear crystal will be determined below from the condition of the optimum phase matching. In the case $z_2 = 0$, the input part of the scheme performs the

Fourier transform of the input wavefront from the object plane P_1 into the input plane of the nonlinear crystal of length l playing the role of a traveling-wave OPA. Parametric interaction takes place between the signal wave of frequency ω, carrying the input image, and the plane monochromatic pump wave of double frequency, $\omega_p = 2\omega$, illuminating the input plane of the crystal. At the output plane P_3 of the nonlinear crystal there is a pupil of area S_P. We

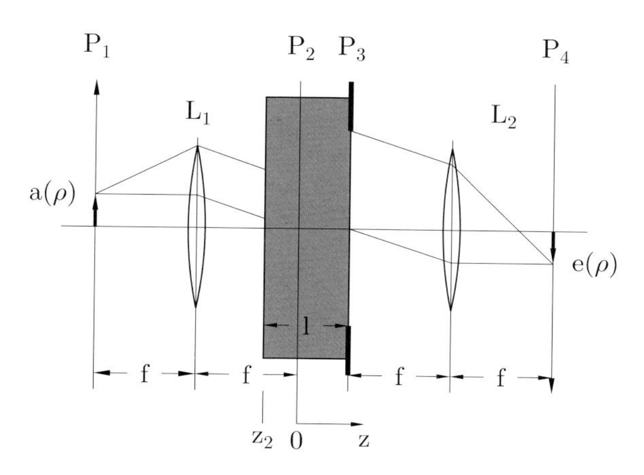

Fig. 7.1. Schematic of a traveling-wave optical image amplifier.

consider the case when the wave pattern of the object wave contains a finite range of transverse spatial frequencies. Input waves with the highest spatial frequencies are in the vacuum state. After passing the lens L_1 they illuminate the crystal at largest distances from the longitudinal axes z and stimulate only generation of spontaneous parametric down-conversion photons. The pupil cuts off these noise photons, which do not contribute to the signal part of the amplified wavefront in the image plane P_4.

The back Fourier transform of the amplified object wave is performed by the lens L_2. With the axis orientation as in Fig. 7.1, the imaging magnification of the scheme is +1. The amplified image is detected by a dense array of small photodetectors (pixels) in the image plane P_4.

In Ref. [6] it was shown that for an infinitely large pupil, $S_P \to \infty$, the photon annihilation operator $\hat{e}(\boldsymbol{\rho}, t)$ in the image plane is related to the photon annihilation operator $\hat{a}(\boldsymbol{\rho}, t)$ in the object plane by the following unitary transformation,

$$\hat{e}(\boldsymbol{\rho}, \Omega) = u(\boldsymbol{\rho}, \Omega)\hat{a}(\boldsymbol{\rho}, \Omega) + v(\boldsymbol{\rho}, \Omega)\hat{a}^\dagger(-\boldsymbol{\rho}, -\Omega). \tag{7.1}$$

Here $\hat{e}(\boldsymbol{\rho}, \Omega), \hat{a}(\boldsymbol{\rho}, \Omega)$ are the temporal Fourier transforms of $\hat{e}(\boldsymbol{\rho}, t), \hat{a}(\boldsymbol{\rho}, t)$. The coefficients $u(\boldsymbol{\rho}, \Omega)$ and $v(\boldsymbol{\rho}, \Omega)$ are given by

$$u(\boldsymbol{\rho},\Omega) = \exp\left\{i\Big[\Omega n/c - (2k-k_p)/2\Big]l\right\}\Big[\cosh\Gamma(\boldsymbol{\rho},\Omega) + \frac{i\delta(\boldsymbol{\rho},\Omega)}{2\Gamma(\boldsymbol{\rho},\Omega)}\sinh\Gamma(\boldsymbol{\rho},\Omega)\Big],$$

$$v(\boldsymbol{\rho},\Omega) = \exp\left\{i\Big[\Omega n/c - (2k-k_p)/2\Big]l\right\}\frac{g}{\Gamma(\boldsymbol{\rho},\Omega)}\sinh\Gamma(\boldsymbol{\rho},\Omega). \quad (7.2)$$

Here $c/n = 1/k'_\Omega$ is the velocity of light in the crystal. The dimensionless mismatch function $\delta(\boldsymbol{\rho},0)$ for $\Omega = 0$, which we shall call the local mismatch, is given by

$$\delta(\boldsymbol{\rho},0) = \delta_0 - \rho^2/\rho_0^2, \qquad \delta_0 = (2k-k_p)l, \quad (7.3)$$

where ρ_0 is defined as

$$\rho_0 = f\sqrt{\frac{1}{kl}}, \quad (7.4)$$

and

$$\Gamma = \sqrt{g^2 - \delta^2/4}. \quad (7.5)$$

An important case of degenerate phase matching in the crystal corresponds to $\delta_0 = 0$. In this case the maximum parametric amplification takes place for the waves, propagating in the z-direction. This implies the maximum amplification of the area around the optical axis of the scheme, $\boldsymbol{\rho} = 0$. The spatial scale ρ_0 determines the linear dimension of the effectively amplified area in the object plane.

In the case of nondegenerate phase matching with positive δ_0, $\delta_0 > 0$, the dimensionless mismatch (7.3) is equal to zero for waves with $q \neq 0$. This corresponds to effective parametric amplification of a region in the object plane, which has the shape of a ring.

In the case of a pupil of finite size Eq. (7.1) becomes

$$e(\boldsymbol{\rho},\Omega) = \frac{1}{\lambda f}\int d\boldsymbol{\rho}' p(\boldsymbol{\rho}-\boldsymbol{\rho}')\left[u(\boldsymbol{\rho}',\Omega)a(\boldsymbol{\rho}',\Omega) + v(\boldsymbol{\rho}',\Omega)a^\dagger(-\boldsymbol{\rho}',-\Omega)\right]. \quad (7.6)$$

The function $p(\boldsymbol{\rho})$ is related to the pupil frame function as

$$p(\boldsymbol{\rho}) = \frac{1}{\lambda f}\int d\boldsymbol{\xi}\, P(\boldsymbol{\xi})\exp\left[i\frac{2\pi}{\lambda f}\boldsymbol{\rho}\cdot\boldsymbol{\xi}\right], \quad (7.7)$$

and is called the *impulse response* of the optical system.

The pupil cuts off the photons of spontaneous parametric down-conversion, emitted from a peripheral region of nonlinear crystal. The wavepackets of these photons pass lens L_2 and come to the detection plane at strongly inclined directions with respect to the axes z. These wavepackets do not contribute to the amplified image, because their emission is stimulated by the vacuum fluctuations and not by the signal waves from the object plane. But the self-beats of these noise waves on the surface of the elementary detector (pixel) produce additional photocurrent noise and degrade the signal-to-noise ratio.

The optimum choice of the pupil area S_P is done as follows. Let the input image of the area S_O have details or *image elements* of a certain area $S_{\mathrm{el}} \ll S_O$. The size of the image element determines the diffraction spread of the wave pattern between the object plane and the input face of the crystal, that is, the size of the spot illuminated by this element at the input face of the crystal. The area of this spot is of the order of $(f\lambda)^2/S_{\mathrm{el}}$. The pupil must be large enough to let the light within this spot pass and to cut off the spontaneous photons:

$$S_P \geq \frac{(f\lambda)^2}{S_{\mathrm{el}}}. \tag{7.8}$$

In the presence of the pupil the field operator $e(\boldsymbol{\rho}, t)$ (7.6) is given by an integral over the diffraction area S_{diff} in the object plane. With the choice (7.8) of the pupil area S_P this diffraction spread is of the order of or slightly less than the area of the image element:

$$S_{\mathrm{diff}} \simeq \frac{(f\lambda)^2}{S_P} \leq S_{\mathrm{el}}. \tag{7.9}$$

In our investigation we shall assume that the squeezing coefficients $u(\boldsymbol{\rho}, \Omega)$ and $v(\boldsymbol{\rho}, \Omega)$ do not significantly vary within the diffraction area S_{diff} and can be considered as constants in diffraction integrals. This assumption means that the area of the effectively amplified object wave is much larger than the size of the image element; that is, the image consists of many elements. For this reason we discuss the phase properties of Eq. (7.1) for the field evolution in our optical scheme as if there were no pupil.

7.3 Optimum Phase Matching for Parametric Amplification

As already mentioned above, we shall assume the temporal evolution of the image to be slow and put $\Omega \to 0$ in Eq. (7.1) for calculation of the amplified signal. Because the output field $e(\boldsymbol{\rho}, \Omega)$ is the sum of the contributions coming from symmetric points $\boldsymbol{\rho}$ and $-\boldsymbol{\rho}$ of the object plane, it is natural to consider such input signals that are even functions of $\boldsymbol{\rho}$. We assume that the input field is in a coherent state with complex amplitude $s(\boldsymbol{\rho})$, where

$$s(\boldsymbol{\rho}) = s(-\boldsymbol{\rho}). \tag{7.10}$$

In the case when $s(\boldsymbol{\rho})$ has the same phase for all $\boldsymbol{\rho}$ (e.g., $s(\boldsymbol{\rho})$ is real), the wavefront of the input signal in the object plane is flat and, therefore, the classical part of the input signal is in one and the same quadrature component for all $\boldsymbol{\rho}$.

To gain more physical insight into the squeezing transformation (7.1), let us consider what happens under this transformation with a classical complex field amplitude α taken for the moment as constant over the whole object plane. Denoting the output classical field amplitude ϵ, we can write

$$\epsilon = u\alpha + v\alpha^\star = e^{i\psi}\left\{|u|e^{-i\phi}\alpha + |v|e^{i\phi}\alpha^\star\right\}, \tag{7.11}$$

where we have introduced the angles ψ and ϕ as

$$\psi = \frac{1}{2}(\arg u + \arg v), \qquad \phi = \frac{1}{2}(\arg v - \arg u). \tag{7.12}$$

Let us define the quadrature components of the field on the input and the output of the crystal in their eigen coordinate systems determined by the angles ϕ and ψ as follows,

$$\alpha = e^{i\phi}(\alpha_1 + i\alpha_2), \tag{7.13}$$

$$\epsilon = e^{i\psi}(\epsilon_1 + i\epsilon_2). \tag{7.14}$$

It is easy to see from Eq. (7.1) that these eigen quadrature components are related by the following transformation:

$$\epsilon_1 = e^r\alpha_1, \qquad \epsilon_2 = e^{-r}\alpha_2, \tag{7.15}$$

where we have introduced the *squeezing parameter* r:

$$e^{\pm r} = |u| \pm |v|. \tag{7.16}$$

Therefore, the results of the squeezing transformation (7.1) can be summarized as follows.

(a) In the input to the system the complex field amplitude together with its uncertainty region must be decomposed into the eigen quadrature components defined by (7.13) in the coordinate system rotated by the angle ϕ given by (7.12); this is illustrated in Fig. 2(a) for a coherent input state with real amplitude;
(b) The quadrature component $e^{i\phi}a_1$ is rotated by the angle $-\phi$, that is, brought to the real axis, and stretched by the factor e^r; the quadrature component $ie^{i\phi}a_2$ is brought to the imaginary axis and squeezed by the factor e^{-r}, (see Fig. 2(b));
(c) The resulting complex field amplitude together with its uncertainty region is rotated by the angle ψ given by (7.12) (see Fig. 7.2(c)); in the chosen example the final state of the system represents a squeezed state, because its uncertainty region became an ellipse with unequal dispersion of different quadratures.

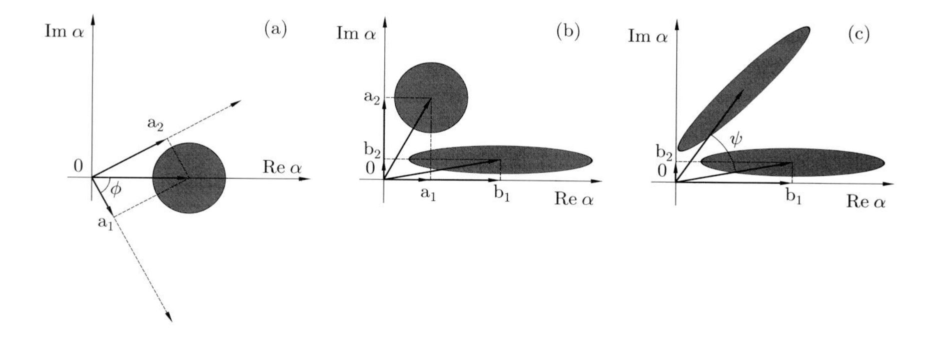

Fig. 7.2. Graphic illustration of the squeezing transformation described by Eq. (7.1).

In our case the coefficients u and v depend on $\boldsymbol{\rho}$. Therefore, the phase $\phi(\boldsymbol{\rho})$ that determines the amplified quadrature component is a function of the transverse coordinate $\boldsymbol{\rho}$. As a consequence, we obtain that the input classical signal $s(\boldsymbol{\rho})$ with constant phase for all $\boldsymbol{\rho}$ does not match the condition of maximum amplification everywhere.

There are two possibilities to gain the maximum parametric amplification for all points of the signal wave-front in the object plane. The first one is to introduce a $\boldsymbol{\rho}$-dependent phase shift into the incident object wave. In the paraxial approximation this phase shift must be quadratic in ρ; that is, the wave-front must have a certain curvature.

Another possibility is to adjust the optical scheme such as to make the phase $\phi(\boldsymbol{\rho})$ independent of $\boldsymbol{\rho}$. It follows from Eqs. (7.2) and (7.12) that

$$\phi(\boldsymbol{\rho}) = -\frac{1}{2}\arg\left\{\cosh\Gamma + \frac{i\delta}{2\Gamma}\sinh\Gamma\right\} = -\frac{1}{2}\tan^{-1}\left(\frac{\delta}{2\Gamma}\frac{\sinh\Gamma}{\cosh\Gamma}\right). \quad (7.17)$$

In the case of interest, when the coupling constant is large, $g \gg 1$, and the dimensionless mismatch is relatively small, $\delta \ll g$, we can approximate $\phi(\boldsymbol{\rho})$ as

$$\phi(\boldsymbol{\rho}) \approx -\frac{\delta(\boldsymbol{\rho},0)}{4g} = \frac{\rho^2/\rho_0^2 - \delta_0}{4g}. \quad (7.18)$$

Let us shift the crystal together with the lens L_2 and the detection plane P_4 at a certain distance so that the input face of the crystal is now located at $z_2 \neq 0$. As shown in Ref. [6], the phase ϕ that determines the amplified and attenuated quadrature components of the signal changes to $\tilde{\phi}$ according to

$$\tilde{\phi} = \phi - \varphi_2(z_2, \boldsymbol{\rho}), \tag{7.19}$$

with

$$\varphi_2(z_2, \boldsymbol{\rho}) = -\rho^2 \frac{k}{2f^2} z_2. \tag{7.20}$$

Physically speaking, the signal wave emitted at the point $\boldsymbol{\rho}$ of the object plane propagates after the lens L_1 as inclined plane wave with $\boldsymbol{q} = -(k/f)\boldsymbol{\rho}$. If the crystal and the rest of the scheme are shifted, the additional optical phase of this wave on the input face of the crystal depends on $\boldsymbol{q}$ or $\boldsymbol{\rho}$ quadratically, as seen from Eq. (7.20). Therefore, by appropriate choice of the distance z_2 one can eliminate the ρ-dependence in $\tilde{\phi}(\boldsymbol{\rho})$. Taking into account Eqs. (7.18), (7.19), and (7.20), we obtain that the optimum choice of z_2 is given by

$$z_2^{(\mathrm{opt})} = -\frac{2f^2}{k\rho_0^2 4g} = -\frac{l}{2g} = -\frac{l_{\mathrm{amp}}}{2}. \tag{7.21}$$

Here $l_{\mathrm{amp}} = l/g$ is the characteristic length of parametric amplification in the case of large coupling constant g. The optimum adjustment of the optical scheme is achieved when the plane P_2 of the first Fourier transform is located at the distance $l_{\mathrm{amp}}/2$ behind the input face of the crystal.

If z_2 is chosen according to Eq. (7.21), the amplified and attenuated quadratures of the signal field in the object plane are defined by Eq. (7.13) with

$$\tilde{\phi} = -\frac{\delta_0}{4g}. \tag{7.22}$$

In the case of degenerate phase matching, $\delta_0 = 0$, the input signal with a real classical amplitude $s(\boldsymbol{\rho})$ plane is uniformly amplified for all $\boldsymbol{\rho}$, at least in the paraxial approximation.

Let us consider nondegenerate phase matching with $\delta_0 > 0$. Assume that the input image is complex,

$$s(\boldsymbol{\rho}) = e^{i\varphi(\boldsymbol{\rho})}|s(\boldsymbol{\rho})|, \tag{7.23}$$

with a constant phase $\varphi(\boldsymbol{\rho}) = \tilde{\phi}$ given by Eq. (7.22). As follows from Eq. (7.13), such a signal has optimally matched quadrature components. Because the phase $\tilde{\phi}$ in Eq. (7.22) does not depend on transverse coordinate $\boldsymbol{\rho}$, this matching can be achieved for the whole image.

If the input signal is in a coherent state, both quadrature components of the quantum fluctuation of the signal field have the same mean square value. For this reason the amplified component of the fluctuation is always present. Its mean square value in the plane of photodetection does not depend on the above-described phase matching. On the contrary, if there is no optimum phase matching of the signal field at all points of the input image, the effective parametric gain of the signal is less than the gain of the quantum fluctuations. This can degrade the signal-to-noise ratio for some parts of the amplified image.

7.4 Quantum Fluctuations in the Amplified Image and Conditions for Noiseless Amplification

In this section we shall give the results from Ref. [6] for the mean number of photoelectrons $\langle N_I(\boldsymbol{\rho}, t)\rangle$ collected from a pixel at a point $\boldsymbol{\rho}$ in the image plane, and its variance $\langle \Delta N_I^2(\boldsymbol{\rho}, t)\rangle$. Using these quantities we shall find the signal-to-noise R_I ratio in the image plane, defined as [18, 19]

$$R_I = \langle N_I(\boldsymbol{\rho}, t)\rangle^2 / \langle \Delta N_I^2(\boldsymbol{\rho}, t)\rangle, \tag{7.24}$$

and compare it with the corresponding quantity R_O in the object plane. Our goal is to find conditions for noiseless operation of the scheme when these two signal-to-noise ratios are equal.

To simplify the results we shall use two approximations. First, we assume that both the diffraction area S_{diff} and the pixel area S_d are small compared to the image area S_O and that of the image element S_{el},

$$S_{\text{diff}}, S_d \leq S_{\text{el}} \ll S_O. \tag{7.25}$$

This is a natural choice in order to resolve the details of the image. The weight of the noise terms depends on the relation between the pixel and the diffraction areas. In what follows we shall evaluate two situations: $S_d \geq S_{\text{diff}}$, and $S_d \ll S_{\text{diff}}$, and show that in both cases the overall procedure (i.e., amplification and detection) can be noiseless.

Second, we shall consider the situation when the signal is slow and the observation time T_d can be taken long compared to the inverse bandwidth of parametric amplification $1/\Omega_p$, which is the characteristic temporal scale of the amplifier. With these approximations we obtain for the mean number of photoelectrons collected from a pixel at the point $\boldsymbol{\rho}$:

$$\langle N_I(\boldsymbol{\rho}, t)\rangle = \eta S_d T_d |s(\boldsymbol{\rho})|^2 G_\phi(\boldsymbol{\rho}) + \eta \frac{S_d}{S_{\text{diff}}} \frac{T_d}{2\pi} \int d\Omega |v(\boldsymbol{\rho}, \Omega)|^2. \tag{7.26}$$

Here $G_\phi(\boldsymbol{\rho})$ is the phase-sensitive gain of the amplifier,

$$G_\phi(\boldsymbol{\rho}) = \Big| \exp\big[i\varphi(\boldsymbol{\rho})\big] u(\boldsymbol{\rho}, 0) + \exp\big[-i\varphi(\boldsymbol{\rho})\big] v(\boldsymbol{\rho}, 0) \Big|^2. \tag{7.27}$$

Introducing the relative angle $\Delta\phi(\boldsymbol{\rho})$ between the amplified quadrature and the complex signal field in the object plane,

$$\Delta\phi(\boldsymbol{\rho}) = \phi(\boldsymbol{\rho}) - \varphi(\boldsymbol{\rho}), \tag{7.28}$$

and the squeezing parameter $r(\boldsymbol{\rho})$,

$$\exp[\pm r(\boldsymbol{\rho})] = |u(\boldsymbol{\rho}, 0)| \pm |v(\boldsymbol{\rho}, 0)|, \tag{7.29}$$

we can write the phase-sensitive gain $G_\phi(\boldsymbol{\rho})$ as follows,

$$G_\phi(\boldsymbol{\rho}) = \cos^2\left[\Delta\phi(\boldsymbol{\rho})\right]\exp[2r(\boldsymbol{\rho})] + \sin^2\left[\Delta\phi(\boldsymbol{\rho})\right]\exp[-2r(\boldsymbol{\rho})]. \tag{7.30}$$

The maximum gain $G = G_\phi(\boldsymbol{R})$ is attained for such $\boldsymbol{\rho} = \boldsymbol{R}$, where the local mismatch $\delta(\boldsymbol{R}, 0)$ vanishes: $\delta(\boldsymbol{R}, 0) = \delta_0 - (R/\rho_0)^2 = 0$. For degenerate phase matching $\delta_0 = 0$ this maximum is located at the optical axis of the system $\boldsymbol{R} = \boldsymbol{0}$. For nondegenerate matching with $\delta_0 > 0$ the maximum gain is reached along the ring of radius R given by

$$R = \rho_0 \delta_0^{1/2}. \tag{7.31}$$

From Eqs. (7.2) and (7.29) it follows that for both degenerate and nondegenerate matching the maximum gain G is equal to

$$G = \exp[2g]. \tag{7.32}$$

Figures 7.3a,b show the phase-sensitive gain $G_\phi(\boldsymbol{\rho})$ as a function of transverse distance for degenerate and nondegenerate matching. The dashed lines correspond to the gain profile without optimization of the phase matching discussed in Section 7.3, the solid lines, with optimization. We observe from Fig. 7.3 that in the area close to the maximum of the gain, the optimization has no effect. However, when departing from $\boldsymbol{\rho} = \boldsymbol{R}$, the gain curves with optimization become flatter and wider in both cases of degenerate and nondegenerate matching. We consider first the case $S_d \geq S_{\text{diff}}$. In this limit the diffraction spread of the image is small and the impulse response function $p(\boldsymbol{\rho})$ can be approximated by the delta-function:

$$p(\boldsymbol{\rho}) = \lambda f \delta(\boldsymbol{\rho}). \tag{7.33}$$

Using this approximation, we arrive at the following result,

$$\begin{aligned} \langle \Delta N_I^2(\boldsymbol{\rho}, t)\rangle &= \eta S_d T_d |s(\boldsymbol{\rho})|^2 G_\phi(\boldsymbol{\rho})\Big[1 - \eta + \eta S_\theta(\boldsymbol{\rho})\Big] \\ &+ \eta \frac{S_d}{S_{\text{diff}}}\frac{T_d}{2\pi}\int d\Omega |v(\boldsymbol{\rho}, \Omega)|^2\Big[1 + \eta + 2\eta|v(\boldsymbol{\rho}, \Omega)|^2\Big], \end{aligned} \tag{7.34}$$

where we have introduced the orientation angle $\theta(\boldsymbol{\rho})$ of the squeezing ellipse relative to the complex amplitude of the signal in the image plane,

$$\theta(\boldsymbol{\rho}) = \psi(\boldsymbol{\rho}) - \arg\Big(\exp[i\varphi(\boldsymbol{\rho})]u(\boldsymbol{\rho}, 0) + \exp[-i\varphi(\boldsymbol{\rho})]v(\boldsymbol{\rho}, 0)\Big), \tag{7.35}$$

and the squeezing function $S_\theta(\boldsymbol{\rho})$,

$$S_\theta(\boldsymbol{\rho}) = \cos^2\theta(\boldsymbol{\rho})\exp[2r(\boldsymbol{\rho})] + \sin^2\theta(\boldsymbol{\rho})\exp[-2r(\boldsymbol{\rho}). \tag{7.36}$$

There are two shot-noise contributions in Eq. (7.34), proportional to the mean intensity of the amplified image and that of the parametric down-conversion. The term proportional to $\eta^2|s(\boldsymbol{\rho})|^2 G_\phi(\boldsymbol{\rho})$ stems from the interference of the

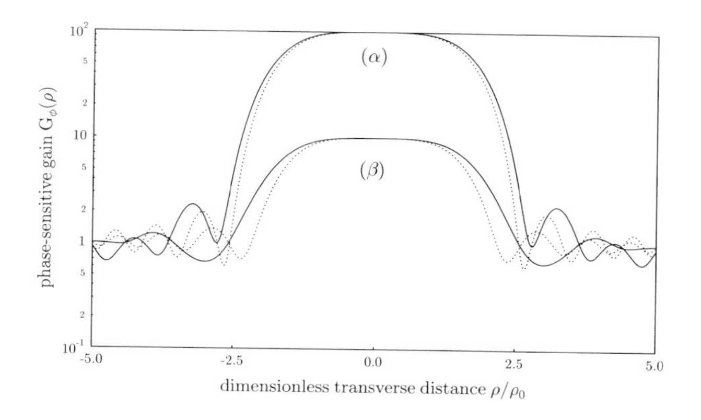

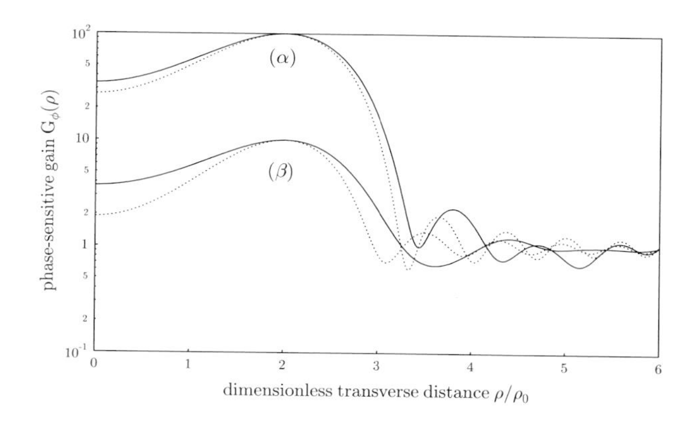

Fig. 7.3. Phase-sensitive gain $G_\phi(\rho)$ as a function of dimensionless transverse distance ρ/ρ_0 for degenerate (a) and nondegenerate (b) phase matching with $\delta_0 = 4$. Curves α correspond to the maximum gain $G = 100$, curves β to $G = 10$. Dotted lines are obtained without phase correction, solid: with correction.

amplified signal with noise and the other quadratic in the η term from the self-interference of the noise. This self-interference determines the inherent noise of the amplifier present even without the signal on its input.

As follows from Eq. (7.26), the mean number of detected photoelectrons, $\langle N_I(\boldsymbol{\rho}, t)\rangle$, contains two contributions. The first one is proportional to intensity $|s(\boldsymbol{\rho})|^2$ of the input image at the point $\boldsymbol{\rho}$ and constitutes the amplified image. The second term exists even when no signal is present at the input, and represents the inherent noise of the amplifier. Its physical origin is in the phenomenon of spontaneous parametric down-conversion. This noise term determines the ultimate lower limit for the input signal $s(\boldsymbol{\rho})$ that can be amplified without adding noise. Indeed, one can neglect the inherent noise of the amplifier compared to the amplified signal if

$$|s(\boldsymbol{\rho})|^2 G_\phi(\boldsymbol{\rho}) \gg \frac{1}{S_{\text{diff}}} \frac{1}{2\pi} \int d\Omega |v(\boldsymbol{\rho}, \Omega)|^2. \tag{7.37}$$

In the case of high gain, $G \gg 1$, we have $|u| \simeq |v| \gg 1$, and $G_\phi \simeq 4|v|^2$. Assessing the integral in the right-hand side of Eq. (7.37) as $\Omega_p|v(\boldsymbol{\rho}, 0)|^2$, where Ω_p is the bandwidth of parametric down-conversion, we can rewrite this condition as

$$|s(\boldsymbol{\rho})|^2 S_{\text{diff}} T_{\text{amp}} \gg 1/4. \tag{7.38}$$

Here we have introduced the typical temporal scale of the amplifier as $T_{\text{amp}} = 2\pi/\Omega_p$. The left-hand side of Eq. (7.38) gives the degeneracy parameter of the input field, that is, the mean number of photons in an elementary volume $cS_{\text{diff}}T_{\text{amp}}$, determined by the geometry of the amplifier and its response time. The input field in such an elementary volume is amplified by our system as an independent degree of freedom. The criterion (7.38) is met for degenerate input signals with relatively small quantum fluctuations.

Another condition for noiseless amplification arises from Eq. (7.34) for dispersion of the observed number of photoelectrons. Namely, the self-interference term in Eq. (7.34), given by the last integral, must be small compared to the term due to the interference of the amplified signal with noise, that is, the term proportional to $\eta^2|s(\boldsymbol{\rho})|^2 G_\phi(\boldsymbol{\rho})$. This gives:

$$|s(\boldsymbol{\rho})|^2 S_{\text{diff}} T_{\text{amp}} \gg 1/8, \tag{7.39}$$

which by the order of magnitude is equivalent to condition (7.38).

Now let us turn to the case of small pixels in the photodetection array $S_d \ll S_{\text{diff}}$. In the case of a symmetrical pupil frame function $P(\boldsymbol{\xi}) = P(-\boldsymbol{\xi})$, we obtain instead of (7.34):

$$\begin{aligned}\langle \Delta N_I^2(\boldsymbol{\rho}, t)\rangle &= \eta S_d T_d \, |s(\boldsymbol{\rho})|^2 G_\phi(\boldsymbol{\rho}) \Big[1 - \eta' + \eta' S_\theta(\boldsymbol{\rho})\Big] \\ &+ \eta' \frac{T_d}{2\pi} \int d\Omega |v(\boldsymbol{\rho}, \Omega)|^2 \Big[1 + \eta' + 2\eta' |v(\boldsymbol{\rho}, \Omega)|^2\Big].\end{aligned} \tag{7.40}$$

Here we have introduced a shorthand $\eta' = \eta S_d/S_{\text{diff}}$. It follows from Eq. (7.40) that the condition (7.39) for noiseless amplification holds true also in the case of small pixels.

To study quantitatively the noise performance of the amplifier we introduce the noise figure F' as

$$F' = \frac{R_O(\eta = 1)}{R_I}. \tag{7.41}$$

Because a linear amplifier cannot improve the signal-to-noise ratio in the input image, the noise figure F' is always not smaller than unity. We refer to the case $F' = 1$ as noiseless amplification. Notice, that in Eq. (7.41) the input signal-to-noise ratio refers to the ideal photodetection in the input plane. As explained in Ref. [4], this correction is necessary to obtain the noise figure that characterizes the noise added by the amplifier but not the noise due to imperfections in the preamplification apparatus (such as nonideal photodetection array in the object plane). Without such a correction (i.e., allowing for nonideal photodetection in the object plane with $\eta < 1$ one could obtain the noise figure smaller than unity. The physical feature of this phenomenon lies in the fact that the noiseless amplifier can compensate for the imperfection of the photodetection scheme at the preamplification stage and formally "improve" the signal-to-noise ratio of the input image degraded by the nonideal photodetection array.

Let us start the investigation of the noise figure with the case $S_d \geq S_{\text{diff}}$. The signal-to-noise ratio in the object plane is defined in analogy to (7.24). Because

$$\langle N_O(\boldsymbol{\rho}, t)\rangle = \eta S_d T_d |s(\boldsymbol{\rho})|^2, \qquad \langle \Delta N_O^2(\boldsymbol{\rho}, t)\rangle = \eta S_d T_d |s(\boldsymbol{\rho})|^2, \tag{7.42}$$

we find

$$R_O = \langle N_O(\boldsymbol{\rho}, t)\rangle^2 / \langle \Delta N_O^2(\boldsymbol{\rho}, t)\rangle = \eta S_d T_d |s(\boldsymbol{\rho})|^2. \tag{7.43}$$

Using the definition (7.24) of the signal-to-noise ratio in the image plane and Eq. (7.34) we obtain

$$F' = \frac{1}{\eta}\frac{1 - \eta + \eta S_\theta(\boldsymbol{\rho})}{G_\phi(\boldsymbol{\rho})}. \tag{7.44}$$

It is immediately seen from Fig. 7.2(b) that $|\theta(\boldsymbol{\rho})| \leq |\Delta\phi(\boldsymbol{\rho})|$. There $\Delta\phi$ is the angle between the input signal and the real axis, and θ is the angle between the amplified (stretched and squeezed) signal and the major axis of the squeezing ellipse (i.e., the real axis again). Because $|\theta(\boldsymbol{\rho})| \leq |\Delta\phi(\boldsymbol{\rho})|$, one can easily show that the noise figure $F'(\boldsymbol{\rho})$ is never smaller than unity. In the case of optimum phase matching, when $|\Delta\phi(\boldsymbol{\rho})| \to 0$, and high gain $G \gg 1$, the noise figure approaches unity. This is illustrated in Fig. 7.4 for $G = 100$. Figure 7.4(a) refers to degenerate, and Fig. 7.4(b) to nondegenerate phase matching. In both cases we have also shown the improvement in the noise figure, achieved by the phase correction as discussed in Section 7.3. Solid curves refer to noise figures with such phase correction, whereas the dotted lines are obtained without it. In our numerical calculations we introduce the optimum phase matching by taking the input signal in the form (7.23), where the signal phase $\varphi(\boldsymbol{\rho})$ is taken equal to the phase $\phi(\boldsymbol{\rho})$ of the amplified quadrature. One can see that with phase correction the spatial region of noiseless amplification becomes much larger. The large peaks of excess noise in the peripheral regions in dotted curves almost disappear in solid ones. Therefore, the effect of optimum phase matching both on gain profile (see Fig. 7.3) and noise figure is more pronounced in the peripheral region of the image, where the dephasing between the signal field and the amplified field quadrature is significant.

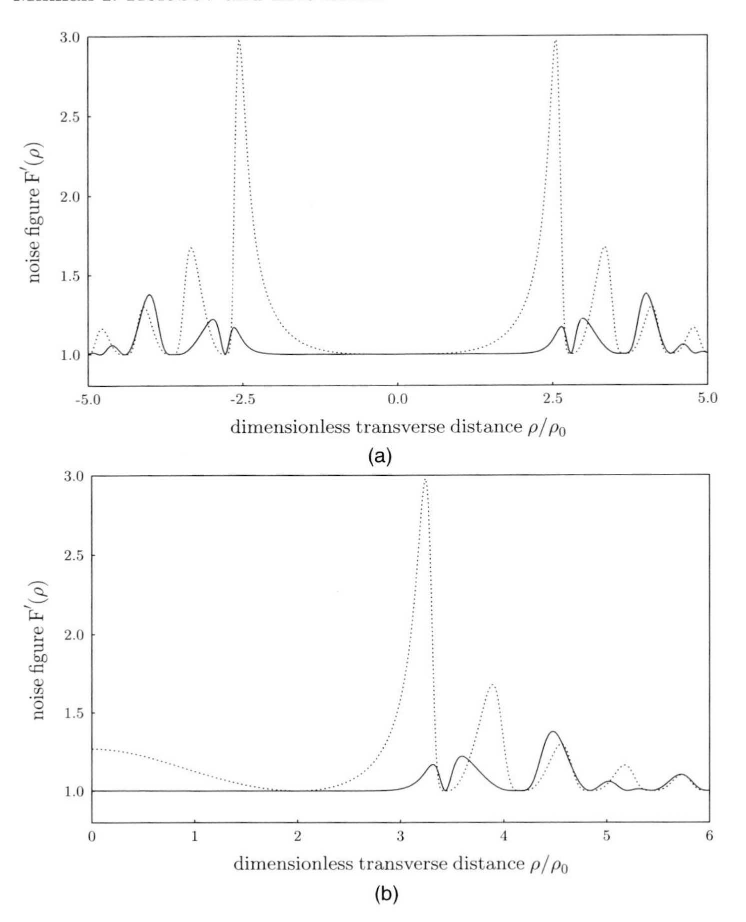

Fig. 7.4. Noise figure $F'(\rho)$ as a function of dimensionless transverse distance ρ/ρ_0 for degenerate (a) and nondegenerate (b) phase matching with $\delta_0 = 4$. The maximum gain $G = 100$. Dotted lines are obtained without phase correction; solid, with correction.

In the case of small pixels, $S_d \ll S_{\rm diff}$, the noise figure is given by

$$F' = \frac{1}{\eta}\frac{1-\eta' + \eta' S_\theta(\boldsymbol{\rho})}{G_\phi(\boldsymbol{\rho})}. \tag{7.45}$$

When the phase-sensitive gain G is so high that $\eta G S_d/S_{\rm diff} \gg 1$, Eq. (7.45) gives the noise figure as small as $S_d/S_{\rm diff} \ll 1$. That is, the signal-to-noise ratio is improved by the scheme with a high-gain parametric amplifier and small photodetectors. This situation is similar to the case of observation in the object plane with nonideal photodetectors, discussed in Ref. [4].

Both effects, namely, detection by pixels with low quantum efficiency $\eta < 1$ and detection by pixels of small area $S_d \ll S_{\text{diff}}$ lead to losses of light from an elementary volume $cS_{\text{diff}}T_{\text{amp}}$ which in our case plays the role of a single degree of freedom of the system. These losses degrade the signal-to-noise ratio of the input image because this image has Poissonian quantum fluctuations. Parametric amplification converts the Poissonian quantum noise into excess super-Poissonian fluctuations that are much less sensitive to the losses of light due to nonideal photodetection. Indeed, it is known that non-ideal photodetection reduces both signal and the excess noise at the same degree, not changing the signal-to-noise ratio. Therefore, one obtains improvement of the signal-to-noise ratio in the image plane corrupted by the losses in the object plane.

One can also say that due to diffraction in the amplifier, a detector of small area S_d in the image plane collects the information about the quantum state of a portion of light, coming from a diffraction area S_{diff} in the object plane. The important practical implication of Eq. (7.45) is that one can decrease the area of the pixel without deteriorating the signal-to-noise ratio as long as $\eta G S_d/S_{\text{diff}} \gg 1$. This conclusion can be very important for such applications where one seeks to reduce the amount of information without significant deterioration of the signal-to-noise ratio, such as in machine vision, for example.

To summarize, the optimum choice for noiseless amplification with sufficient spatial resolution is $S_O \gg S_{\text{el}} \geq S_d \geq S_{\text{diff}}$. Under such a choice, the number of image elements resolved by the amplifier can be assessed as

$$N_{\text{el}} \leq \frac{S_O}{S_{\text{diff}}} \simeq \frac{S_P}{S_{\text{coh}}}, \tag{7.46}$$

where $S_{\text{coh}} = (2\pi/q_p)^2$ is the coherence area of parametric down-conversion at the output plane of the parametric crystal, q_p is the spatial-frequency bandwidth of the crystal [20]. We have used here (7.9) and the estimate $S_O \simeq (fq_p/k)^2$. Equation (7.46) says that there are two ways to evaluate the number of spatial degrees of freedom of our amplifier. In the object plane of area S_O a degree of freedom is specified by S_{diff}, and in the transverse section of parametric crystal of area S_P by S_{coh}.

7.5 Experimental Demonstration of Temporally Noiseless Image Amplification

The first experiment that demonstrated noiseless image amplification considered temporal fluctuations that affect a spatial pattern [16]. The experimental set-up, shown in Fig. 7.5, is a traveling-wave OPA with the center of the crystal in the image plane conjugated with the object (a resolution chart). The noiseless properties of such a scheme have been theoretically assessed in

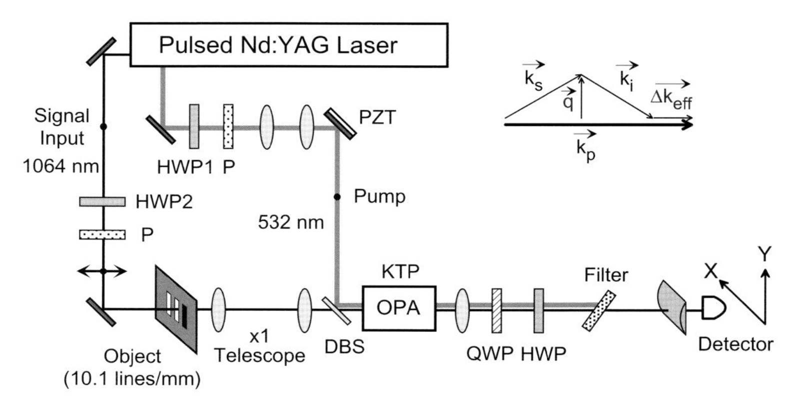

Fig. 7.5. Schematic of the experimental setup to measure 1D spatial profiles of the intensity and noise power of an image amplified by the traveling-wave OPA. A feedback signal to the PZT is used to lock for maximum amplification in the phase-sensitive configuration. The phase-matching diagram in terms of the spatial frequency $\boldsymbol{q}$ is shown in the upper-right corner.

Ref. [7] and are similar to that with the crystal in the Fourier plane described in the preceding sections. The signal and pump pulses are provided, respectively, by the fundamental (1064 nm) and the second harmonic (532 nm) of a Q-switched mode-locked Nd:YAG laser at a repetition rate of 1 kHz. The resulting Q-switch envelopes of the pump and signal pulses are $\simeq$145 and $\simeq$200 ns in duration, respectively. The mode-locked pump and signal pulses underneath these Q-switch envelopes are estimated to be $\simeq$85 and $\simeq$120 ps, respectively. The amplification is performed in a KTP crystal, with a length of either 3.25 or 5.21 mm. The signal is polarized at 45° of the crystal neutral axes, in order to inject both the signal and the idler waves into the amplifier and obtain phase-sensitive amplification. The phase of the pump beam relative to that of the signal beam is locked to maximize the parametric gain, by means of a feedback loop that drives the piezo-electric transducer (PZT). The object illuminated by the signal beam consists of two vertical lines of a USAF test pattern. These lines are imaged with a unity magnification into the center of the crystal by a telescope. The spatial frequency corresponding to these lines, 10.1 lines/mm, is chosen sufficiently low to lie well within the spatial bandwidth of the amplifier [9], and sufficiently high to ensure uniform illumination by the pump beam. The amplified image is magnified 24 times after the crystal in order to be spatially resolved when scanned by an InGaAs detector of 300 μm diameter. In the temporal domain, the bandwidth of the parametric amplifier is much larger than that of the detector, ensuring an identical gain for dc and for 27 MHz photocurrents. Hence the experimental noise figure can be simply determined as

$$\mathrm{NF} = \frac{\text{Noise-power gain}}{\eta\,(\text{Mean-intensity gain})^2}\,, \tag{7.47}$$

where the photocurrent noise power is recorded at 27 MHz. The spatial image profiles of the signal and noise scanned by the photodetector are shown in Fig. 7.6. To minimize the effect of spatial averaging caused by the finite

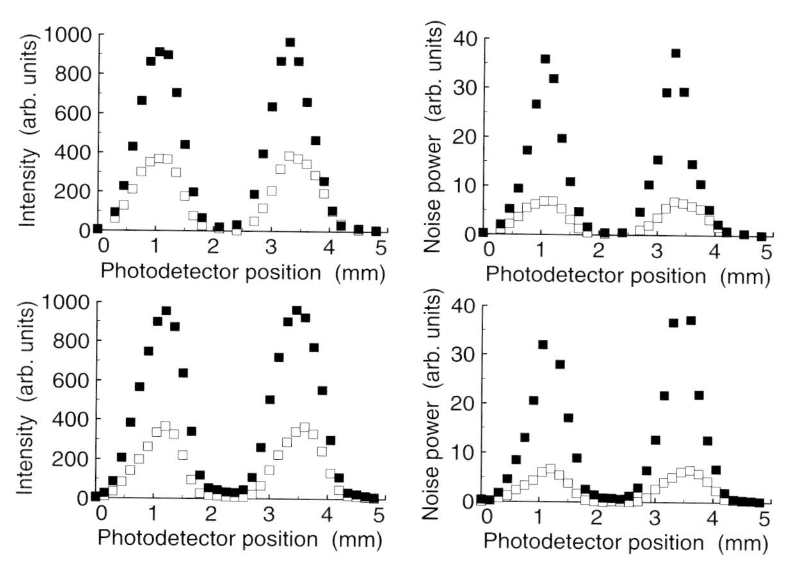

Fig. 7.6. 1D spatial profiles of the intensity (left) and the noise power (right) for a 3.25 mm long (top) and 5.21 mm long (bottom) KTP crystals. Profiles of both the unamplified (empty squares) and the amplified (full squares) images are shown.

size of the photodetector the noise figure was characterized using experimental values of gains measured at the peaks of the spatial profile. Intensity gains $G \simeq 2.5$ were obtained at the peaks, with a noise power gain lower than the intensity gain to the square, because the amplified image is less degraded by the detection than the shot-noise limited input. By taking into account the measured overall detection efficiency $\eta = 0.82$ and using Eq. (7.47), the following values of the total NF of the optical amplifier were obtained: (0.2 ± 0.6) dB at $G = 2.5$ for the 3.25 mm KTP crystal and (0.4 ± 0.6) dB at $G = 2.6$ for the 5.21 mm KTP crystal. These values agree with the theoretical values predicted for a phase-sensitive amplifier. They show clearly the improvement due to preamplification when compared with the NF = 0.86 dB of the detector and are almost 2 dB lower than the quantum limit of an ideal phase-insensitive amplifier of the same gain.

7.6 Experiment on Spatially Noiseless Amplification of Images

The experiment described in the previous section shows that a phase-sensitive scheme allows the signal-to-noise ratio to be unmodified over an entire image, where the noise is recorded at a frequency of 27 MHz by a photodiode. As the photodiode scanned the image, this result proves that phase-sensitive amplification improves the regularity in time of the distribution of photons for each point of the image but, because only fluctuations in the time domain were recorded, it does not directly show a regularity in space. However, patterns in an image are pure spatial information, without any time aspect, that are ultimately degraded by spatial fluctuations of quantum origin for very weak images. The experiment described in this section [17] was designed to assess these spatial fluctuations.

The quantum noise properties of an optical parametric amplifier are described by the noise figure $\mathrm{NF} = \mathrm{SNR}_{\mathrm{in}}/\mathrm{SNR}_{\mathrm{out}}$. However, if $\mathrm{SNR}_{\mathrm{out}}$ is the output signal-to-noise ratio after detection with a quantum efficiency $\eta < 1$, the input signal-to-noise ratio $\mathrm{SNR}_{\mathrm{in}}$ of a Poissonian beam is not a directly measurable quantity. What can be effectively measured is the signal-to-noise ratio after detection without amplification, resulting in multiplying both the input signal-to-noise ratio and the noise figure by the global quantum efficiency of the system η_{tot}. Unlike the "theoretical" noise figure F' defined in Eq. (7.44), the experimental ratio $R = \eta_{\mathrm{tot}} \times F'$ can be smaller than unity, meaning that the amplified image is less degraded by the detection than the shot-noise limited input (see discussion before Eq. (7.42)). Moreover, using a detector with a pixel size S_d much smaller than the coherence area S_{diff} in the amplified image is equivalent to multiplying the quantum efficiency by the ratio S_d/S_{diff} (see Eq. (7.45)). Note that such degradation of the signal-to-noise ratio due to very small pixels can be effectively overcome by the amplification for the temporal fluctuations, whereas this improvement is an artifact of spatial signals, because the OPA itself rejects both signal and noise of high spatial frequencies. The experimental set-up is a traveling-wave OPA similar to that described in [9] (see Fig. 7.7). The signal and the pump pulses are provided, respectively, by the the second harmonic (1.2 ps FWHM duration FWHM at 527.5 nm) and the fourth harmonic (0.93 ps duration at 263.7 nm) of a Q-switched mode-locked Nd:Glass laser (Twinkle laser by Light Conversion) at a repetition rate of 33 Hz. The amplification is performed in a beta-barium-borate (BBO) crystal whose transverse area, $7 \times 7\ \mathrm{mm}^2$, is chosen in order to obtain a sufficient number of resolution cells in the amplified image so as to perform valid statistics. The crystal length, 4 mm, is limited by the group-velocity difference between the UV pump and the green signal. Because of the high dispersion of the crystal in the UV, only type-I amplification is possible for this couple of wavelengths. Hence, collinear interaction is phase sensitive, and phase-insensitive amplification is obtained by a slight angular shift between the pump and the signal beams

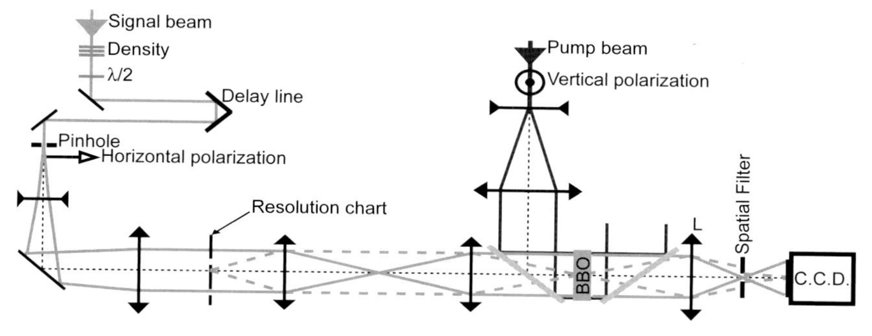

Fig. 7.7. Experimental set-up for spatially noiseless amplification.

as shown in Fig. 7.8. The signal beam is widened by a telescope and illuminates a resolution chart. A line of this chart is imaged on the input face of the crystal by a second telescope and then onto the CCD camera by the lens L (scheme of Ref. [7]). The CCD camera includes a back-illuminated thin silicon array cooled at $-40°$C to ensure a negligible dark current and a low read-out noise. To match the phase-fronts, the beam waists of the signal and the pump are superimposed within the crystal. A filtering aperture, placed in the Fourier plane and centered around the zero spatial frequency of the signal image, limits the detected intensity of spontaneous parametric down-conversion (SPDC) and ensures the elimination of the idler in the PIA scheme. However, the spatial spectral bandwidth in the detected image is reduced by this aperture and, in practice, the size of S_{diff} is no longer determined by the phase matching conditions, but rather by the diameter of the aperture, giving a transverse size of S_{diff} equal to 7.3 pixels on the camera. In the following, the results will be presented for different groupings of the pixels (achieved by software) in order to consider effective detector areas S_d smaller or greater than S_{diff}. The NF depends on the resulting total quantum efficiency η_{tot} that can be computed as the product of the quantum efficiency (η_{CCD}) of the CCD camera by the transmission (η_{opt}) of the optical elements after the crystal,

$$\eta_{\mathrm{tot}} = \eta_{\mathrm{opt}} \times \eta_{\mathrm{CCD}} = 0.69 \times 0.9 = 0.62 \pm 0.10. \tag{7.48}$$

The uncertainty comes from the evaluation of η_{opt}, and η_{CCD} is given by the manufacturer.

The experimental procedure to measure R is achieved in three main steps that are identical in the PIA and PSA schemes. The first step consists in measuring the signal-to-noise ratio without amplification (i.e., $\eta_{\mathrm{tot}} \times \mathrm{SNR}_{\mathrm{in}}$) along with a statistical verification of the Poissonian hypothesis. In the second step, the intensity of the SPDC is measured and its level is subtracted from the amplified images. The last step consists in measuring $\mathrm{SNR}_{\mathrm{out}}$. For the first step, the shot-to-shot stability of the laser is sufficient to estimate the

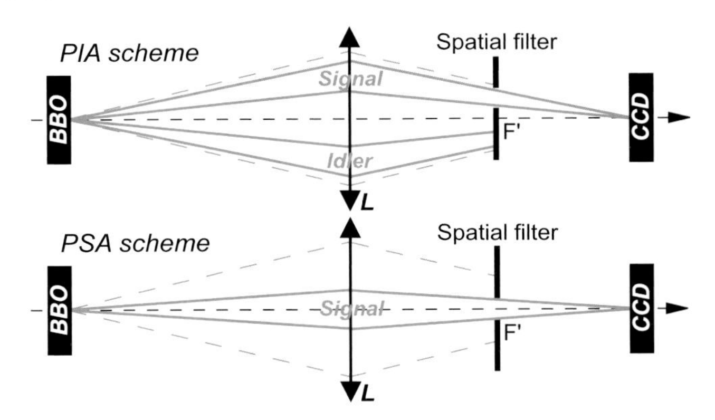

Fig. 7.8. Phase-matching conditions. Top: noncollinear for the PIA scheme. Bottom: collinear for the PSA scheme. Dotted line shows spatial spectral bandwidth of the OPA. Full line indicates the resolution range.

level of the input images by recording a set of about 20 nonamplified images. Two sets with different intensity levels were recorded in the PIA scheme. For each set, we have verified that a nongrouped image is well described by a Poissonian distribution, whereas grouping degrades this distribution because of residual deterministic defects. Nevertheless, the experimental statistics on the difference between two images of a set remains Poissonian because the subtraction of images eliminates deterministic structures,

$$\eta_{\rm tot} \times {\rm SNR}_{\rm in} = \langle n \rangle_p \times S_d = \langle n \rangle, \tag{7.49}$$

where S_d is the detector area after grouping, expressed in pixels. In Eq. (7.49) and in all the following $\langle n \rangle_p$ is the mean number of photoelectrons per pixel, and $\langle n \rangle$ designates the number of photoelectrons on S_d, obtained by summing the gray levels on the pixels and multiplying by the appropriate scaling factor.

To estimate the SPDC level, 20 images were recorded by injecting only the pump. Because type-I phase matching is noncritical in wavelength [11], the SPDC is strongly temporally multimode. These modes add incoherently and each temporal mode is described by a thermal statistic [21]. The number of temporal modes is experimentally assessed as

$$M_t = \frac{(\langle n \rangle_p^{\rm SPDC} / \eta_{\rm tot}) \times S_{\rm diff}}{G - 1}, \tag{7.50}$$

where $\langle n \rangle_p^{\rm SPDC}$ is the mean level of SPDC per pixel before grouping and G the gain of the OPA. Consequently, the variance of the SPDC on a grouped pixel is given by

$$\langle (\Delta n)^2 \rangle_{\rm SPDC} = \frac{\langle n \rangle_{\rm SPDC}^2}{M_t} + \langle n \rangle_{\rm SPDC}, \tag{7.51}$$

where the first term in the right part of Eq. (7.51) describes the classical fluctuations and the second term the shot noise. Equation (7.51) does not take into account deterministic variations of SPDC due, for example, to imperfections of the pump profile, that are not negligible for large grouping when directly measuring the SPDC variance. Because these variations do not affect difference images, Eq. (7.51) is used rather than a direct measurement to calculate the SPDC variance.

The measurement of $\mathrm{SNR}_{\mathrm{out}}$ is performed by using single-shot images, in order to take into account the strong variations of the gain from one shot to another in the phase-sensitive scheme because of the noncontrolled variations of the relative phase between the signal and the pump. Such a control is made difficult by the low repetition rate of both the laser source (33 Hz) and the camera (<1 Hz). Classical noise is predominant in the amplified images, because of deterministic imperfections of the system (pump beam, lenses, etc.) and good results have been obtained only by performing differences of images, in order to eliminate the spatial defects that are reproducible from one shot to another. $\mathrm{SNR}_{\mathrm{out}}$ is measured as follows: (i) an area with approximately constant mean intensity is selected in the amplified image; (ii) the measurement of the signal-to-noise ratio on all amplified images without subtraction allows the selection of images with the highest signal-to-noise ratios; (iii) pairs of images are defined by all permutations between the selected images; (iv) the mean and the variance are calculated for each pair and each grouping: the mean is calculated as the mean of the two images corrected by subtracting the electronic background $\langle n\rangle_b$ and converted in photoelectrons (pe^-). Finally, the mean of the SPDC is subtracted. The whole calculation is summarized as

$$\langle n\rangle = g \times (\langle n\rangle_{gl} - \langle n\rangle_b) - \langle n\rangle_{\mathrm{SPDC}}, \tag{7.52}$$

where $g = 0.97\ (pe^-) \times (gl)^{-1}$ converts the gray level (gl) into pe^-. The variance in the amplified image is computed as half the variance of the difference of the images, with subtraction of the variances of the read-out noise $\langle(\Delta n)^2\rangle_{\mathrm{subread}}$) and of the SPDC:

$$\langle(\Delta n)^2\rangle = \frac{1}{2} \times \{[\langle(\Delta n)^2\rangle_{\mathrm{sub}} - \langle(\Delta n)^2\rangle_{\mathrm{subread}}] - 2 \times \langle(\Delta n)^2\rangle_{\mathrm{SPDC}}\}. \tag{7.53}$$

$\mathrm{SNR}_{\mathrm{out}}$ is computed for each value of the detector area as

$$\mathrm{SNR}_{\mathrm{out}} = \frac{\langle n\rangle^2}{\langle(\Delta n)^2\rangle}. \tag{7.54}$$

From this value and the corresponding value $\eta_{\mathrm{tot}} \times \mathrm{SNR}_{\mathrm{in}}$, the ratio R is determined for each pair of images and the mean ratio is assessed from all pairs. The experimental error bars are finally determined as twice the standard deviation divided by the square root of the number of pairs of selected images. In the PIA scheme, about 100 amplified images were recorded for

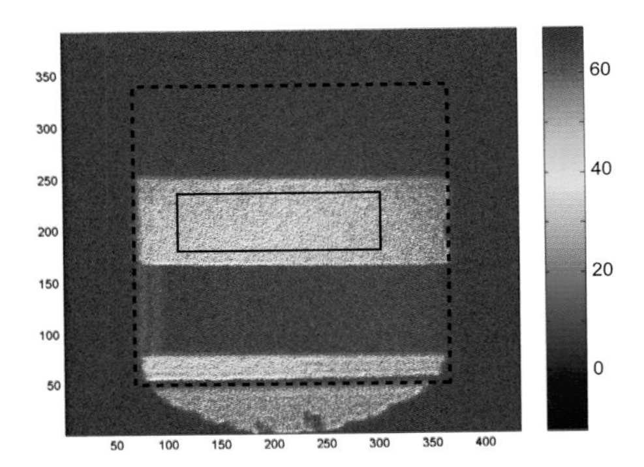

Fig. 7.9. Example of amplified image in the PIA scheme. Dashed line: edges of the crystal. Solid line: limits of the area used for the statistics (8241 pixels).

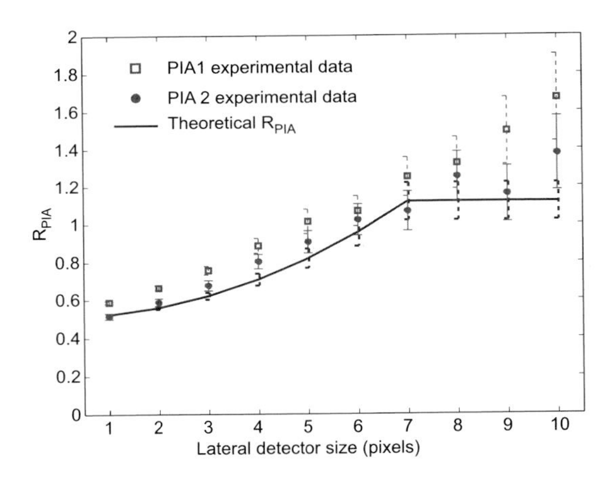

Fig. 7.10. NF after detection in the PIA scheme versus the detector size. Squares: experimental data of PIA series 1 (dotted error bars). Circles: experimental data of PIA series 2 (solid line error bars). Line: theoretical curve (heavy dotted error bars).

each set of nonamplified images and the gain was estimated for each image from the mean in the corresponding set of nonamplified images. The gain variations are due to the shot-to-shot fluctuations of the laser. The average gain is $G_{\mathrm{PIA}} = 1.6 \pm 0.5$. Ten images have been selected with a constant gain for each set. Figure 7.9 shows an example of a selected image. Figure 7.10 reports the NF after detection for the PIA scheme versus the detector size. For $S_d \geq S_{\mathrm{diff}}$, the theoretical R is $R_{\mathrm{PIA}} = 1.1 \pm 0.1$. For $S_d < S_{\mathrm{diff}}$, $\eta_{\mathrm{tot}} \times (S_d/S_{\mathrm{diff}})$ is used as in Eq. (7.40) although this assumption is only correct when $S_d \ll S_{\mathrm{diff}}$ as explained previously. The uncertainty is defined by the uncertainty on η_{tot} given in Eq. (7.48). The experimental data and the theoretical curve are in good agreement. The differences between the two sets, realized in equivalent conditions, remain in the uncertainty range due to the random character of fluctuations. In the PSA scheme about 500 images

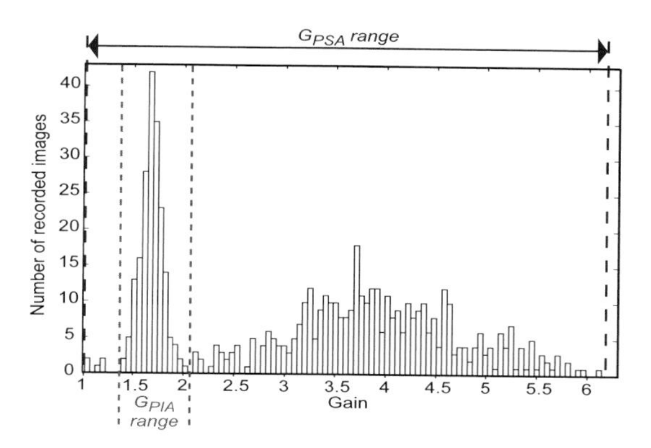

Fig. 7.11. Histogram of the gain for both the PIA and PSA schemes.

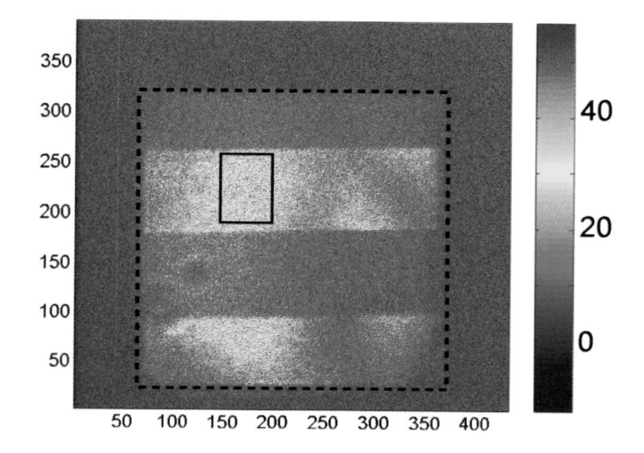

Fig. 7.12. Example of amplified image in the PSA scheme. Area of 3266 pixels.

were recorded and the gain was measured for each image with a range from 1 to 6, as shown in Fig. 7.11. Because the relative phase is not controlled, it is more difficult to find pairs of images that correspond both to the same (maximum) gain and the same phase. Nevertheless, the criteria based on the highest signal-to-noise ratio allow the selection of five images that were amplified in the same conditions. Figure 7.12 shows an example of a selected

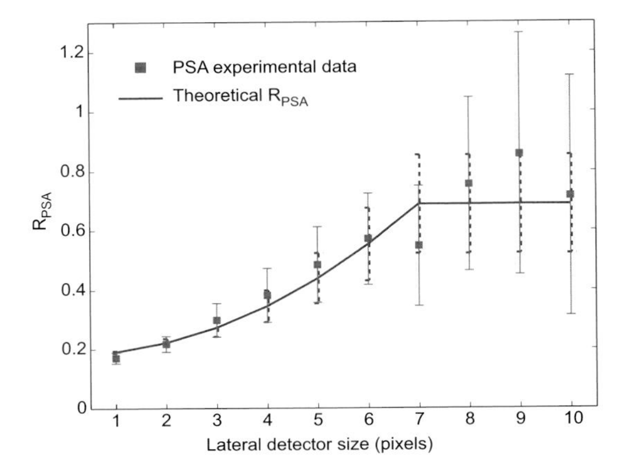

Fig. 7.13. NF after detection in the PSA scheme versus the lateral detector size. Squares: experimental data (solid line error bars). Line: theoretical curve (heavy dotted error bars).

image. The gain is clearly nonhomogeneous along the line because of residual variations of the relative phase. Therefore, the used area where the statistics can be assumed stationary, is smaller than in the PIA case. Figure 7.13 shows the evolution of the NF after detection for the PSA scheme versus the detector size. The theoretical curve is calculated from Eq. (7.45) as in the PIA scheme giving $R_{\mathrm{PSA}} = 0.7 \pm 0.1$ when $S_d \geq S_{\mathrm{diff}}$. The agreement between the experimental data and the theoretical curve is good and proves the noiseless character of the PSA scheme.

In conclusion, the experimental demonstration of purely spatial noiseless amplification of images was achieved in a PSA scheme and the results were compared with the PIA scheme. The obtained results are in satisfactory agreement with the theory. As expected, the PSA does not add noise, whereas PIA leads to the expected 3 dB degradation of the signal-to-noise ratio.

References

1. C. M. Caves, Phys. Rev. D **26**, 1817 (1982).
2. Z. Y. Ou, S. F. Pereira, and H. J. Kimble, Phys. Rev. Lett. **70**, 3239 (1993).
3. J. A. Levenson, I. Abram, T. Rivera, P. Fayolle, J. C. Garreau, and P. Grangier, Phys. Rev. Lett. **70**, 267 (1993).
4. M. I. Kolobov and L. A. Lugiato, Phys. Rev. A **52**, 4930 (1995).
5. S. Mancini, A. Gatti, and L. A. Lugiato, Eur. Phys. J. D **12**, 499 (2000).
6. I. V. Sokolov, M. I. Kolobov, and L. A. Lugiato, Phys. Rev. A **60**, 2420 (1999).
7. K. Wang, G. Yang, A. Gatti, and L. A. Lugiato, J. Opt. B: Quantum Semiclass. Opt. **5**, S535 (2003).

8. F. Devaux, E. Lantz, A. Lacourt, D. Gindre, H. Maillotte, P. A. Doreau, and T. Laurent, Nonlin. Opt. **11**, 25 (1995).
9. F. Devaux and E. Lantz, Opt. Commun. **114**, 295 (1995).
10. F. Devaux and E. Lantz, Opt. Commun. **118**, 25 (1995).
11. F. Devaux and E. Lantz, J. Opt. Soc. Am. B **12**, 2245 (1995).
12. A. Gavrielides, P. Peterson, and D. Gardimona, J. Appl. Phys. **62**, 2640 (1987).
13. C. J. Wetterer, L. P. Schelonka, and M. A. Kramer, J. Appl. Phys. **65**, 3347, (1989).
14. S. M. Cameron, D. F. Bliss, and M. W. Kimmel, Proc. SPIE **2679**, 195 (1996).
15. D. Guthals and D. Sox, in *Proceedings of International Conference on Lasers '89*, eds. D. G. Harris and T. M. Shay (STS, Mclean, VA, 1990), 808–815.
16. S.-K. Choi, M. Vasilyev, and P. Kumar, Phys. Rev. Lett. **83**, 1938 (1999).
17. A. Mosset, F. Devaux, and E. Lantz, Phys. Rev. Lett. **94**, 223603 (2005).
18. J. B. Thomas, *An Introduction to Statistical Communication Theory* (Wiley, New York, 1969).
19. J. W. Goodman, *Statistical Optics* (Wiley, New York, 1985).
20. M. I. Kolobov, Rev. Mod. Phys. **71**, 1539 (1999).
21. A. Mosset, F. Devaux, G. Fanjoux, and E. Lantz, Eur. Phys. J. D **28**, 447 (2004).

8 Optical Image Processing in Second-Harmonic Generation

Pierre Scotto, Pere Colet, Adrian Jacobo, and Maxi San Miguel

Instituto Mediterraneo de Estudios Avanzados, IMEDEA (CSIC-UIB), Campus Universitat de les Illes Balears, E-07122 Palma de Mallorca, Spain
pere@imedea.uib.es

8.1 Introduction

Although the processing of an image by all-optical means is quite less common than the well-developed techniques for digital image processing [1], it has nevertheless been around for quite some time. At a classical level early works demonstrated frequency transfer of an optical image from the infrared to the visible domain [2, 3], and later from the visible to the UV domain [4, 5], as well as parametric amplification of an UV image [6, 7], and contrast inversion [8]. In these schemes, an optical image at a frequency ω is directly injected into a nonlinear crystal illuminated with a strong monochromatic pump wave at frequency ω_p and the processed image is formed in the output plane. As a result of the nonlinearity of the crystal, the input image will be, depending on some phase-matching condition, either transferred to a higher frequency $\omega + \omega_p$ by simple frequency addition [2–4], or amplified by photon down-conversion [6–8]. In the latter case the amplification is accompanied by the formation of a phase conjugated (idler) image at the complementary frequency $\omega_p - \omega$. Considering the spatial dependence of the image-processing mechanism on the position of the object in the transverse plane, the phase-matching condition will determine whether image-processing will be efficient either on a disk centered on the optical main axis of the system, or on a ring of finite width. This latter regime is also useful for selectively amplifying some Fourier components of a given image, leading to contrast enhancement or inversion. A quite significant amount of work in all-optical image-processing operations has been performed in photorefractive media [9] including edge enhancement [10–12], image inversion, division, differentiation and deblurring [13–16], noise suppression [17], and contrast enhancement [18].

More recently, image processing has been considered also on a quantum level, including the investigation of the properties of the quantum fluctuations in the output image. The crucial prediction, which gave rise to rapid developments in the emerging field of quantum imaging [19, 20], was made in the context of image amplification: whereas quantum mechanics imposes that the phase-insensitive amplification of an image is always accompanied by an addition of at least 3 dB extra noise to the output image [21], a phase-sensitive amplifier has much better noise performance [22, 23]. Even more, noiseless image amplification (i.e., an amplification that preserves the signal-to-noise ratio during processing) was shown, first theoretically [24, 25], and then experimentally [26], to be possible. This technique may be applicable to situations in which a faint coherent signal must be amplified prior to detection. In the case of weak signals, the degradation of the signal-to-noise ratio predicted by quantum mechanics in the case of phase-insensitive amplification might irremediably destroy the information encoded in the image.

In this chapter we consider the use of the second-harmonic generation (SHG) for all-optical processing of images. From the point of view of crystal nonlinearities one distinguishes between type-I and type-II second-harmonic generation. In the simplest situation, type-I refers to the case where two fields with the same polarization and the same fundamental frequency ω combine to yield a second-harmonic field at frequency 2ω. In type-II two linearly orthogonally polarized fields with fundamental frequency ω are injected in the nonlinear crystal leading to a second-harmonic field at frequency 2ω.

We will address first the situation in which the nonlinear crystal is placed inside an optical cavity. The fundamental difference of the cavity-based geometry from the traveling-wave case is the existence of instability thresholds which, if used appropriately, allow for nonlinear processing of the image. For example, considering type-II second-harmonic generation inside a planar cavity where all the fields are resonant, it is possible to selectively enhance the contrast of part of an image or to detect its contour [27]. This phenomenon will be discussed in Section 8.2. In Sections 8.3 and 8.4 we will consider the quantum imaging properties of the second-harmonic generation in the traveling-wave configuration. The first of these sections is devoted to a type-I case, and the second one is devoted to a type-II situation where the polarization degree of freedom allows for a larger variety of possible operations.

8.2 Image Processing in Second-Harmonic Generation at a Classical Level

In this section we consider a crystal with a $\chi^{(2)}$ nonlinearity enclosed in an optical cavity, taken ideally to be a planar cavity and we will assume type-II phase matching as sketched in Fig. 8.1. A second-harmonic (SH) field will be generated if the cavity is pumped at two orthogonal polarizations x and y. In the paraxial and mean-field approximation the system can be described by the following set of equations [27–31],

$$\partial_t B = -(1 + i\delta_B)B + \frac{i}{2}\nabla^2 B + iA_x A_y, \tag{8.1}$$

$$\partial_t A_x = -(1 + i\delta_A)A_x + i\nabla^2 A_x - iA_y^* B + E_x, \tag{8.2}$$

$$\partial_t A_y = -(1 + i\delta_A)A_y + i\nabla^2 A_y - iA_x^* B + E_y, \tag{8.3}$$

which govern the temporal evolution of the intracavity field envelopes A_x and A_y at the fundamental frequency ω with linear polarizations x and y and B

at the second-harmonic frequency 2ω, polarized along the y axis; δ_A and δ_B are the detunings at the fundamental and second-harmonic frequencies, respectively. Times are expressed in units of the cavity decay time and lengths in units of the diffraction length. Diffraction is taken into account through the transverse Laplacian $\nabla^2 = \partial^2/\partial x^2 + \partial^2/\partial y^2$. The pumping amplitudes E_x and E_y in each linear polarization state are chosen such that an image is injected with the x-polarization and a homogeneous field is inserted in the y-polarization. The study of the steady-state solution of Eqs. (8.1)–(8.3) for

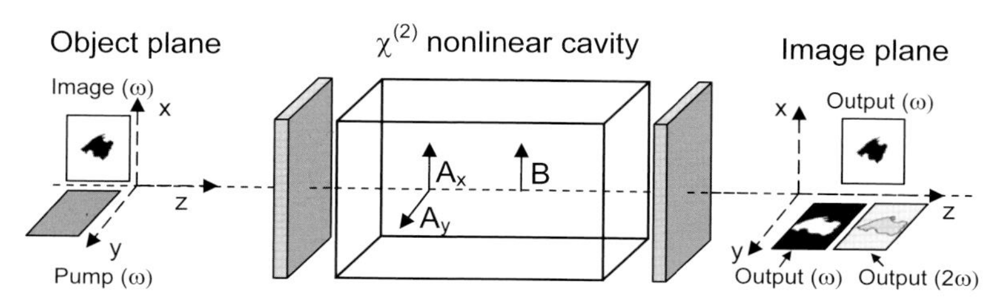

Fig. 8.1. Scheme of an optical device based on intracavity type-II second-harmonic generation. The nonlinear crystal is enclosed in a plane-mirror cavity.

homogeneous pumps provides valuable insight into the relevant properties of the system that will be used for image processing. For the pump waves homogeneous in the transverse plane, E_x and E_y can be taken as real without loss of generality. Typically, Eqs. (8.1)–(8.3) have been considered for the case of symmetrical pumping $E_x = E_y$, which maximizes the production of the second harmonic. In such a case the homogeneous stationary steady state becomes unstable for a pump above the critical value [28–32]:

$$|E_{as}|^2 = 2(1+\delta_B^2)^{1/2}(1+\delta_A^2)^{3/2} + 2(1+\delta_A^2)(1-\delta_A\delta_B). \tag{8.4}$$

The system evolves to a homogeneous state for which $|A_x|$ and $|A_y|$ are different, so the intracavity field polarization is no longer the same as the pump (polarization instability). Because of the symmetry of the system two equivalent but different states can exist, with large value of $|A_x|$ and small value of $|A_y|$ and vice versa [27–29]. For asymmetric homogeneous pumping $E_x \neq E_y$, the homogeneous steady state for $|A_y|$ is given by the solution of the polynomial

$$\begin{aligned}\Delta_A|A_y|^{10} &+ [4(1-\delta_b)\Delta_A - |E_y|^2]|A_y|^8 \\ &+ 2[\Delta_A Q + \Delta_{AB}(|E_x|^2 - 2|E_y|^2)]|A_y|^6 \\ &+ 2[2\Delta_A^2\Delta_B^2\Delta_{AB} - Q|E_y|^2 - 2\Delta_{AB}^2|E_x|^2]|A_y|^4 \\ &+ [\Delta_A^3\Delta_B^2 + 2\Delta_A\Delta_B\Delta_{AB}(|E_x|^2 - 2|E_y|^2)]|A_y|^2 \\ &- \Delta_A^2\Delta_B^2|E_y|^2 = 0,\end{aligned} \tag{8.5}$$

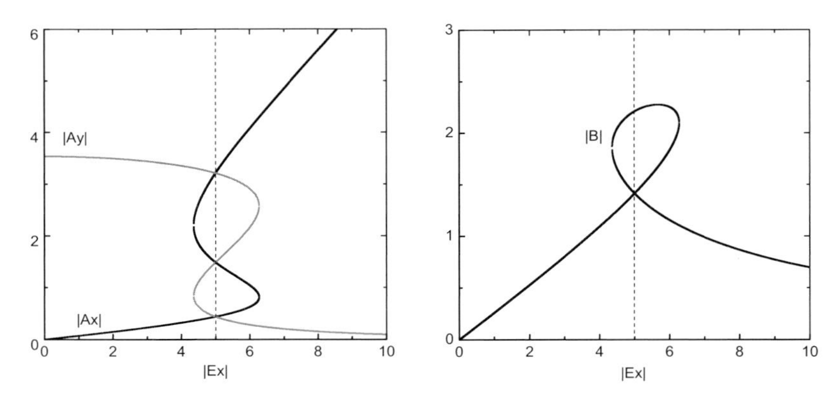

Fig. 8.2. Second-harmonic generation for asymmetric pumping. Steady-state intracavity field amplitudes as a function of E_x, for $E_y = 5$, $\delta_A = 1$, and $\delta_B = 0$. The vertical dashed line corresponds to the symmetric pumping $E_x = E_y$.

where $\Delta_A = 1 + \delta_A^2$, $\Delta_B = 1 + \delta_B^2$, $\Delta_{AB} = 1 - \delta_A \delta_B$, and $Q = (\delta_A + \delta_B)^2 + 3\Delta_{AB}^2$. Once $|A_y|$ is known, $|A_x|$ and $|B|$ are given by

$$\begin{aligned} |A_x|^2 &= \frac{\Delta_B |E_x|^2}{|A_y|^2 + 2\Delta_{AB}|A_y| + \Delta_A \Delta_B}, \\ |B|^2 &= \frac{|A_x|^2 |A_y|^2}{\Delta_B}. \end{aligned} \tag{8.6}$$

Figure 8.2 shows a typical dependence of the stationary solutions for the intracavity fields $|A_x|$, $|A_y|$, and $|B|$ on $|E_x|$ when $E_y = 5$. For small $|E_x|$, the functions $A_x(E_x)$ and $B(E_x)$ take small values, whereas $A_y(E_x)$ is large and close to $E_y/(1 + i\delta_A)$. All of them are single-valued. When $|E_x|$ approaches $|E_y|$, the system displays bistability: $A_x(E_x)$ and $A_y(E_x)$ become S-shaped and $B(E_x)$ closes over itself. For large $|E_x|$ all the functions become again single-valued but now $A_x(E_x) \gg A_y(E_x)$. The existence of three steady-state solutions of Eqs. (8.1)–(8.3) in a region of finite width centered on $|E_x| = |E_y|$ is closely related to the polarization instability occurring in the symmetrical pumping case. If fact, this S-shape can be observed only if $|E_y| > |E_{as}|$.

Below we will consider the effects produced on an image inserted in the system as spatial variations in the intensity of the x-polarized pump field along with a homogeneous pump E_y. Varying the amplitude of the homogeneous pump it is possible to achieve different regimes of operation. In a first regime the image can be transferred from the fundamental to the second-harmonic. In a second regime it is possible to enhance its contrast and to detect the contour of the image [27]. Furthermore, it is also possible to filter noise eventually present in the image. For simplicity E_x and E_y are taken as real except when noise is considered. Here we will consider only the case of an ideal cavity with flat mirrors that is resonant with both fundamental fields and with the second-harmonic field. Similar operations can be performed using cavities with spherical mirrors or where only the fundamental fields are resonant, as discussed in Ref. [33].

8.2.1 Frequency Up-Conversion of an Image

We consider the injection of an image, that is, the amplitude of the x-polarized signal $|E_x(x)|$ is a function of the transverse coordinate x. At a given position x the intracavity fields $A_{x,y}(x)$ and $B(x)$ tend to take the stationary values shown in Fig. 8.2 as if the pumps were homogeneous, despite the spatial coupling caused by diffraction. Diffraction becomes relevant for image details on the scale of the diffraction length. Figure 8.3 shows a scheme for a very simple one-dimensional image where $|E_x|$ takes only two values. If $|E_x(x)|$ remains well below $|E_y|$, $A_x(x)$ never leaves the lower branch of the curve $A_x(E_x)$, so $|A_x|$ reproduces the spatial distribution of the input image $|E_x|(x)$. The output at the SH frequency $B(x)$ also reproduces $|E_x(x)|$.

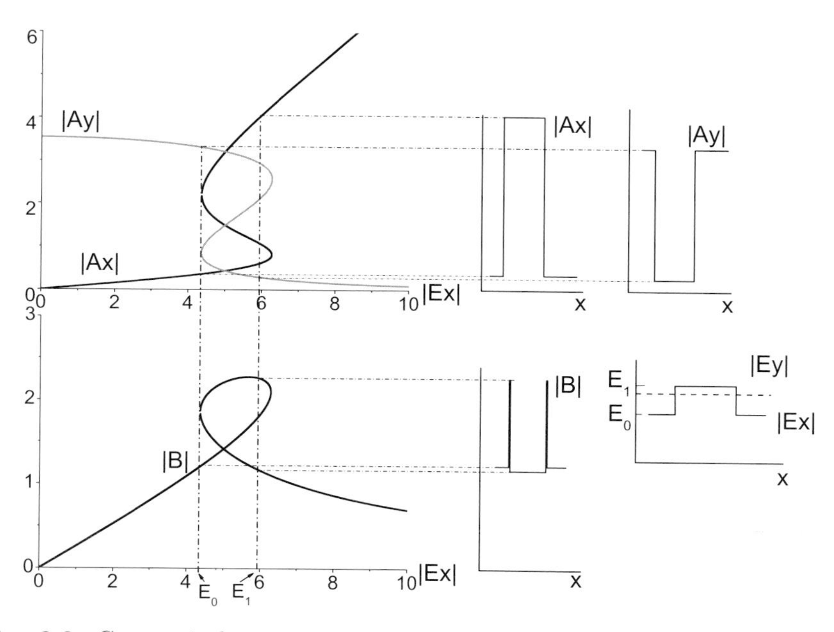

Fig. 8.3. Geometrical construction to illustrate the frequency transfer regime. On the left we plot the stationary amplitude of the intracavity fields for homogeneous asymmetric pumping as a function of E_x for $E_y = 5$ ($\delta_B = 0$, $\delta_A = 1$, and $E_{as} = 3.10755$). On the right we plot the response of the system to a simple image (sketched on the far right) where E_x takes only the values E_0 and E_1 with $E_0 < E_1 < E_y$.

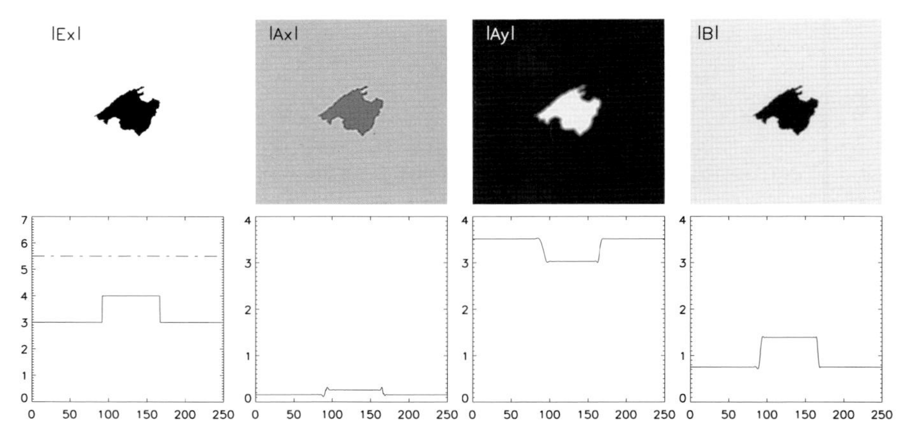

Fig. 8.4. Frequency transfer. The left column shows (from top to bottom) the spatial distribution of the input field E_x, and of the intracavity fields $|A_x|$, $|A_y|$ and $|B|$. In all the figures of this chapter, except if otherwise noted, the grey scale varies from the minimum (white) to the maximum (black) of each field. The right column shows a transversal cut of the fields along the dashed line on the top left panel. We have considered $E_y = 5.5$ (shown as dashed line on the top right panel).

Therefore, intracavity type-II SHG allows for an input image to be transferred from the fundamental to the SH frequency. In addition, polarization switching is performed, because the image encoded in $|E_x|$ and the SH field B has orthogonal polarizations. As a side effect, the input image appears in negative as a weak modulation of A_y around $E_y/(1+i\delta_A)$. This procedure is illustrated in Fig. 8.4 in a more realistic two-dimensional image. One can see that because of diffraction the edges of the image are softened. The image in the intracavity and the second-harmonic fields can be considered as a union of two different stationary states and the oscillatory tails of the front connecting the two states induce some distortion near the border. Nevertheless, the image is quite well reproduced as follows from Fig. 8.4.

8.2.2 Contrast Enhancement and Contour Recognition

Now we will consider the case in which the amplitude of the signal locally exceeds $|E_y|$. In this case the multivalued dependence of $A_x(E_x)$, $A_y(E_x)$, and $B(E_x)$ comes into play as is shown in Fig. 8.5. If E_1 is larger than the upper end of the hysteresis cycle and E_0 is smaller than the lower end, $A_x(E_x(x))$ has to jump from the lower to the upper branch, and $A_y(E_x(x))$ has to jump from the upper to the lower branch. This will give rise to a sharp spatial variation of A_x and A_y. In fact for vanishing intracavity fields as an initial condition, it is not necessary to fully cross the hysteresis cycle to have a jump. With those initial conditions where $|E_x| < |E_y|$ the system locally selects the steady-state solution with a small value for $|A_x|$ and a large value

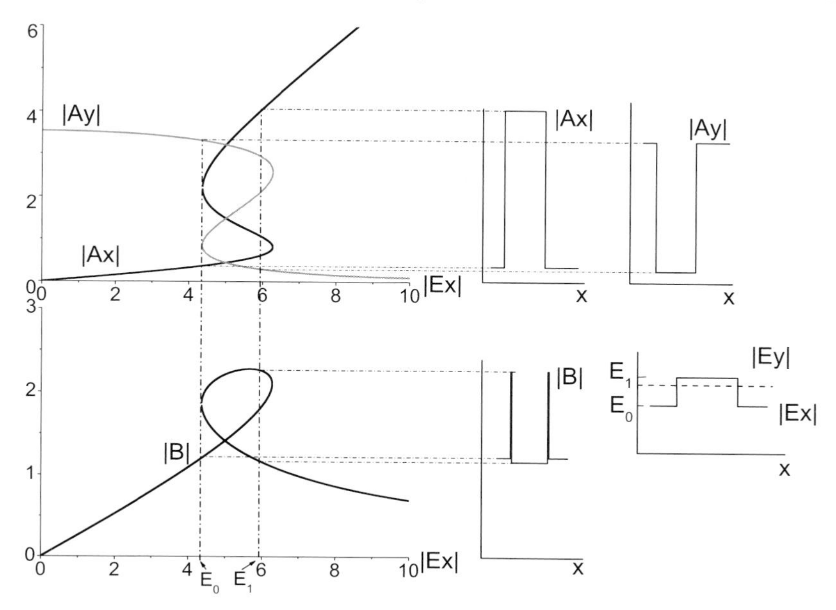

Fig. 8.5. Geometrical construction to illustrate the contrast enhancement and contour recognition regime similar to Fig. 8.3 but with $E_0 < E_y < E_1$.

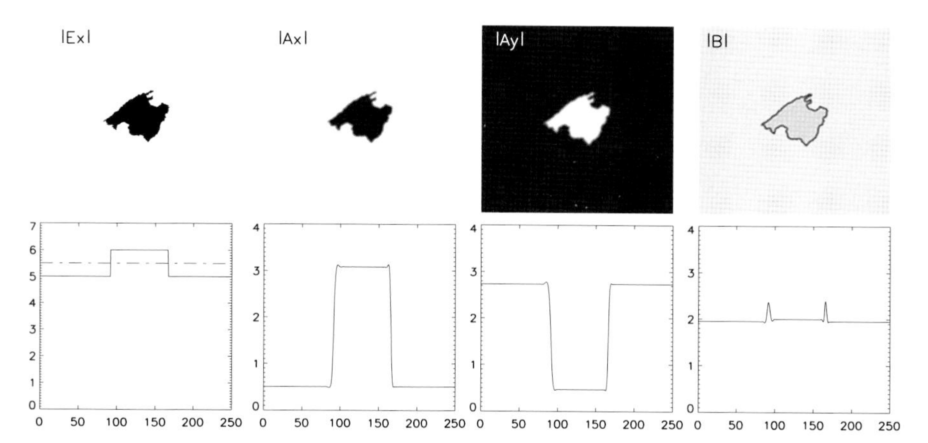

Fig. 8.6. Contrast enhancement and contour recognition. The left column shows (from top to bottom) the spatial distribution of the amplitude of the input image $|E_x|$, and the amplitude of the intracavity fields $|A_x|$, $|A_y|$, and $|B|$. The right column shows a transversal cut of the fields. We have considered $|E_y| = 5$.

for $|A_y|$, and it selects the steady state with large $|A_x|$ and small $|A_y|$ where $|E_x| > |E_y|$. Therefore, the reference value $|E_y|$ plays in fact the role of an effective threshold and the jump already occurs if $|E_x|$ crosses $|E_y|$ as sketched in Fig. 8.5. In the region where $|E_x(x)|$ is larger than the reference level $|E_y|$, $|A_x|$ has a large value compared with the zones where $|E_x(x)| < |E_y|$; so the contrast in this field appears enhanced with respect to the contrast in the input fields (see Fig. 8.5). The amplitude $|A_y|$ takes lower values where $|E_x| < |E_y|$ leading to an image that is inverted with respect to the input. At the border between the regions $|E_x| > |E_y|$ and $|E_x| < |E_y|$ the second harmonic field B displays a sharp peak, because locally $|A_x| \simeq |A_y|$; that is, the system goes through the symmetric steady-state solution characterized by a higher intracavity second-harmonic field than the asymmetric stable ones. As a consequence, the second-harmonic field displays the contour of the input image Fig. 8.5. These effects are shown for a two-dimensional input image in Fig. 8.6. Image processing is slightly affected by diffraction effects in two dimensions, which tends to smooth out sharp angles in the input image and sets a minimum contrast below which no contrast enhancement can occur. It should be emphasized that the previous results show that for a given image different processing capabilities are possible by tuning the amplitude of the homogeneous field $|E_y|$. This is even more interesting when considering images that are composed of many levels of intensity, as in a gray-scale image. In that sense, if the homogeneous pump $|E_y|$ is set to a value larger than $|E_x(x)|$ for any x then the frequency transfer process will take place and the whole image will be displayed by the second-harmonic frequency field $B(x)$. If $|E_y|$ is decreased, then the parts of the image where $|E_x(x)| > |E_y|$ will undergo a contrast enhancement process.

8.2.3 Noise Filtering Properties

Another interesting effect arises when the inserted image is superimposed with a complex random field, creating a noisy image both in intensity and phase. In this case the system shows noise filtering properties, and the images at the fundamental and the second-harmonic fields have a lower noise level than the input image. The noise filtering effect arises as an interplay between the diffraction and the nonlinear interaction that filters out all the high spatial frequency components of the input image. Therefore, the small-scale fluctuations are effectively removed [27]. It appears both in the frequency transfer regime and in the contour recognition regime, but it is more effective in the second case, when the nonlinearities play a more important role and the contrast of the image is enhanced, as can be seen from Figs. 8.7 and 8.8.

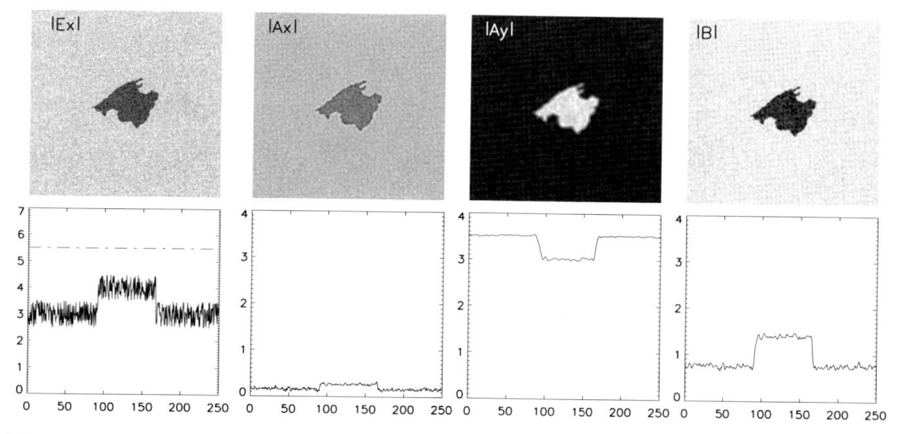

Fig. 8.7. Noise filtering in the frequency transfer regime. The left column shows (from top to bottom) the spatial distribution of the amplitude of the input image $|E_x|$, and the amplitude of the intracavity fields $|A_x|$, $|A_y|$, and $|B|$. The right column shows a transversal cut of the fields. We have considered $|E_y| = 5$.

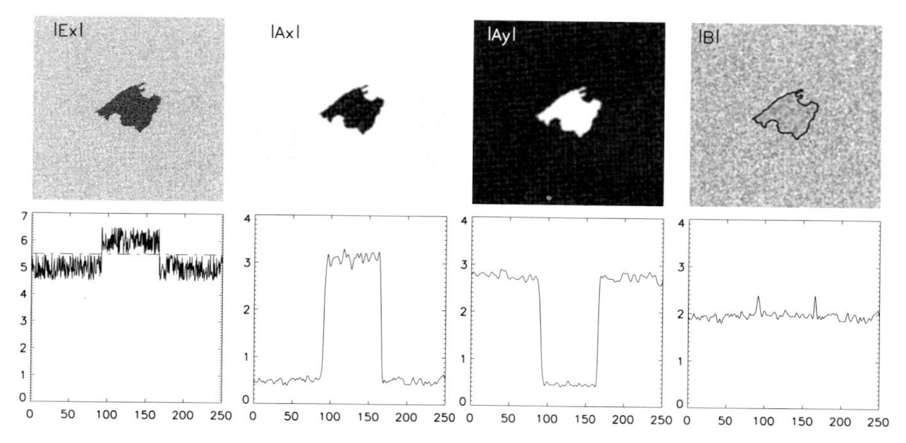

Fig. 8.8. Noise filtering in the contrast enhancement regime. The left column shows (from top to bottom) the spatial distribution of the amplitude of the input image $|E_x|$, and the amplitude of the intracavity fields $|A_x|$, $|A_y|$, and $|B|$. The right column shows a transversal cut of the fields. We have considered $|E_y| = 5.5$.

8.3 Quantum Image Processing in Type-I Second-Harmonic Generation

The treatment in the previous section was fully classical. In this and the next section we will explore the possibilities offered by second-harmonic generation for image processing from a quantum point of view. We will consider a traveling-wave configuration because in this case we will be able to separate the classically strong fields from the (weak) quantum fluctuations for which we will obtain linearized propagation equations. In this section we will consider the simpler case of a type-I interaction where the fundamental field has only one relevant polarization direction and in the next section we will address the type-II case where the polarization degree of freedom of the fundamental fields plays an important role. In the following subsections we address the dynamics of the field operators, obtain the propagation equations, and explore the results for different configurations.

8.3.1 Field-Operator Dynamics

We follow a procedure similar to the one used in Ref. [25] for an OPA. The main difference is that for an OPA it is generally justified to work in a classical approximation for the pump field because it is undepleted. In SHG, pump depletion cannot be neglected and both the fundamental and the second-harmonic fields have to be treated simultaneously as quantum fields [34]. This section is divided in two parts: we first deduce the nonlinear propagation equations for the operators associated with the fundamental and the second-harmonic fields and then we linearize the quantum fluctuations around the nonlinear classical fields.

A. Propagation Equations

We will begin by defining the slowly varying photon annihilation operators for the fundamental and the second-harmonic fields $\hat{A}_F(z,\boldsymbol{\rho},t)$ and $\hat{A}_S(z,\boldsymbol{\rho},t)$ from the positive-frequency part of the electric field,

$$\hat{E}_i^{(+)}(z,\boldsymbol{\rho},t) = i\xi_i\sqrt{\frac{\hbar\omega_i}{2\epsilon_0 c}}\exp[i(k_i z-\omega_i t)]\hat{A}_i(z,\boldsymbol{\rho},t)\,, \tag{8.7}$$

for $i = F, S$. The wave-numbers of the fundamental and the second-harmonic waves in the nonlinear medium, k_F and k_S, depend on the wave frequency through the dispersion relation $\omega = \omega(k)$. In Eq. (8.7) the prefactors

$$\xi_i = \frac{u(k_i)v(k_i)}{c^2\cos\rho(k_i)} \tag{8.8}$$

involve the group velocity $u(k_i)$, the phase velocity $v(k_i)$, and some generalized anisotropy angle $\rho(k_i)$. They describe the strength of the electric field in the medium, as compared to that in the vacuum; z is the coordinate on the longitudinal axis, which is defined as the beam axis, and $\boldsymbol{\rho} = (x,y)$ is the two-dimensional coordinate vector in the transverse plane.

The dynamics of these two fields in a $\chi^{(2)}$ nonlinear crystal is described by the Hamiltonian operator [25,34]:

$$\hat{H}_{\mathrm{int}} = \hat{H}_{0,F} + \hat{H}_{0,S} + \hat{H}_{\mathrm{int}}, \tag{8.9}$$

in which $\hat{H}_{0,F}$ and $\hat{H}_{0,S}$ are the free-field Hamiltonians for the fundamental and the second-harmonic field in the medium, whereas $\hat{H}_{\text{int}}$ describes the interaction between the two fields generated by the nonlinearity of the crystal. In terms of the slowly varying operators $\hat{\mathcal{A}}_i(z,\boldsymbol{\rho},t)$, the free-field Hamiltonians are given by [25]

$$\hat{H}_{0,i} = \frac{\hbar\omega_i}{c}\int_V dz\, d^2\rho \hat{\mathcal{A}}_i^\dagger(z,\boldsymbol{\rho},t)\hat{\mathcal{A}}_i(z,\boldsymbol{\rho},t), \tag{8.10}$$

where space integration is extended to the whole volume of the crystal. The expectation value $\langle\hat{\mathcal{A}}_i^\dagger(z,\boldsymbol{\rho},t)\hat{\mathcal{A}}_i(z,\boldsymbol{\rho},t)\rangle$ can be interpreted as the energy density per unit volume, scaled by a factor $\hbar\omega_i/c$. The interaction part $\hat{H}_{\text{int}}$ describes the three-wave interaction, which, under the usual assumption of an instantaneous and local nonlinear response of the medium [35], is given in terms of the slowly varying field operators by

$$\begin{aligned}\hat{H}_{\text{int}} = i\hbar\lambda\int_V dz\, d^2\rho \quad & \Big[e^{i\Delta kz}\hat{\mathcal{A}}_S^\dagger(z,\boldsymbol{\rho},t)\hat{\mathcal{A}}_F^2(z,\boldsymbol{\rho},t) \\ & -e^{-i\Delta kz}\hat{\mathcal{A}}_S(z,\boldsymbol{\rho},t)\hat{\mathcal{A}}_F^{\dagger 2}(z,\boldsymbol{\rho},t)\Big],\end{aligned} \tag{8.11}$$

where $\hbar\lambda = \chi^{(2)}(\hbar/2\epsilon_0 c)^{3/2}\xi_F^2\xi_S\sqrt{\omega_F^2\omega_S}$ and $\Delta k = 2k_F - k_S$ is the collinear phase mismatch. The Hamiltonian $\hat{H}_{\text{int}}$ is the sum of two contributions: the first term in Eq. (8.11) is responsible for second-harmonic generation and the second for the down-conversion. The dynamics of the two field operators is described by the Heisenberg equations, which for the Hamiltonian defined by Eqs. (8.10) and (8.11) are:

$$\begin{aligned}\partial_t\hat{\mathcal{A}}_F(z,\boldsymbol{\rho},t) = {} & i\omega_F\hat{\mathcal{A}}_F(z,\boldsymbol{\rho},t) \\ & -i\omega_F\int_V dz'd^2\rho' G_F(z-z',\boldsymbol{\rho}-\boldsymbol{\rho}')\hat{\mathcal{A}}_F(z',\boldsymbol{\rho}',t) \\ & -2c\lambda\int_V dz'd^2\rho' G_F(z-z',\boldsymbol{\rho}-\boldsymbol{\rho}')e^{-i\Delta kz'}\hat{\mathcal{A}}_S(z',\boldsymbol{\rho}',t)\hat{\mathcal{A}}_F^\dagger(z',\boldsymbol{\rho}',t),\end{aligned} \tag{8.12}$$

$$\begin{aligned}\partial_t\hat{\mathcal{A}}_S(z,\boldsymbol{\rho},t) = {} & i\omega_S\hat{\mathcal{A}}_S(z,\boldsymbol{\rho},t) \\ & -i\omega_S\int_V dz'd^2\rho' G_S(z-z',\boldsymbol{\rho}-\boldsymbol{\rho}')\hat{\mathcal{A}}_S(z',\boldsymbol{\rho}',t) \\ & -2c\lambda\int_V dz'd^2\rho' G_F(z-z',\boldsymbol{\rho}-\boldsymbol{\rho}')e^{-i\Delta kz'}\hat{\mathcal{A}}_F^2(z',\boldsymbol{\rho}',t),\end{aligned} \tag{8.13}$$

where

$$G_i(z-z',\boldsymbol{\rho}-\boldsymbol{\rho}') = \int\frac{dk_z d^2q}{(2\pi)^3}\,\frac{\omega(\sqrt{k_z^2+\boldsymbol{q}^2})}{\omega_i}\,e^{i(k_z-k_i)(z-z')+i\boldsymbol{q}\cdot(\boldsymbol{\rho}-\boldsymbol{\rho}')}. \tag{8.14}$$

It is helpful to work with operators in the Fourier space rather than in real space,

$$\hat{\mathcal{A}}_\sigma(z, \boldsymbol{q}, \Omega) = \int d^2\rho e^{-i\boldsymbol{q}\cdot\boldsymbol{\rho}} \int dt e^{i\Omega t} \hat{A}_\sigma(z, \boldsymbol{\rho}, t). \tag{8.15}$$

To separate the effects of the free propagation through the crystal from the nonlinear effects it is convenient to define for each field a propagation-corrected Fourier amplitude,

$$\hat{A}_i(z, \boldsymbol{q}, \Omega) = \xi_i \sqrt{n_i} \exp\{-i[k_i^z(\boldsymbol{q}, \Omega) - k_i]z\} \hat{\mathcal{A}}_i(z, \boldsymbol{q}, \Omega), \tag{8.16}$$

where $k_i^z(\boldsymbol{q}, \Omega) = \sqrt{k(\omega_i + \Omega)^2 - \boldsymbol{q}^2}$ is the longitudinal wave-number of a wave with frequency $\omega_i + \Omega$ and the transverse vector $\boldsymbol{q}$. The exponential phase factor in Eq. (8.15) is chosen to absorb in the free-propagation case the exact z-dependence of the wave associated with the field operator $\hat{A}_i(z, \boldsymbol{q}, \Omega)$. The prefactor $\xi_i \sqrt{n_i} = \sqrt{u_i/c}$, with u_i defined as the group velocity of a wave with frequency ω_i, allows us to identify $\langle \hat{A}_i^\dagger(z, \boldsymbol{\rho}, t) \hat{A}_i(z, \boldsymbol{\rho}, t) \rangle$ with the mean photon flux density in the medium (photons cm^{-2} sec^{-1}).

In the standard paraxial ($|\boldsymbol{q}| \ll k_\sigma^z(\boldsymbol{q}, \Omega)$) and quasi-monochromatic ($\Omega \ll \omega_\sigma$) approximation and under the assumption of slow z-dependence of the field operators, it can be shown that the propagation-corrected Fourier amplitudes obey the following set of propagation equations [34],

$$\frac{\partial}{\partial z} \hat{A}_F(z, \boldsymbol{q}, \Omega) = -2K \int d^2q' d\Omega' \hat{A}_F^\dagger(z, \boldsymbol{q}', \Omega') \hat{A}_S(z, \boldsymbol{q} + \boldsymbol{q}', \Omega + \Omega')$$
$$\times \exp\{i[k_S^z(\boldsymbol{q} + \boldsymbol{q}', \Omega + \Omega') - k_F^z(\boldsymbol{q}, \Omega) - k_F^z(\boldsymbol{q}', \Omega')]z\}, \tag{8.17}$$
$$\frac{\partial}{\partial z} \hat{A}_S(z, \boldsymbol{q}, \Omega) = +K \int d^2q' d\Omega' \hat{A}_F(z, \boldsymbol{q}', \Omega') \hat{A}_F(z, \boldsymbol{q} - \boldsymbol{q}', \Omega - \Omega')$$
$$\times \exp\{i[k_F^z(\boldsymbol{q}', \Omega') + k_F^z(\boldsymbol{q} - \boldsymbol{q}', \Omega - \Omega') - k_S^z(\boldsymbol{q}, \Omega)]z\}, \tag{8.18}$$

where $K = (2\pi)^{-3} \sqrt{c^3/u_F^2 u_S} \lambda$ is the coupling constant of the interaction. These coupled differential operator equations describe the propagation of the fundamental and the second-harmonic field through the nonlinear medium. The right-hand side represents a sum over all three wave processes that are able to generate a fundamental and a second-harmonic wave, respectively, with transverse wave-vector $\boldsymbol{q}$ and frequency Ω under the physical constraints of momentum and energy conservation. These equations generalize the propagation equations derived in [36] and [37] for a single-mode case.

By solving Eqs. (8.17) and (8.18) one can obtain the functional dependence between the fields at the output plane and the fields at the input one. In principle, this allows us to calculate the output of the system for an arbitrary quantum-mechanical state of the electromagnetic field illuminating the crystal. However, due to the nonlinear character of the equations some approximations are needed to solve them, as described bellow.

B. Two-Field Input-Output Relations

We consider now a situation suitable for image-processing problems: we assume the field distribution in the input plane of the crystal as given by a superposition of a strong homogeneous pump field at frequency ω and a weak coherent signal at 2ω with some spatiotemporal distribution corresponding to an input image. We assume that at any point inside the crystal the fundamental field generated by the input signal remains weak with respect to the pump field. Following [36] and [37] we write the propagation-corrected field operators associated with the fundamental and the second-harmonic field as

$$\hat{A}_F(z, \boldsymbol{q}, \Omega) = \tilde{c}_F(z)\delta^{(2)}(\boldsymbol{q})\delta(\Omega) + \hat{a}_F(z, \boldsymbol{q}, \Omega), \tag{8.19}$$

$$\hat{A}_S(z, \boldsymbol{q}, \Omega) = \tilde{c}_S(z)\delta^{(2)}(\boldsymbol{q})\delta(\Omega) + \hat{a}_S(z, \boldsymbol{q}, \Omega), \tag{8.20}$$

where $\tilde{c}_F(z)$ and $\tilde{c}_S(z)$ are the amplitudes of the strong monochromatic waves at frequencies ω and 2ω generated by the pump inside the crystal which, for simplification, are considered in the plane-wave approximation, and $\hat{a}_F(z, \boldsymbol{q}, \Omega)$ and $\hat{a}_S(z, \boldsymbol{q}, \Omega)$ are the quantum field operators associated with the two fields. These representations take into account the propagation of any field distribution injected into the crystal in addition to the strong pump field. In particular, they encode the propagation of the vacuum fluctuations entering the crystal through the input plane, which are responsible for the quantum fluctuations in the output fields, as analyzed in [36, 37]. Substituting Eqs. (8.19) and (8.20) in Eqs. (8.17) and (8.18) we obtain at zero order

$$\frac{d}{dz}\tilde{c}_F(z) = -2K\tilde{c}_F^*(z)\tilde{c}_S(z)e^{-i\Delta kz}, \tag{8.21}$$

$$\frac{d}{dz}\tilde{c}_S(z) = K\tilde{c}_F^2(z)e^{i\Delta kz}, \tag{8.22}$$

which are the classical propagation equations of nonlinear optics. The total power W is conserved $|\tilde{c}_F(z)|^2 + 2|\tilde{c}_S(z)|^2 = |\tilde{c}_F(0)|^2 + 2|\tilde{c}_S(0)|^2 = W$ in correspondence with the conservation of energy flow in the lossless crystal (Manley–Rowe relation). Introducing the dimensionless characteristic interaction length $z_0 = 1/\sqrt{2WK}$ and scaling the space as $\zeta = z/z_0$ and the field amplitudes as $c_F(z) = \tilde{c}_F(z)/\sqrt{W}$ and $c_S(z) = \tilde{c}_S(z)/\sqrt{W/2}$, we arrive at

$$\frac{d}{d\zeta}c_F(\zeta) = -c_F^*(\zeta)c_S(\zeta)e^{-i\Delta s\zeta}, \tag{8.23}$$

$$\frac{d}{d\zeta}c_S(\zeta) = c_F^2(\zeta)e^{i\Delta s\zeta}, \tag{8.24}$$

where $\Delta s = \Delta k z_0$. These equations can be solved analytically [46]. At first order we obtain

$$\frac{\partial}{\partial\zeta}\hat{a}_F(\zeta, \boldsymbol{q}, \Omega) = -c_S(\zeta)\hat{a}_F^\dagger(\zeta, -\boldsymbol{q}, -\Omega)e^{-i\Delta(\boldsymbol{q},\Omega)\zeta}$$
$$-\sqrt{2}c_F^*(\zeta)\hat{a}_S(\zeta, \boldsymbol{q}, \Omega)e^{iD(\boldsymbol{q},\Omega)\zeta}, \tag{8.25}$$

$$\frac{\partial}{\partial\zeta}\hat{a}_S(\zeta, \boldsymbol{q}, \Omega) = \sqrt{2}c_F(\zeta)\hat{a}_F(\zeta, \boldsymbol{q}, \Omega)e^{iD(\boldsymbol{q},\Omega)\zeta}. \tag{8.26}$$

Equations (8.25) and (8.26) involve two different dimensionless phase-mismatch functions,

$$\Delta(\boldsymbol{q},\Omega) = [k_F^z(\boldsymbol{q},\Omega) + k_F^z(-\boldsymbol{q},-\Omega) - k_S]z_0, \tag{8.27}$$

$$D(\boldsymbol{q},\Omega) = [k_F^z(\boldsymbol{q},\Omega) + k_F - k_S^z(\boldsymbol{q},\Omega)]z_0. \tag{8.28}$$

We should notice that, in the perfect phase-matched case the linearized-fluctuation analysis predicts that for large interaction lengths the fundamental field should evolve to a perfect squeezed vacuum [37]. This is in contradiction to the linearization requirement that at frequency ω the amplitude of the fluctuations should be smaller than the mean value of the field [38]. Comparing the predictions of the linearized analysis for traveling-wave SHG with the stochastic integration of the full nonlinear propagation equations obtained in the positive P representation [38–40], the approximation can be considered valid for interaction lengths $\zeta < 4$ [34].

The different terms in Eqs. (8.25) and (8.26) have a clear physical interpretation: the first term in the right-hand side of Eq. (8.25) reflects the transformation inside the crystal of photons of the strong homogeneous second-harmonic wave generated by the pump into two fundamental photons with opposite frequency offsets Ω and $-\Omega$ and transverse wave-vectors $\boldsymbol{q}$ and $-\boldsymbol{q}$. This process (which we call Process I) generates a coupling between the amplitudes $\hat{a}_F(z,\boldsymbol{q},\Omega)$ and $\hat{a}_F^\dagger(z,\boldsymbol{q},\Omega)$. The second term in the right-hand side of Eq. (8.25) describes the frequency down-conversion of a second-harmonic wave with $(\boldsymbol{q},\Omega)$, into a fundamental wave with $(\boldsymbol{q},\Omega)$, which translates into a coupling between the field operators $\hat{a}_F(z,\boldsymbol{q},\Omega)$ and $\hat{a}_S(z,\boldsymbol{q},\Omega)$ (Process II). Energy conservation implies that this frequency-changing process occurs under the radiation of a fundamental pump photon. The reverse process (Process III) acts as a source of second-harmonic photons and corresponds to the right-hand side of Eq. (8.26). A large phase mismatch results in fast spatial oscillations of the source term, which reduces the efficiency of a particular process. Therefore, Process I will be efficient for $\Delta(\boldsymbol{q},\Omega)\zeta \ll 1$, whereas processes II and III will be efficient for $D(\boldsymbol{q},\Omega)\zeta \ll 1$. In the paraxial and monochromatic approximation the longitudinal wave number can be written as

$$k_i^z(\boldsymbol{q},\Omega) = k_i + \left[\frac{\omega_i n_i'}{c} + \frac{k_i}{\omega_i}\right]\Omega + \frac{k_i''}{2}\Omega^2 - \frac{q^2}{2k_i}\,, \tag{8.29}$$

where $k_i'' = \partial^2 k/\partial\omega^2$ and $n_i' = \partial n/\partial\omega$ evaluated at $\omega = \omega_i$. Then [45]

$$\Delta(\boldsymbol{q},\Omega) = \Delta s + \mathrm{sign}(k_F'')\frac{\Omega^2}{\Omega_2^2} - \frac{q^2}{q_2^2}\,, \tag{8.30}$$

$$D(\boldsymbol{q},\Omega) = \Delta s - \frac{\Omega}{\Omega_1} + \mathrm{sign}(k_F'')\frac{\Omega^2}{4\Omega_2^2} - \frac{1}{4}\left(1 - \frac{\Delta k}{2k_F}\right)\frac{q^2}{q_2^2}\,, \tag{8.31}$$

where $q_2 = \sqrt{k_F/z_0}$, $\Omega_1 = (\omega/c(2n_S' - n_F')z_0)^{-1}$ and $\Omega_2 = (\mid k_F'' \mid z_0)^{-1/2}$.

It is convenient to introduce a vectorlike notation by defining [45]

$$\hat{\boldsymbol{a}}(z,\boldsymbol{q},\Omega) = \begin{pmatrix} \hat{a}_F(z,\boldsymbol{q},\Omega) \\ \hat{a}_S(z,\boldsymbol{q},\Omega) \end{pmatrix}; \qquad \hat{\boldsymbol{a}}^\dagger(z,\boldsymbol{q},\Omega) = \begin{pmatrix} \hat{a}_F^\dagger(z,\boldsymbol{q},\Omega) \\ \hat{a}_S^\dagger(z,\boldsymbol{q},\Omega) \end{pmatrix}. \tag{8.32}$$

The solution of Eqs. (8.25) and (8.26) can be expressed in the form of a compact input-output transformation that connects the field operators at the exit plane of the crystal with those at the input plane,

$$\hat{a}_i(z,\boldsymbol{q},\Omega) = \boldsymbol{U}_i(z,\boldsymbol{q},\Omega)\cdot\hat{\boldsymbol{a}}(0,\boldsymbol{q},\Omega) + \boldsymbol{V}_i(z,\boldsymbol{q},\Omega)\cdot\hat{\boldsymbol{a}}^\dagger(0,-\boldsymbol{q},-\Omega). \tag{8.33}$$

This transformation involves eight complex coefficients: $\boldsymbol{U}_F = (U_{FF}, U_{FS})$, $\boldsymbol{U}_S = (U_{SF}, U_{SS})$, $\boldsymbol{V}_F = (V_{FF}, V_{FS})$ and $\boldsymbol{V}_S = (V_{SF}, V_{SS})$. They can be determined by solving the propagation equations.

Analytical expressions for the coefficients of the input-output transformation Eq. (8.33) were obtained in [37] for $q = 0$ and $\Omega = 0$. In the general case, however, no analytical solution is known, so the coefficients have to be determined by numerical integration of Eqs. (8.25) and (8.26)

8.3.2 Quantum Image Processing

We consider an optical device represented in Fig. 8.9: a $\chi^{(2)}$-nonlinear crystal pumped at a frequency ω enclosed in a two-lens telescopic system. The field distribution injected into the nonlinear crystal will be the spatial Fourier transform of the original image. After processing, another lens will perform the back-transformation into the real space. This two-lens imaging configuration maps any point of the object and the image plane onto a plane wave with a given transverse wave-vector. This configuration is similar to the one based on parametric down-conversion used in [41,42]. In terms of image processing, we consider an input image at the second-harmonic frequency 2ω. This optical device is expected to deliver a pair of symmetric amplifieiesd versions of the input image at both fundamental and second-harmonic frequencies [34]. We will discuss the results in terms of plane waves with given wave-vectors, because the telescopic system converts these wave-vectors into positions in the transverse plane. Furthermore, we will assume the temporal evolution of the input image to be slow and put $\Omega \to 0$ for the calculation of the output images.

The nonlinear crystal is pumped only at a frequency ω. Considering a vanishing collinear phase mismatch $\Delta k = 0$, Eqs. (8.23) and (8.24) lead to

$$\widetilde{c}_F(\zeta) = e^{i\phi_F^{(0)}}\,\mathrm{sech}(\zeta), \tag{8.34}$$

$$\widetilde{c}_S(\zeta) = e^{i\phi_F^{(0)}}\,\tanh(\zeta), \tag{8.35}$$

where $\phi_F^{(0)}$ is the phase of the pump field.

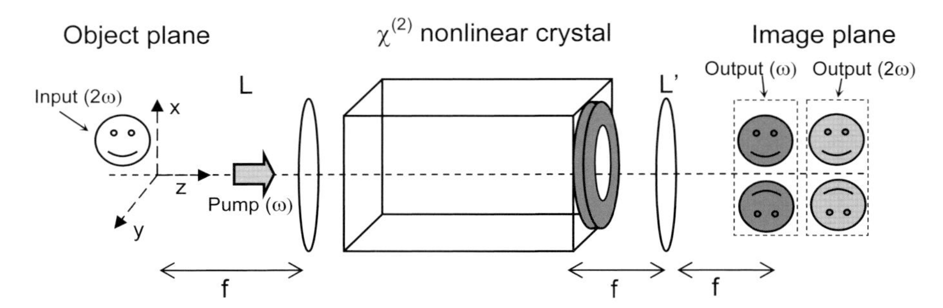

Fig. 8.9. Scheme of an optical device based on second-harmonic generation. A nonlinear crystal, pumped at frequency ω, is enclosed in a two-lens telescopic system. In the output plane of the crystal, a pupil of finite width represents the finite spatial bandwidth of the system.

The input signal at frequency 2ω is described by a coherent state $|\alpha_{\text{in}}\rangle$ characterized by a complex amplitude $\alpha_{\text{in}}(\boldsymbol{q}, \Omega)$. With respect to the frequency ω, $|\alpha_{\text{in}}\rangle$ is assumed to be in the vacuum state. Therefore we have

$$\begin{aligned} \hat{a}_S(0, \boldsymbol{q}, \Omega)|\alpha_{\text{in}}\rangle &= \alpha_{\text{in}}(\boldsymbol{q}, \Omega)|\alpha_{\text{in}}\rangle, \\ \hat{a}_F(0, \boldsymbol{q}, \Omega)|\alpha_{\text{in}}\rangle &= 0. \end{aligned} \tag{8.36}$$

In parametric down-conversion according to the symmetry properties of $\alpha_{\text{in}}(\boldsymbol{q}, \Omega)$ one distinguishes between a phase-insensitive regime, which corresponds to an input signal confined to one half of the object plane ($\alpha_{\text{in}}(\boldsymbol{q}, \Omega) = 0$ for $q_y < 0$) [19], and the phase-sensitive regime, which refers to the case of symmetrical input signal, that is, when $\alpha_{\text{in}}(\boldsymbol{q}, \Omega) = \alpha_{\text{in}}(-\boldsymbol{q}, -\Omega)$ [42]. In the following we discuss these two regimes for the SHG system.

A. Phase-Insensitive Configuration

For the input signal defined by Eq. (8.36) using the input-output transformations (8.33) one obtains that the fundamental output field is given by

$$\begin{aligned} \langle \hat{a}_F^\dagger(\zeta, \boldsymbol{q}, \Omega)\hat{a}_F(\zeta, \boldsymbol{q}, \Omega)\rangle &= (2\pi)^3\delta^{(3)}(0)(|V_{FF}(\zeta, \boldsymbol{q}, \Omega)|^2 + |V_{FS}(\zeta, \boldsymbol{q}, \Omega)|^2) \\ &+ |U_{FS}(\zeta, \boldsymbol{q}, \Omega)|^2|\alpha_{\text{in}}(\boldsymbol{q}, \Omega)|^2 + |V_{FS}(\zeta, \boldsymbol{q}, \Omega)|^2|\alpha_{\text{in}}(-\boldsymbol{q}, -\Omega)|^2, \end{aligned} \tag{8.37}$$

which indeed is independent of the phase of the input signal. Four different contributions can be distinguished: the first two terms on the right-hand side of Eq. (8.37) are independent of the strength of the input wave and correspond to spontaneous parametric fluorescence, which takes place in the crystal even in the absence of any coherent input signal. The two other contributions are proportional to the intensity of the input wave at $(\boldsymbol{q}, \Omega)$ and $(-\boldsymbol{q}, -\Omega)$, respectively. Because of the particular injection scheme considered here, for a given wave-vector $\boldsymbol{q}$ at which the output is considered, only one of these two terms is nonvanishing. For an object confined to the upper part of the object plane, at the output one obtains two output images. One confined to the upper output plane is an amplified version of the input image and has the intensity given by $|U_{FS}(\zeta, \boldsymbol{q}, \Omega)|^2 |\alpha_{\rm in}(\boldsymbol{q}, \Omega)|^2$. The other is confined to the lower output plane with the intensity given by $|V_{FS}(\zeta, \boldsymbol{q}, \Omega)|^2 |\alpha_{\rm in}(-\boldsymbol{q}, -\Omega)|^2$ and corresponds to a reversed amplified version of the input image. The physical underlying mechanism is Process I; second-harmonic photons generated by the strong pump wave inside the crystal are converted into pairs of fundamental twin photons propagating in opposite directions [19].

The second-harmonic output intensity is given by a similar expression,

$$\begin{aligned} \langle \hat{a}_S^\dagger(\zeta, \boldsymbol{q}, \Omega) \hat{a}_S(\zeta, \boldsymbol{q}, \Omega) \rangle &= (2\pi)^3 \delta^{(3)}(0) (|V_{SF}(\zeta, \boldsymbol{q}, \Omega)|^2 + |V_{SS}(\zeta, \boldsymbol{q}, \Omega)|^2) \\ &+ |U_{SS}(\zeta, \boldsymbol{q}, \Omega)|^2 |\alpha_{\rm in}(\boldsymbol{q}, \Omega)|^2 + |V_{SS}(\zeta, \boldsymbol{q}, \Omega)|^2 |\alpha_{\rm in}(-\boldsymbol{q}, -\Omega)|^2 . \end{aligned} \quad (8.38)$$

The second-harmonic output also displays both amplified and phase-conjugated amplified versions of the input image. Here the underlying mechanism is different; it is not the simultaneous generation of two second-harmonic waves with opposite wave-vectors and frequency offsets, but rather the frequency up-conversion of two fundamental photons (Process III).

The efficiency of these mechanisms can be quantified, defining for each of the four output images a local phase-insensitive gain as the ratio of the intensity of the output wave to the intensity of the input wave $|\alpha_{\rm in}(\boldsymbol{q}, \Omega)|^2$. Considering a pupil of finite aperture and adequate dimensions located at the output plane of the crystal then the output intensities due to spontaneous processes can be neglected [41]. In this situation the gains are given by

$$G_F(\zeta, \boldsymbol{q}, \Omega) = \frac{\langle \hat{a}_F^\dagger(\zeta, \boldsymbol{q}, \Omega) \hat{a}_F(\zeta, \boldsymbol{q}, \Omega) \rangle}{|\alpha_{\rm in}(\boldsymbol{q}, \Omega)|^2} = |U_{FS}(\zeta, \boldsymbol{q}, \Omega)|^2 , \quad (8.39)$$

$$\begin{aligned} G_F(\zeta, -\boldsymbol{q}, -\Omega) &= \frac{\langle \hat{a}_F^\dagger(\zeta, -\boldsymbol{q}, -\Omega) \hat{a}_F(\zeta, -\boldsymbol{q}, -\Omega) \rangle}{|\alpha_{\rm in}(\boldsymbol{q}, \Omega)|^2} \\ &= |V_{FS}(\zeta, \boldsymbol{q}, \Omega)|^2 , \end{aligned} \quad (8.40)$$

$$G_S(\zeta, \boldsymbol{q}, \Omega) = \frac{\langle \hat{a}_S^\dagger(\zeta, \boldsymbol{q}, \Omega) \hat{a}_S(\zeta, \boldsymbol{q}, \Omega) \rangle}{|\alpha_{\rm in}(\boldsymbol{q}, \Omega)|^2} = |U_{SS}(\zeta, \boldsymbol{q}, \Omega)|^2 , \quad (8.41)$$

$$\begin{aligned} G_S(\zeta, -\boldsymbol{q}, -\Omega) &= \frac{\langle \hat{a}_S^\dagger(\zeta, -\boldsymbol{q}, -\Omega) \hat{a}_S(\zeta, -\boldsymbol{q}, -\Omega) \rangle}{|\alpha_{\rm in}(\boldsymbol{q}, \Omega)|^2} \\ &= |V_{SS}(\zeta, \boldsymbol{q}, \Omega)|^2 . \end{aligned} \quad (8.42)$$

Figure 8.10 shows the phase-insensitive gains as functions of the interaction length inside the crystal. At the input plane we have $G_F(\zeta = 0, \boldsymbol{q}, \Omega) = G_F(\zeta = 0, -\boldsymbol{q}, -\Omega) = G_S(\zeta = 0, -\boldsymbol{q}, -\Omega) = 0$ and $G_S(\zeta = 0, \boldsymbol{q}, \Omega) = 1$, which simply identifies the chosen input. Increasing the interaction length, $G_S(\zeta, \boldsymbol{q}, \Omega)$ decreases and $G_F(\zeta, \boldsymbol{q}, \Omega)$ increases whereas $G_F(\zeta, -\boldsymbol{q}, -\Omega)$ and $G_S(\zeta, -\boldsymbol{q}, -\Omega)$ remain very small. At small interaction length Process II is dominant. At $\zeta \simeq 1.4$, the injected second-harmonic signal is fully down-converted. Further increase of the interaction length leads to a second step in the signal processing, which, for small wave numbers (solid lines in Fig. 8.10), is characterized by a rapid and symmetric growth of both the fundamental gains at $\boldsymbol{q}$ and $-\boldsymbol{q}$. This is the manifestation of the down-conversion Process I. This region of the crystal acts mainly as an OPA with a z-dependent pump. However, the presence of a weak residual pump field at frequency ω allows a partial frequency up-conversion of the amplified waves at fundamental frequency through Process III. This mechanism is responsible for a slow increase of the gains at $\boldsymbol{q}$ and $-\boldsymbol{q}$ when increasing interaction length, and leads to formation of the two phase-conjugate output images at the second-harmonic frequency.

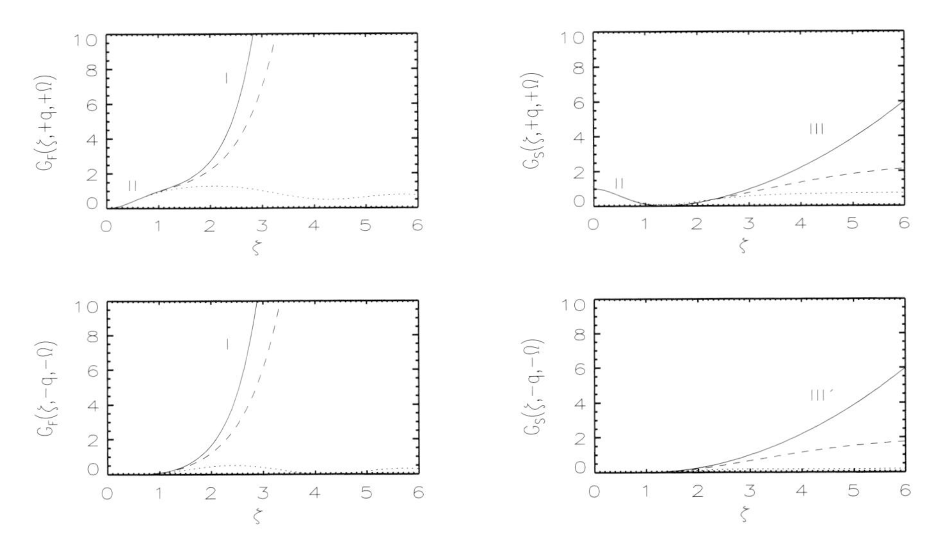

Fig. 8.10. Phase-insensitive gains as a function of the interaction length ζ for waves with $\Omega = 0$ and different wave-numbers: $q = 0.5$ (solid line), $q = 1.2$ (dashed line), and $q = 1.6$ (dotted line) (in units of $q_2 = \sqrt{k_F/z_0}$. The symbols I, II, and III refer to the corresponding dominant elementary process. Process III′ is the same as process III but with $\boldsymbol{q}$ replaced by $-\boldsymbol{q}$.

Due to diffraction, which appears in the dependence of the phase mismatch functions $D(\boldsymbol{q},\Omega)$ and $\Delta(\boldsymbol{q},\Omega)$ on $\boldsymbol{q}$ and Ω, the gains at high transverse wave-number will be reduced and for $q > \sqrt{2}$ the input signal is no longer amplified; rather it has an oscillatory behavior as a function of the interaction length [34, 43]. Therefore, only a finite disk-shaped portion of the input image centered on the beam axis will be efficiently processed [34] similarly to what happens in a perfect-matched OPA [19].

B. Phase-Sensitive Configuration

We consider here a symmetrical input image $\alpha_{\rm in}(-\boldsymbol{q},-\Omega)=\alpha_{\rm in}(\boldsymbol{q},\Omega)$. For static signals, this translates into a symmetry of the input image with respect to the beam axis. In the case of the OPA, it is well known for these symmetric images that the output is the result of the coherent superposition of both twin waves produced in an elementary down-conversion process [44]. The amplification is, therefore, phase sensitive, which is one of the requisites for amplifying an image without deteriorating its signal-to-noise ratio [21,24,42].

When the image inserted in the SHG system is symmetric with respect to the beam axis, the output images at each frequency will display the same symmetry. Under the same assumptions as for the phase-insensitive case, the ratio of the intensity in a given portion of each output image to that in the corresponding part of the input image defines the gains

$$G_F^{(\phi_{\rm in})}(\zeta,\boldsymbol{q},\Omega)=|U_{FS}(\zeta,\boldsymbol{q},\Omega)e^{i\phi_{\rm in}}+V_{FS}(\zeta,\boldsymbol{q},\Omega)e^{-i\phi_{\rm in}}|^2\,, \tag{8.43}$$

$$G_S^{(\phi_{\rm in})}(\zeta,\boldsymbol{q},\Omega)=|U_{SS}(\zeta,\boldsymbol{q},\Omega)e^{i\phi_{\rm in}}+V_{SS}(\zeta,\boldsymbol{q},\Omega)e^{-i\phi_{\rm in}}|^2\,, \tag{8.44}$$

which both depend on the phase of the input signal $\phi_{\rm in}$. For simplicity, we only consider the input images with a homogeneous phase $\alpha_{\rm in}(\boldsymbol{q},\Omega)=|\alpha_{\rm in}(\boldsymbol{q},\Omega)|e^{i\phi_{\rm in}}$.

Figure 8.11 shows the phase dependence of the gains for different transverse wave-numbers which in the telescopic system correspond to different regions of the transverse plane. On the optical axis ($\boldsymbol{q}=0$), both gains reach maximum values for $\phi_{\rm in}=\pi/2+n\pi$ and minimum at $\phi_{\rm in}=n\pi$ [34]. For off-axis regions of the transverse plane, one observes a shift in the position of the maximum and minimum gains (dashed and dotted lines in Fig. 8.11). This implies that for an input image with a homogeneous phase, the maximal gain condition can only be satisfied at one point of the transverse plane. However, it should be possible to compensate this through a displacement of the nonlinear crystal with respect to the lenses, which amounts to superposing a parabolic phase profile to the overall phase of the input image, as was considered for the OPA in Ref. [42].

The effect of diffraction is similar to the phase-insensitive case: the image processing will be efficient within a region of finite width centered on the beam axis, whereas outside this region, the nonlinear crystal will behave as a transparent medium [34].

Now we address the noise properties of the system, which are determined by the quantum fluctuations of the output fields. We first define for the fundamental and the second-harmonic fields the following quadrature operators,

$$\hat{x}_i^{\phi_{LO}}(\zeta,\boldsymbol{q},\Omega)=\frac{1}{2}[\hat{\bar{a}}_i(\zeta,\boldsymbol{q},\Omega)e^{-i\phi_{LO}}+\hat{\bar{a}}_i^{\dagger}(\zeta,-\boldsymbol{q},-\Omega)e^{i\phi_{LO}}], \tag{8.45}$$

which involves the field amplitude operator $\hat{\bar{a}}_i(\zeta,\boldsymbol{q},\Omega)$ related to the propagation-corrected amplitude $\hat{a}_i(\zeta,\boldsymbol{q},\Omega)$ as follows.

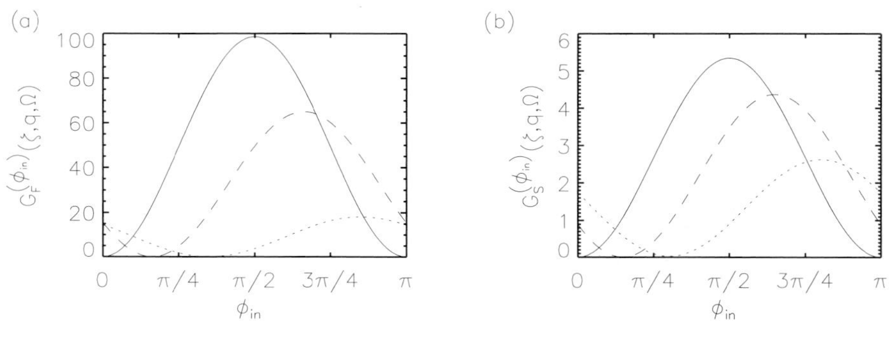

Fig. 8.11. Phase-sensitive gains for fundamental (a) and second-harmonic (b) fields as a function of the phase of the input signal (in radians) for different values of the transverse wave-number $q = 0$ (solid line), $q = 1$ (dashed line), and $q = 1.4$ (dotted line). Interaction length $\zeta = 3.32$.

$$\hat{\hat{a}}_i(\zeta, \boldsymbol{q}, \Omega) = \exp\{i[k_i^z(\boldsymbol{q}, \Omega) - k_i]z_0\zeta\}\hat{a}_i(\zeta, \boldsymbol{q}, \Omega)\,. \tag{8.46}$$

The exponential factor simply restores the phase accumulated during the propagation, which had been factored out in Eq. (8.16) for technical reasons. Unlike in the quantities considered before, now this phase factor is important as illustrated in [25].

The correlation function of the quadrature fluctuation $\delta\hat{x}_i^{(\phi_{LO})}(\zeta, \boldsymbol{q}, \Omega)$ defines the spectrum of squeezing $S_i^{(\phi_{LO})}(\zeta, \boldsymbol{q}, \Omega)$ through the relation

$$\begin{aligned} &\langle \delta\hat{x}_i^{(\phi_{LO})}(\zeta, \boldsymbol{q}, \Omega)\delta\hat{x}_i^{(\phi_{LO})}(\zeta, \boldsymbol{q}', \Omega')\rangle = \\ &\frac{1}{4}\delta^{(2)}(\boldsymbol{q} + \boldsymbol{q}')\delta(\Omega + \Omega')S_i^{(\phi_{LO})}(\zeta, \boldsymbol{q}, \Omega). \end{aligned} \tag{8.47}$$

In the case of photodetectors with perfect quantum efficiency, the spectrum of squeezing coincides with the spectral density for photocurrent fluctuations, normalized to the shot-noise level, as measured in a homodyne detection scheme. The phase ϕ_{LO} represents the phase of the local oscillator used in this detection setup.

Figure 8.12 shows $S_\sigma^{(\phi_{LO})}(\zeta, \boldsymbol{q}, \Omega)$ as a function of the local oscillator phase. As it happens for a single-mode squeezing transformation, changing the local oscillator phase allows one to explore the shape of the uncertainty region covered by the quantum fluctuations of the field. The maximum of $S_\sigma^{(\phi_{LO})}(\zeta, \boldsymbol{q}, \Omega)$ corresponds to the local oscillator pointing along the quadrature with stretched fluctuations, whereas the minimum indicates the direction of the squeezed quadrature. The effects of diffraction can be analyzed considering different values of $\boldsymbol{q}$. It is clear that the maximum and minimum values of the spectrum of squeezing are shifted for different values of $\boldsymbol{q}$. This shift can be interpreted as a rotation of the axis of the uncertainty region, as in the OPA [25]. Simultaneously, reduction of the amplitude of the oscillations of the spectrum of squeezing indicates reduction of the squeezing effect with larger $\boldsymbol{q}$, with the consequence that the uncertainty region recovers more and more the circular shape characteristic for a coherent state.

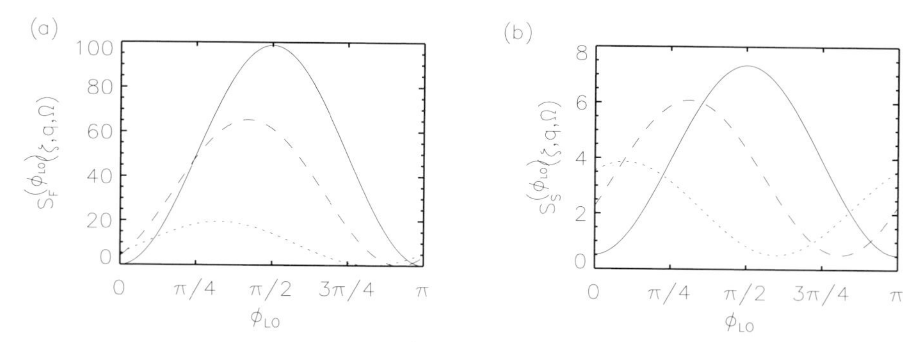

Fig. 8.12. Spectrum of squeezing $S_\sigma^{(\phi_{LO})}(\zeta, \boldsymbol{q}, \Omega)$ for fundamental (a) and second-harmonic (b) fields as a function of the phase of the local oscillator (in radians) for $\Omega = 0$ and different values of the transverse wave vector number $q = 0$ (solid line), $q = 1$ (dashed line), and $q = 1.4$ (dotted line). Interaction length $\zeta = 3.32$.

To appreciate the performances of the SHG device with respect to noiseless signal processing, it is convenient to consider the same detection scheme as for noiseless amplification in the OPA [19]. It consists in measuring the sum of the photocurrents from two symmetric pixels in the output plane. For such a device, the noise figure is given by the ratio of the intensity-squeezing spectrum to the phase-sensitive gain [34]

$$F_i(q) = \frac{S_i^{\phi_\sigma^{\text{out}}}(\zeta, \boldsymbol{q}, \Omega)}{G_i^{\phi_{\text{in}}}(\zeta, \boldsymbol{q}, \Omega)}, \tag{8.48}$$

where

$$\begin{aligned} \phi_F^{\text{out}} &= \arg[U_{FS}(\zeta, \boldsymbol{q}, \Omega)e^{i\phi_{\text{in}}} + V_{FS}(\zeta, \boldsymbol{q}, \Omega)e^{-i\phi_{\text{in}}} \\ &\quad + [k_F^z(\boldsymbol{q}, \Omega) - k_F]z_0\zeta, \end{aligned} \tag{8.49}$$

$$\begin{aligned} \phi_S^{\text{out}} &= \arg[U_{SS}(\zeta, \boldsymbol{q}, \Omega)e^{i\phi_{\text{in}}} + V_{SS}(\zeta, \boldsymbol{q}, \Omega)e^{-i\phi_{\text{in}}} \\ &\quad + [k_S^z(\boldsymbol{q}, \Omega) - k_S]z_0\zeta. \end{aligned} \tag{8.50}$$

Figure 8.13 shows the noise figure for the fundamental and the second-harmonic outputs choosing the phase of the input signal such that the phase-sensitive gains are maximal at $q = 0$. Because we are considering linearized propagation equations for the field operators, the noise figure can be never less than the unity (dotted line) [21] which corresponds to a noiseless operation.

In the domain of wave-numbers for which image processing is efficient we find that the fundamental output shows the same level of noise as the

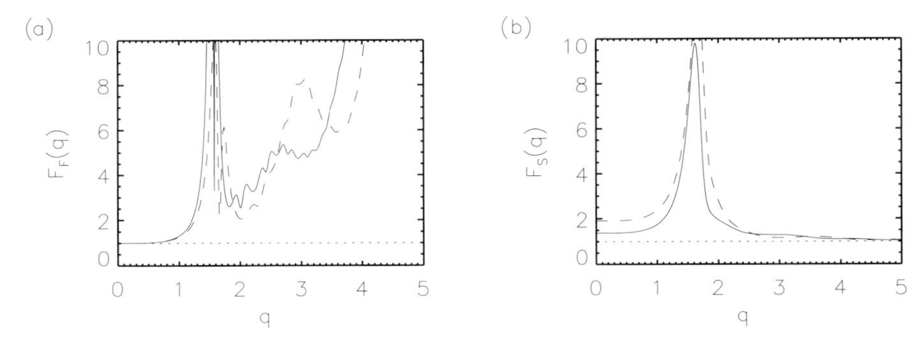

Fig. 8.13. Noise figure for the fundamental (a) and the second-harmonic (b) fields as a function of the transverse wave-number for two different values of the interaction length $\zeta = 3.32$ (solid line) and $\zeta = 2.5$ (dashed line). Dotted line shows the smallest possible value $F_i(q) = 1$ for a linear system.

input image ($F_S \approx 1$), so the SHG device operates without adding noise to the signal. For the second-harmonic field the noise figure is slightly above unity, implying a degradation of the signal-to-noise ratio in the output image. However, increasing the interaction length, the noise figure approaches unity for the spatial frequencies inside the bandwidth for image processing. Finally, one notes that, for large transverse wave-numbers the noise figure for the second-harmonic frequency is equal to one, because the input signal is unaffected by the system in this region, whereas for the noise figure the fundamental frequency diverges, as a consequence of a vanishing output intensity at ω in the limit $|\boldsymbol{q}| \to \infty$.

8.4 Quantum Image Processing in Type-II Second-Harmonic Generation

In this section we will explore the possibilities offered by type-II second-harmonic generation for image processing, where the polarization degree of freedom at the fundamental frequency plays a very important role, as discussed in Section 8.2 at the classical level. Here we will consider the role of quantum fluctuations in the telescopic traveling-wave configuration shown in Fig. 8.14 which is similar to the one considered in Section 8.3. However, although in that case the image was introduced as a second-harmonic signal, here, because we have two orthogonally polarized fundamental fields, we introduce the image on the fundamental fields, as in Section 8.2. We will show that with traveling-wave type-II SHG it is possible to noiselessly up-convert the part of an image with a given polarization while the part with orthogonal polarization is noiselessly amplified. In the following subsections we will first present the propagation equations for this system, as an extension of the equations presented in Subsection 8.3.2. Then we will discuss the imaging properties of this system in two different configurations.

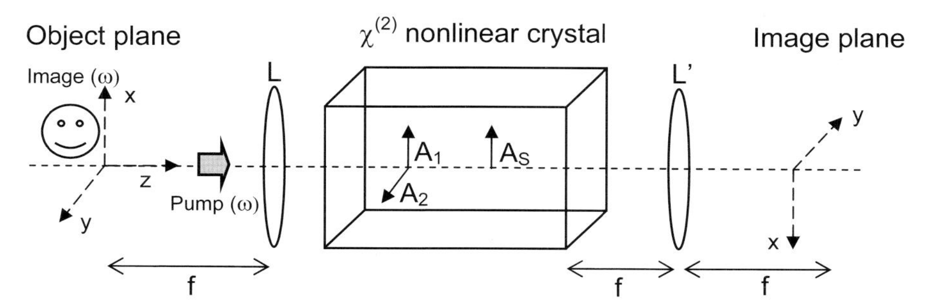

Fig. 8.14. Scheme of an optical device based on type-II second-harmonic generation. A nonlinear crystal, pumped at frequency ω, is enclosed in a two-lens telescopic system.

8.4.1 Propagation Equations

Following the same steps as in Subsection 8.3.2, the equations for the propagation-corrected Fourier amplitudes are

$$\frac{\partial}{\partial z}\hat{A}_S(z,\boldsymbol{q},\Omega) = K\int d^2q'd\Omega'\hat{A}_1(z,\boldsymbol{q}',\Omega')\hat{A}_2(z,\boldsymbol{q}-\boldsymbol{q}',\Omega-\Omega')$$
$$\times\exp\{i[k_1^z(\boldsymbol{q}',\Omega')+k_2^z(\boldsymbol{q}-\boldsymbol{q}',\Omega-\Omega')-k_S^z(\boldsymbol{q},\Omega)]z\}, \quad (8.51)$$

$$\frac{\partial}{\partial z}\hat{A}_1(z,\boldsymbol{q},\Omega) = -K\int d^2q'd\Omega'\hat{A}_2^\dagger(z,\boldsymbol{q}',\Omega')\hat{A}_S(z,\boldsymbol{q}+\boldsymbol{q}',\Omega+\Omega')$$
$$\times\exp\{i[k_S^z(\boldsymbol{q}+\boldsymbol{q}',\Omega+\Omega')-k_1^z(\boldsymbol{q},\Omega)-k_2^z(\boldsymbol{q}',\Omega')]z\}, \quad (8.52)$$

$$\frac{\partial}{\partial z}\hat{A}_2(z,\boldsymbol{q},\Omega) = -K\int d^2q'd\Omega'\hat{A}_1^\dagger(z,\boldsymbol{q}',\Omega')\hat{A}_S(z,\boldsymbol{q}+\boldsymbol{q}',\Omega+\Omega')$$
$$\times\exp\{i(k_S^z(\boldsymbol{q}+\boldsymbol{q}',\Omega+\Omega')-k_2^z(\boldsymbol{q},\Omega)-k_1^z(\boldsymbol{q}',\Omega')]z\}, \quad (8.53)$$

where the indices 1,2, and S correspond to the x-polarized fundamental field, the y-polarized fundamental field, and the second-harmonic field, respectively. These equations have a simple physical interpretation: the generation of a given field mode is seen as the result of all possible three-wave processes that fulfill the energy and the transverse momentum conservation lows.

Now we use a linearization approach as in the previous section. Assuming that the pump field is a strong classical field, which, inside the crystal will eventually produce a strong classical field at the second-harmonic frequency, we have

$$\hat{A}_i(z,\boldsymbol{q},\Omega) = \tilde{c}_i(z)\delta^{(2)}(\boldsymbol{q})\delta(\Omega) + \hat{a}_i(z,\boldsymbol{q},\Omega)\,, \quad (8.54)$$

where i stands for 1,2,S and $\tilde{c}_i(z)$ are the amplitudes of the classical strong monochromatic fields, and $\hat{a}_i(z, \boldsymbol{q}, \Omega)$ are the quantum field operators. The total power $W = |\tilde{c}_1(z)|^2 + |\tilde{c}_2(z)|^2 + 2|\tilde{c}_S(z)|^2$ is conserved as a consequence of the conservation of the energy flux in the lossless crystal. Introducing the dimensionless characteristic interaction length $z_0 = 1/\sqrt{2WK}$, scaling the space as $\zeta = z/z_0$ and the field amplitudes as $c_1(z) = \tilde{c}_1(z)/\sqrt{W}$, $c_2(z) = \tilde{c}_2(z)/\sqrt{W}$, and $c_S(z) = \tilde{c}_S(z)/\sqrt{W/2}$, and substituting Eqs. (8.54) in Eqs. (8.51)–(8.53), we obtain at zero order the classical equations of nonlinear optics,

$$\frac{d}{d\zeta}c_1(\zeta) = -c_2^*(\zeta)c_S(\zeta)e^{-i\Delta s\zeta}, \tag{8.55}$$

$$\frac{d}{d\zeta}c_2(\zeta) = -c_1^*(\zeta)c_S(\zeta)e^{-i\Delta s\zeta}, \tag{8.56}$$

$$\frac{d}{d\zeta}c_S(\zeta) = +2c_1(\zeta)c_2(\zeta)e^{i\Delta s\zeta}, \tag{8.57}$$

where $\Delta s = \Delta k z_0$.

At first order one obtains the propagation equations for the quantum operators associated with the fundamental and the second-harmonic fields

$$\frac{\partial}{\partial\zeta}\hat{a}_1(\zeta,\boldsymbol{q},\Omega) = -c_S(\zeta)e^{-i\Delta_1(\boldsymbol{q},\Omega)z}\hat{a}_2^\dagger(\zeta,-\boldsymbol{q},-\Omega)$$
$$-\sqrt{2}c_2^*(\zeta)e^{-iD_1(\boldsymbol{q},\Omega)z}\hat{a}_S(\zeta,\boldsymbol{q},\Omega), \tag{8.58}$$

$$\frac{\partial}{\partial\zeta}\hat{a}_2(\zeta,\boldsymbol{q},\Omega) = -c_S(\zeta)e^{i\Delta_1(-\boldsymbol{q},-\Omega)z}\hat{a}_1^\dagger(\zeta,-\boldsymbol{q},-\Omega)$$
$$-\sqrt{2}c_1^*(\zeta)e^{-iD_2(\boldsymbol{q},\Omega)z}\hat{a}_S(\zeta,\boldsymbol{q},\Omega), \tag{8.59}$$

$$\frac{\partial}{\partial\zeta}\hat{a}_S(\zeta,\boldsymbol{q},\Omega) = +\sqrt{2}c_2(\zeta)e^{iD_1(\boldsymbol{q},\Omega)z}\hat{a}_1(\zeta,\boldsymbol{q},\Omega)$$
$$+\sqrt{2}c_1(\zeta)e^{iD_2(\boldsymbol{q},\Omega)z}\hat{a}_2(\zeta,\boldsymbol{q},\Omega). \tag{8.60}$$

The phase factors involved in the equations are

$$\Delta_1(\boldsymbol{q},\Omega) = z_0(k_1^z(\boldsymbol{q},\Omega) + k_2^z(-\boldsymbol{q},-\Omega) - k_S), \tag{8.61}$$

$$D_1(\boldsymbol{q},\Omega) = z_0(k_1^z(\boldsymbol{q},\Omega) + k_2 - k_S^z(\boldsymbol{q},\Omega)), \tag{8.62}$$

$$D_2(\boldsymbol{q},\Omega) = z_0(k_1 + k_2^z(\boldsymbol{q},\Omega) - k_S^z(\boldsymbol{q},\Omega)). \tag{8.63}$$

Using the vectorlike notation [45]

$$\hat{\boldsymbol{a}}(z,\boldsymbol{q},\Omega) = \begin{pmatrix} \hat{a}_1(z,\boldsymbol{q},\Omega) \\ \hat{a}_2(z,\boldsymbol{q},\Omega) \\ \hat{a}_S(z,\boldsymbol{q},\Omega) \end{pmatrix}, \qquad \hat{\boldsymbol{a}}^\dagger(z,\boldsymbol{q},\Omega) = \begin{pmatrix} \hat{a}_1^\dagger(z,\boldsymbol{q},\Omega) \\ \hat{a}_2^\dagger(z,\boldsymbol{q},\Omega) \\ \hat{a}_S^\dagger(z,\boldsymbol{q},\Omega) \end{pmatrix}, \tag{8.64}$$

the solution of the propagation equations (8.58)–(8.60) can be written as

$$\hat{a}_i(z, \boldsymbol{q}, \Omega) = \boldsymbol{U}_i(z, \boldsymbol{q}, \Omega) \cdot \hat{\boldsymbol{a}}(0, \boldsymbol{q}, \Omega) + \boldsymbol{V}_i(z, \boldsymbol{q}, \Omega) \cdot \hat{\boldsymbol{a}}^\dagger(0, -\boldsymbol{q}, -\Omega), \quad (8.65)$$

which involves 18 complex coefficients that can be determined by solving the propagation equations.

Superposed to the homogeneous pump field, an optical image will be injected into the nonlinear crystal in a coherent state $|\alpha_{\rm in}\rangle$. This quantum state can be written as

$$\hat{\boldsymbol{a}}(0, \boldsymbol{q}, \Omega)|\alpha_{\rm in}\rangle = \boldsymbol{\alpha}(q)\delta(\Omega)|\alpha_{\rm in}\rangle. \quad (8.66)$$

The amplitude $\boldsymbol{\alpha} = (\alpha_1(q), 0, 0)$ encodes, within the vectorial notation defined by Eqs. (8.64), the spatial distribution of the input image at the fundamental frequency with polarization x and y, and at the second-harmonic frequency, respectively. We consider here an image stationary in time and inserted in the x-polarization.

The quantities of interest are the numbers of photons of each field detected by a photodetector located at a given position $\boldsymbol{\rho}$ of the image plane, which in the telescopic configuration is identified with a given transverse number. Assuming a detector whose size σ_d is much smaller than the typical variation scales of the fields and with a perfect quantum efficiency, the average number of photons is given by

$$< \hat{N}_i(\zeta, q) >= \sigma_d < \hat{a}_i^\dagger(\zeta, q, 0)\hat{a}_i(\zeta, q, 0) >= \sigma_d|\alpha^{\rm out}(q)|^2 \,, \quad (8.67)$$

where

$$\alpha_i^{\rm out}(q) = \boldsymbol{U}_i(q, 0) \cdot \boldsymbol{\alpha}(q) + \boldsymbol{V}_i(q, 0) \cdot \boldsymbol{\alpha}^*(-q) \,, \quad (8.68)$$

represents the amplitude of the outgoing wave for the field $i = 1, 2, S$. The photon numbers depend on the amplitude of the input image at the wave number q but also, essentially as a consequence of the down-conversion process in the crystal, on the input wave at $-q$. The photon number variance is given by

$$\langle(\Delta\hat{N}_i(L, q))^2\rangle = \sigma_d \mid \alpha_i^{\rm out}(q) \mid^2 (1 + 2\boldsymbol{V}_i^*(q, 0) \cdot \boldsymbol{V}_i(q, 0)) \,. \quad (8.69)$$

The noise level of the detected image is quantified by the signal-to-noise ratio:

$${\rm SNR}_i = \frac{\langle\hat{N}_i(L, q)\rangle^2}{\langle(\Delta\hat{N}_i(L, q))^2\rangle} \,, \quad (8.70)$$

which has to be compared to the signal-to-noise ratio in the input image, ${\rm SNR}_1^{\rm in} = \sigma_d \mid \alpha_1(q) \mid^2$. The noise figure is the ratio ${\rm SNR}_1^{\rm in}/{\rm SNR}_i$,

$$F_i = \frac{1 + 2\boldsymbol{V}_i^*(q, 0) \cdot \boldsymbol{V}_i(q, 0)}{|\alpha_i^{\rm out}(q)|^2/|\alpha_1(q)|^2}. \quad (8.71)$$

As stated in Subsection 8.3.2, if the noise figure is equal to unity, the image processing is noiseless, and at the level of quantum fluctuations, the quality of the image is preserved. This is the best possible situation in a system like the one we are considering here, but generally the noise figure is larger than unity, therefore a degradation of the image quality occurs which can even lead to a complete loss of information in the case of weak input signals.

8.4.2 Linearly y-Polarized Pump: Frequency Addition Regime

The first case that we will consider is the simplest one: the pump field is taken as linearly polarized along the y-direction. This is expressed in the initial condition

$$c_1(0) = 0; \qquad c_2(0) = 1; \qquad c_3(0) = 0. \tag{8.72}$$

Because no field is pumped in the orthogonal x-polarization, SHG cannot take place and the classical equations of nonlinear optics [27–31, 34] predict that

$$c_1(\zeta) = 0; \qquad c_2(\zeta) = 1; \qquad c_3(\zeta) = 0. \tag{8.73}$$

The linearized propagation equations (8.58)–(8.60) for the quantum operators can be easily solved analytically. One finally obtains the following input-output transformation.

$$\begin{aligned}
\hat{a}_1(\zeta, q, \Omega) &= \left(\cos(\widetilde{D}_1\zeta) + i\frac{D_1}{2\widetilde{D}_1}\sin(\widetilde{D}_1\zeta)\right) e^{-iD_1\zeta/2}\hat{a}_1(0, q, \Omega) \\
&\quad - \frac{\sqrt{2}}{\widetilde{D}_1}\sin(\widetilde{D}_1\zeta)e^{-iD_1\zeta/2}\hat{a}_S(0, q, \Omega), && (8.74) \\
\hat{a}_S(\zeta, q, \Omega) &= \left(\cos(\widetilde{D}_1\zeta) - i\frac{D_1}{2\widetilde{D}_1}\sin(\widetilde{D}_1\zeta)\right) e^{iD_1\zeta/2}\hat{a}_S(0, q, \Omega) \\
&\quad + \frac{\sqrt{2}}{\widetilde{D}_1}\sin(\widetilde{D}_1\zeta)e^{iD_1\zeta/2}\hat{a}_1(0, q, \Omega), && (8.75) \\
\hat{a}_2(\zeta, q, \Omega) &= \hat{a}_2(0, q, \Omega), && (8.76)
\end{aligned}$$

where $\widetilde{D}_1(q, \Omega) = \sqrt{2 + D_1^2(q, \Omega)/4}$. Although the y-polarized image would not be affected by propagation through the nonlinear crystal, this device is able to up-convert an x-polarized input image at frequency ω to frequency 2ω. This effect comes from dependence of the output operator $a_s(\zeta, q, \Omega)$, as well as $a_1(\zeta, q, \Omega)$, on both input operators $a_1(0, q, \Omega)$ and $a_S(0, q, \Omega)$. Because the second-harmonic wave is not created in the crystal, the down-conversion process is impossible, and hence no coupling between the frequencies (q, Ω) and $(-q, -\Omega)$ occurs during propagation. Comparing with the general form of the input-output transformation (8.65), all the coefficients $\boldsymbol{V}_i(q, \Omega)$ vanish in the case under consideration, and as a consequence, the output signal does not depend on the phase of the input signal.

The gain of the up-conversion process is

$$G_S(\zeta, q) \equiv \frac{\langle \hat{N}_S(\zeta, q)\rangle}{\langle \hat{N}_1(0, q)\rangle} = 2\frac{\sin^2(\tilde{D}_1(q, \Omega)z)}{\tilde{D}_1(q, \Omega)^2}. \tag{8.77}$$

Perfect up-conversion, $G_S = 1$, is achieved if $D_1(q, \Omega) = 0$ and $\sin(\tilde{D}_1(q, \Omega)z) = 1$. The first condition, corresponding to the perfect wave-vector matching defines an ensemble of points in the spatiotemporal frequency plane (q, Ω), and, in the case of static images, at most two values of q. The second condition establishes that even in that case maximum efficiency occurs only at given propagation lengths $\zeta_k = (\pi/\sqrt{2})(k + 1/2)$, with $k = 0, 1, \cdots$.

The noise figure coincides with the inverse of the up-conversion rate.

$$F(q) = \frac{1}{R_{\mathrm{up}}(q)}. \tag{8.78}$$

The input-output transformation (8.74)–(8.76) is completely equivalent to the transformation performed by a beamsplitter. In the case of a coherent image, and assuming that the second-harmonic input state is the vacuum, an incomplete up-conversion means that the mean intensity of the output is reduced with respect to the input image, but the amount of noise is conserved, corresponding to the one of a coherent state. As a result, a deterioration of the image takes place and the noise figure for the transmitted image is equal to the inverse of the transmission coefficient of the beamsplitter. The quantum noise present in the up-converted image can be interpreted as the superposition of the noise already present in the input image, which is partially transmitted together with the image itself, and the noise originally present in the second-harmonic modes, partially reflected by the beamsplitter. Therefore, reduction of the noise level in the output image can only be achieved by reducing the fluctuations present in the second-harmonic input field. This can be done by injecting a squeezed vacuum at the second-harmonic frequency instead of a normal vacuum. By properly choosing the squeezed quadrature, this contribution to the total noise level in the image can be reduced, and one may approach, in the limit of a perfect squeezed quadrature, a noise figure of unity, that is, a noiseless image processing. In the next subsection we will consider a slightly different set-up that allows us to achieve a noiseless up-conversion without any additional source of nonclassical light.

8.4.3 45°-Linearly Polarized Pump: Noiseless Up-Conversion and Amplification

We consider the case where the pump field is linearly polarized at 45° with respect to the axis x. The initial condition for the classical strong fields is

$$c_1(0) = c_2(0) = 1/\sqrt{2}; \qquad c_s(0) = 0. \tag{8.79}$$

Because now the fundamental field is pumped at the two orthogonal polarizations, a strong second-harmonic field will be generated inside the crystal. Considering the perfectly matched wave-vector, the solution of the classical equations [27–31, 34] is particularly simple and reads [46]:

$$c_1(\zeta) = c_2(\zeta) = \mathrm{sech}(\zeta)/\sqrt{2}; \qquad c_s(\zeta) = \tanh(\zeta). \tag{8.80}$$

It is useful to introduce a polarization basis rotated by an angle of 45° with respect to the basis (x,y) [36, 37, 45]:

$$\hat{a}_\pm(\zeta, q) = \frac{1}{\sqrt{2}}(\hat{a}_1(\zeta, q) \pm \hat{a}_2(\zeta, q)). \tag{8.81}$$

In this basis $c_+(\zeta) = \mathrm{sech}(\zeta)$ and $c_-(\zeta) = 0$. The propagation equations in this new basis can easily be derived from Eqs. (8.58)–(8.60) and one finds

$$\begin{aligned}\frac{\partial}{\partial\zeta}\ \hat{a}_\pm + (\zeta, \boldsymbol{q}, \Omega) &= -c_S(\zeta)g_\pm(\zeta, \boldsymbol{q}, \Omega)\hat{a}_+^\dagger(\zeta, -\boldsymbol{q}, -\Omega) \\ &+ c_S(\zeta)g_\mp(\zeta, \boldsymbol{q}, \Omega)\hat{a}_-^\dagger(\zeta, -\boldsymbol{q}, -\Omega) \\ &- \sqrt{2}c_+^*(\zeta)h_\pm(\zeta, \boldsymbol{q}, \Omega)\hat{a}_S(\zeta, \boldsymbol{q}, \Omega), \end{aligned}\tag{8.82}$$

$$\begin{aligned}\frac{\partial}{\partial\zeta}\ \hat{a}_S(\zeta, \boldsymbol{q}, \Omega) &= \sqrt{2}c_+^*(\zeta)h_+^*(\zeta, \boldsymbol{q}, \Omega)\hat{a}_+(\zeta, \boldsymbol{q}, \Omega) \\ &+ \sqrt{2}c_+^*(\zeta)h_-^*(\zeta, \boldsymbol{q}, \Omega)\hat{a}_-(\zeta, \boldsymbol{q}, \Omega), \end{aligned}\tag{8.83}$$

where

$$g_\pm(\zeta, \boldsymbol{q}, \Omega) = \frac{e^{-i\Delta_1(\boldsymbol{q},\Omega)\zeta} \pm e^{-i\Delta_1(-\boldsymbol{q},-\Omega)\zeta}}{2}, \tag{8.84}$$

$$h_\pm(\zeta, \boldsymbol{q}, \Omega) = \frac{e^{-iD_1(\boldsymbol{q},\Omega)\zeta} \pm e^{-iD_2(\boldsymbol{q},\Omega)\zeta}}{2}. \tag{8.85}$$

Equations (8.82)–(8.83) can be simplified, making the following additional assumption

$$\Delta_1(q, \Omega) = \Delta_1(q, -\Omega), \qquad D_1(q, \Omega) = D_2(q, -\Omega). \tag{8.86}$$

This assumption is always fulfilled for $q = \Omega = 0$. In any case, we are mainly interested in the region around $q = 0$ because, as in type-I, image processing will be mainly efficient in this region. Inserting Eqs. (8.86) into Eqs. (8.82)–(8.83), the propagation equation for the field $a_-(\zeta, q)$ decouples leading to a type-I OPA equation, where $c_S(\zeta)$ plays the role of a ζ-dependent pump,

$$\frac{\partial}{\partial\zeta}\hat{a}_-(\zeta, \boldsymbol{q}, \Omega) = c_S(\zeta)e^{-i\Delta_1(\boldsymbol{q},\Omega)\zeta}\hat{a}_-^\dagger(\zeta, -\boldsymbol{q}, -\Omega). \tag{8.87}$$

The equations that describe the remaining two fields reduce to the equations of the type-I SHG,

$$\begin{aligned}\frac{\partial}{\partial\zeta}\hat{a}_+(\zeta, \boldsymbol{q}, \Omega) &= -c_S(\zeta)\hat{a}_+^\dagger(\zeta, -\boldsymbol{q}, -\Omega)e^{-i\Delta_1(\boldsymbol{q},\Omega)\zeta} \\ &\qquad - \sqrt{2}c_+^*(\zeta)\hat{a}_S(\zeta, \boldsymbol{q}, \Omega)e^{-iD_1(\boldsymbol{q},\Omega)\zeta}, \end{aligned}\tag{8.88}$$

$$\frac{\partial}{\partial\zeta}\hat{a}_+(\zeta, \boldsymbol{q}, \Omega) = \sqrt{2}c_+(\zeta)\hat{a}_+(\zeta, \boldsymbol{q}, \Omega)e^{iD_1(\boldsymbol{q},\Omega)\zeta}. \tag{8.89}$$

This decomposition of type-II SHG into a type-I SHG and a type-I OPA generalizes to the transverse spatial multimode case the conclusions obtained for a single-mode model in the traveling-wave configuration [37], or in the cavity-based case [47].

To take advantage of the coupling between waves with spatiotemporal frequencies $(\boldsymbol{q}, \Omega)$ and $(-\boldsymbol{q}, -\Omega)$ generated by the down-conversion, we consider a symmetrical input image as in Subsection 8.3.2,

$$\alpha_{\pm}(q) = \alpha_{\pm}(-q). \tag{8.90}$$

The output intensity of the system depends now on the phase ϕ_{in} of the input image. For simplicity we will consider a homogeneous phase for the input image: $\boldsymbol{\alpha} e^{i\phi_{\text{in}}}$, where the elements of $\boldsymbol{\alpha}$ are real. The output intensity is then given by

$$|\alpha_i^{\text{out}}(q)|^2 = |\boldsymbol{U}_i(q,0) \cdot \boldsymbol{\alpha}(q) e^{i\phi_{\text{in}}} + \boldsymbol{V}_i(q,0) \cdot \boldsymbol{\alpha}(q) e^{-i\phi_{\text{in}}}|^2 \, . \tag{8.91}$$

The simplest detection scheme involves one detector located around q. The noise figure is then given by Eq. (8.71) with an extra factor 2 due to the symmetrization of the input image prior to injection,

$$F_i = 2 \frac{1 + 2\boldsymbol{V}_i^*(q,0) \cdot \boldsymbol{V}_i(q,0)}{|\boldsymbol{U}_i(q,0) \cdot \boldsymbol{\alpha} e^{i\phi_{\text{in}}} + \boldsymbol{V}_i(q,0) \cdot \boldsymbol{\alpha} e^{-i\phi_{\text{in}}}|^2 / |\alpha_1(q)|^2} \, . \tag{8.92}$$

For an OPA the noiseless character of image processing is only valid in the limit of large gains [19]. In the up-conversion case considered here the up-conversion rate is never large. It is then crucial from the point of view of quantum imaging to consider the symmetrized detection scheme that involves two detectors located at opposite spatial points (corresponding to the wave-vectors in the telescopic configuration considered here). One measures the sum of the photocurrents:

$$\hat{\mathcal{N}}_i(L,q) = \hat{N}_i(L,q) + \hat{N}_i(L,-q). \tag{8.93}$$

The noise figure for this quantity is given by

$$F_i = \frac{1 + 2\boldsymbol{V}_i^*(q,0) \cdot \boldsymbol{V}_i(q,0) + 2Re(e^{-2i\phi_i^{\text{out}}} \boldsymbol{U}_i(q,0) \cdot \boldsymbol{V}_i(-q,0))}{|\boldsymbol{U}_i(q,0) \cdot \boldsymbol{\alpha} e^{i\phi_i n} + \boldsymbol{V}_i(q,0) \cdot \boldsymbol{\alpha} e^{-i\phi_{\text{in}}}|^2 / |\alpha_1(q)|^2} , \tag{8.94}$$

which differs from the one given in Eq. (8.71) by an additional interference term. ϕ_i^{out} denotes the phase of the output amplitude,

$$\phi_i^{\text{out}} = Arg[\boldsymbol{U}_i(q,0) \cdot \boldsymbol{\alpha} e^{i\phi_{\text{in}}} + \boldsymbol{V}_i(q,0) \cdot \boldsymbol{\alpha} e^{-i\phi_{\text{in}}}] \, . \tag{8.95}$$

For an OPA it has been shown that if the phase of the input image is adapted for maximal amplification, this amplification preserves the signal-to-noise ratio, and hence the operation is noiseless [41].

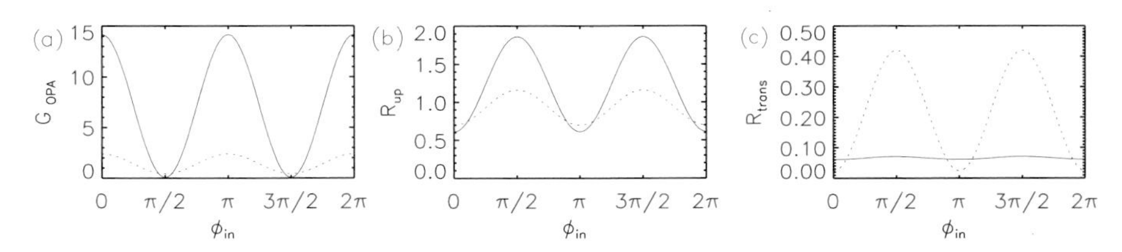

Fig. 8.15. Amplification factor (a), up-conversion rate (b), and transmission rate (c) as a function of the input phase for $\zeta = 1$ (dotted line) and $\zeta = 2$ (dashed line).

As previously stated, the phase-sensitivity of the image-processing scheme is an important requisite for interesting quantum imaging properties. This sensitivity is reflected in several quantities: first, the behavior of the OPA with a z-dependent pump can be characterized by the gain

$$G_{\mathrm{OPA}}(q) = \frac{\langle N_-(L,q)\rangle}{\langle N_-(0,q)\rangle}. \tag{8.96}$$

Figure 8.15a shows $G_{\mathrm{OPA}}(0)$ as a function of the phase of the input image. This gain oscillates between $\cosh(L)$ (maximal amplification) and $\mathrm{sech}(L)$ (minimum amplification). Second, the behavior of the type-I SHG can be characterized by two quantities, the up-conversion and the transmission rate

$$R_{\mathrm{up}}(q) = \frac{\langle N_S(L,q)\rangle}{\langle N_+(0,q)\rangle}, \qquad R_{\mathrm{trans}}(q) = \frac{\langle N_+(L,q)\rangle}{\langle N_+(0,q)\rangle}. \tag{8.97}$$

The phase-dependence of these quantities for $q = 0$ is plotted in Fig. 8.15. In the limit of large propagation lengths, the up-conversion rate oscillates between a minimal value of 1/2 and a maximum value of 2 which can be understood from the energy and phase conservation in SHG. The energy conservation implies that, after a complete conversion, the number of second-harmonic photons has to be half the number of initial fundamental photons, and hence the amplitude is reduced by a factor $\sqrt{2}$. On the contrary, the phase of the output field is, after complete up-conversion, twice the phase of the input fundamental field. In the case of negligible phase mismatch, the maximal up-conversion rate occurs at $\phi_{\mathrm{in}} = \pi/2$ and $3\pi/2$ for all distances. However, we should note that in general the phase for which the up-conversion is maximal may depend on the distance. Also, it is interesting to observe, in Fig. 8.15, that the input signal phase for which the up-conversion is minimal, ensures a maximal amplification of the "-" polarization component.

We consider now the noise behavior of the up-conversion process. We limit ourselves to the case of the input phase adjusted for minimal/maximal up-conversion and plot the up-conversion rate and the noise figure as a function of the propagation length (Fig. 8.16).

We observe that in the regime of minimal up-conversion, the noise figure rapidly drops to unity, and hence image up-conversion is performed without

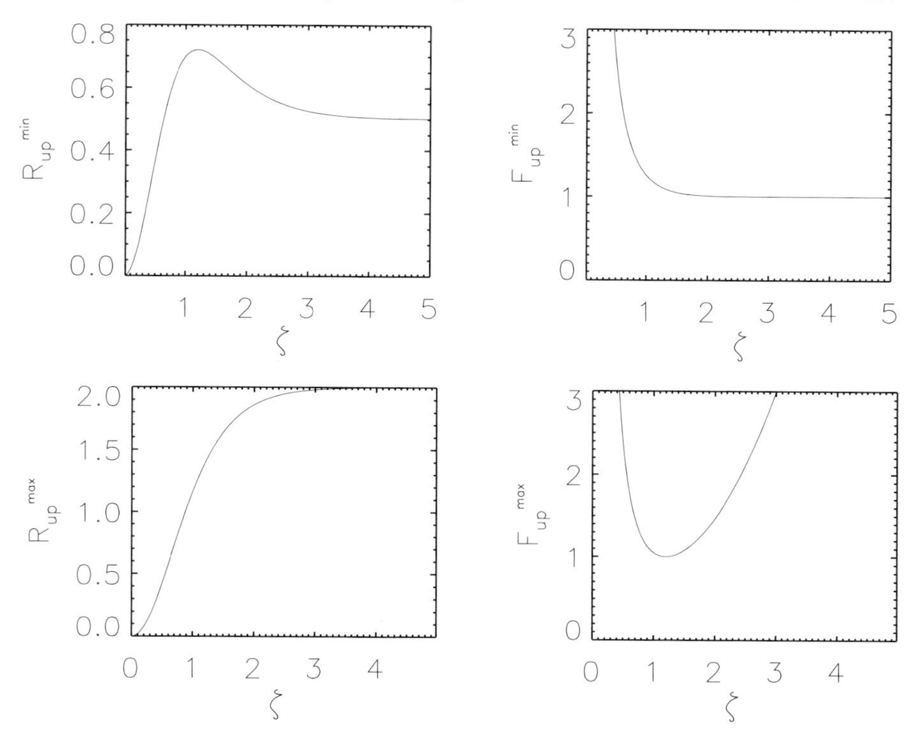

Fig. 8.16. Up-conversion rate and the noise figure at a vanishing transverse wave vector $q = 0$ as a function of the propagation length in the symmetric detection scheme. Top row $\phi_{\rm in} = 0$; bottom row $\phi_{\rm in} = \pi/2$.

noise addition. This can be understood because in the limit of large propagation lengths, the amplitude fluctuations entering the crystal at the second-harmonic frequency are damped during propagation, so that in this limit they do not contribute to the noise in the amplitude of the second-harmonic frequency output. This is not valid for the noise figure in the case of maximal up-conversion. Defining $y_i(\zeta, q, \Omega) = \left(\hat{a}_i(\zeta, q, \Omega) - \hat{a}_i^\dagger(\zeta, -q, -\Omega)\right)/2i$, in the limit of large ζ one obtains

$$y_S(z) = -\sqrt{2}y_+(0) + (1 - z\,\tanh z)y_S(0). \tag{8.98}$$

Therefore, the phase fluctuations of the input second-harmonic field contribute to the total fluctuations and the noise figure is larger than one, as shown in Fig. 8.16. Only for propagation lengths such that

$$1 - \zeta\tanh(\zeta) = 0, \tag{8.99}$$

the second-harmonic input fluctuations do not contribute, and the image processing is noiseless ($F = 1$). Finally, we consider the behavior of this

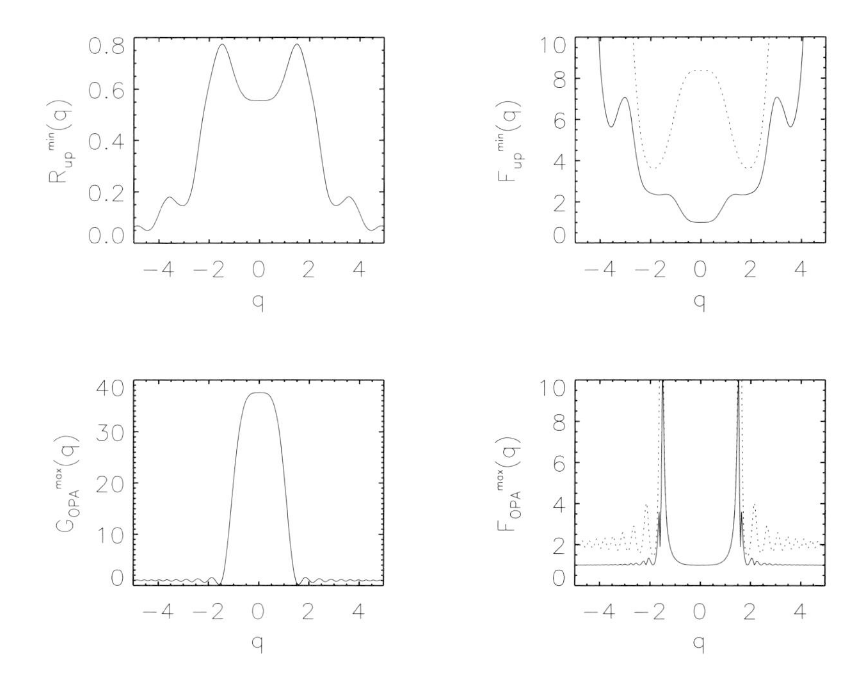

Fig. 8.17. Up-conversion rate (top left) and OPA gain (bottom left) as function of the position in the transverse plane for $\phi_{\text{in}} = 0$. The corresponding noise figures are shown on the right. The dotted line corresponds to the single-detector case and the solid line corresponds to symmetric detection.

image-processing scheme as a function of the position in the transverse plane. Figure 8.17 shows the up-conversion rate and the OPA gain for an input phase corresponding to minimum up-conversion (maximum OPA gain). The corresponding noise figures for the single detector placed at q and the symmetric detector scheme with two detectors at q and $-q$ are shown as well. The up-conversion rate and the OPA gain are roughly constant on a disk of finite diameter centered on the main optical axis of the system ($q = 0$). The noise figure of the up-conversion F_{up} rate in this central region approaches the value of one characterizing a noiseless image processing. However, this interesting property is only true in the symmetric detection scheme. If the output were detected only with one detector, a large amount of excess noise with respect to the noise in the injected image would be observed. For the OPA gain, in the central region of maximal gain, the two detection schemes give exactly the same noise level, as a consequence of the strong nonclassical correlations of the two outputs at q and $-q$. At higher q where almost no amplification takes place, the noise figure F_{OPA} for the symmetric and single-detector case differ by a factor 2 as a consequence of the symmetrization procedure of the input image.

References

1. J. Teuber, *Digital Image Processing* (Prentice-Hall, Englewood Cliffs, NJ, 1993).
2. J. Mitwinter and J. Warner, J. Appl. Phys. **38**, 519 (1967).
3. J. Mitwinter, Appl. Phys. Lett. **12**, 68 (1968).
4. F. Devaux, A. Mosset, E. Lantz, S. Monneret, and H. Le Gall, Appl. Opt. **40**, 4957 (2001).
5. A. H. Firester, J. Appl. Opt. **40**, 4842 (1969); 4849 (1969).
6. F. Devaux and E. Lantz, J. Opt. Soc. Am. B **12**, 2245 (1995).
7. E. Lantz and F. Devaux, Quantum Semiclass. Opt. **9**, 279 (1997).
8. F. Devaux and E. Lantz, Opt. Commun. **114**, 295 (1995).
9. P. Günther and J. P. Huignard (eds.), *Photorefractive Materials and Their Applications II*, Topics in Applied Physics Vol. 62. (Springer Verlag, New York, 1989).
10. J. Feinberg, Opt. Lett. **5**, 330 (1980).
11. J. P. Huignard and J. P. Herriau, Appl. Opt. **17**, 2671 (1978).
12. B. Liang, Z. Wang, J. Guang, G. Mu, and C. M. Cartwright, Opt. Lett. **25**, 1086 (2000).
13. E. Ochoa, L. Hesselink, and W. Goodman, Appl. Opt. **24**, 1826 (1985).
14. Y. H. Ja, Opt. Comm. **44**, 24 (1982).
15. Y. H. Ja, Opt. Quant. El. **15**, 457 (1983).
16. Y. H. Ja, Appl. Phys. B **36**, 21 (1985).
17. H. Rajbenbach, A. Delboublé, J. P. Huignard, Opt. Lett. **14**, 1275 (1989).
18. M. D. Rahn, D. P. West, J. D. Shakos, J. Appl. Phys. **87**, 127 (2000).
19. A. Gatti, E. Brambilla, L. A. Lugiato, and M. I. Kolobov, J. Opt. B: Quantum Semiclassical Opt. **2**, 196 (2000).
20. Eur. Phys. J. D, **22** (2003), special issue "Quantum fluctuations and coherence in optical and atomic structures".
21. C. M. Caves, Phys. Rev. D **26**, 1817 (1982).
22. Z. Y. Ou, S. F. Pereira, and H. J. Kimble, Phys. Rev. Lett. **70**, 3239 (1993).
23. J. A. Levenson, I. Abram, T. Rivera, and P. Grangier, J. Opt. Soc. Am. B **10**, 2233 (1993)..
24. M. I. Kolobov and I. V. Sokolov, Sov. Phys. JETP **69**, 1097 (1989); Phys. Lett. A **140**, 101 (1989).
25. M. I. Kolobov, Rev. Mod. Phys. **71**, 1539 (1999).
26. S. K. Choi, M. Vasilyev, and P. Kumar, Phys. Rev. Lett. **83**, 1938 (1999).
27. P. Scotto, P. Colet, and M. San Miguel, Opt. Lett. **28**, 1695 (2003).
28. U. Peschel, C. Etrich, and F. Lederer, Opt. Lett. **23**, 500 (1998).
29. U. Peschel, C. Etrich, and F. Lederer, Phys. Rev. E, **58**, 4005 (1998).
30. S. Longhi, Opt. Lett. **23**, 346 (1998).
31. S. Longhi, Phys. Rev. A **59**, 346 (1999).
32. Z. Y. Ou, Phys. Rev. A **59**, 4021 (1999).
33. A. Jacobo, P. Colet, P. Scotto, and M. San Miguel, submitted for publication.

34. P. Scotto and M. San Miguel, Phys. Rev. A **65**, 043811 (2002).
35. C. L. Tang and L. K. Cheng, *Fundamentals of Optical Parametric Processes and Oscilators* (Harwood Academic, Amsterdam, 1995), pp. 30-31.
36. R. D. Li and P. Kumar, Phys. Rev. A **49**, 2157 (1994).
37. Z. Y. Ou, Phys. Rev. A **49**, 2106 (1994).
38. M. K. Olsen, R. J. Horowicz, L. I. Plimak, N. Treps, and C. Fabre, Phys. Rev. A **61**, 021803 (2000).
39. M. K. Olsen and R. J. Horowicz, Opt. Commun. **168**, 135 (1999).
40. M. K. Olsen, L. I. Plimak, M. J. Collet, and D. F. Walls, Phys. Rev. A **62**, 023802 (2000).
41. M. I. Kolobov and L. A. Lugiato, Phys. Rev. A **52**, 4930 (1995).
42. I. V. Sokolov, M. I. Kolobov, and L. A. Lugiato, Phys. Rev. A **60**, 2420 (1999).
43. A. Berzanskis, W. Chinaglia, L. A. Lugiato, K.-H. Feller, and P. Di Trapani, Phys. Rev. A **60** 1626 (1999).
44. E. Lantz and F. Devaux, Quantum Semiclassic. Opt. **9**, 279 (1997).
45. P. Scotto, Phys. Rev. A **68**, 033814(2003).
46. J. A. Armstrong, N. Bloembergen, J. Ducuing, and P. S. Pershan, Phys. Rev. **127**, 1918 (1962).
47. Z. Y. Ou, Phys. Rev. A **49**, 4902 (1994).

9 Transverse Distribution of Quantum Fluctuations in Free-Space Spatial Solitons

Eric Lantz[1], Nicolas Treps[2], and Claude Fabre[2]

[1] Laboratoire d'Optique P. M. Duffieux, Université de Franche-Comté, F-25030 Besançon cedex, France
e-mail: elantz@univ-fcomte.fr
[2] Laboratoire Kastler Brossel, Université Pierre et Marie Curie, case 74, F-75252 Paris cedex 05, France

9.1 Introduction

Solitons, whatever the nonlinear phenomenon that produces them, are isolated and long-living entities that have been considered for a long time as potential candidates to store and carry digital information. This is, in particular, the case for temporal solitons, which have been envisioned for high-bit-rate telecommunications in fibers, but also for spatial solitons, which can be used in parallel processing of information.

With the spectacular current development of quantum information, a question arises: can solitons also be of some interest as potential supports not only of classical digital information, but also of quantum information? In view of these applications, it is therefore highly interesting to study the quantum properties of solitons. Whereas this has been the subject of significant research efforts, both theoretical and experimental for temporal solitons [1], the subject is still almost virgin for spatial solitons. The purpose of this chapter is to review the preliminary theoretical studies that have started recently on this subject for the case of solitons freely propagating in nonlinear media. The case of cavity solitons is treated in the next chapter.

Because of an exact balance between diffraction and nonlinearity, spatial solitons propagate without any deformation over long distances, leading to an enhancement of the quantum correlations and also to the local quantum noise reduction effects. Some recent studies [2–4] have considered these quantum effects in spatial solitons. Reference [2] shows that vacuum-induced jitter remains negligible versus the width of a soliton. In Ref. [3], production of sub-Poissonian light was theoretically demonstrated by placing either a stop in the center of the beam in the near field or a lowpass filter in the far field. Quantum fluctuations appear here as weak fluctuations of intense fields that can be treated by linearization methods [5] in the absence of a cavity. On the experimental side, nonclassical temporal solitons, having fluctuations below the standard quantum limit, have been experimentally produced [6,7]. We will see below that there exist some similarities between the quantum properties of temporal solitons and those of spatial solitons, although evience of quantum properties of spatial solitons has still not been found experimentally.

We use in the first part of the chapter, devoted to the free-space solitons, a general method of linearization, called the "Green's function method," that gives numerically the expectation values of quantum moments without averaging of stochastic simulations. This method, which can be applied to any type of soliton [4, 8] and more generally to any weak quantum fluctuations either of an intense field or amplified by a pump field [9], will be described in Section 9.2. We consider scalar and vector $\chi^{(3)}$ as well as $\chi^{(2)}$ spatial solitons. Their main properties are recalled in Section 9.3. Section 9.4 deals with the degree of global squeezing of these solitons, and Section 9.5 presents quantum properties of local fluctuations. In most of the cases, the global squeezing increases with the propagation distance, whereas local squeezing exists only for short distances.

9.2 General Method

We present here the outline of the method allowing us to determine the spatial distribution of quantum fluctuations in the regime of free propagation. We closely follow the presentation in [4]. As we will apply it to different cases of solitons ($\chi^{(2)}$ or $\chi^{(3)}$, scalar or vectorial) we will present the method in the general case of a nonlinear coupling between n monochromatic complex electric fields of frequency ω_i, written as $E_i(\boldsymbol{r}, z)\, e^{i(k_i z - \omega_i t)}$, where z is the main propagation direction, $k_i = n_i \omega_i / c$ the longitudinal k-vector (n_i being the linear index of refraction of the medium at frequency ω_i), and $\boldsymbol{r} = (x, y)$ the position in the transverse plane.

9.2.1 Propagation Equations for the Fluctuations

Within the slowly varying envelope, paraxial and scalar approximations, the coupled propagation equations for the classical field envelopes E_i are [10]

$$2ik_i \frac{\partial}{\partial z} E_i(\boldsymbol{r}, z) + \nabla^2 E_i(\boldsymbol{r}, z) = F_i(E_1, \ldots, E_n), \qquad (9.1)$$

where $\nabla^2 = \partial^2/\partial x^2 + \partial^2/\partial y^2$ is the transverse Laplacian, and $F_i(E_1, \ldots, E_n)$ a term proportional to the nonlinear polarization created in the medium at frequency ω_i by the different Fourier components of the field. We will assume that we are able to determine, by analytical or numerical means, the solutions of Eq. (9.1) for a given input field, that we will call $\bar{E}_i(\boldsymbol{r}, z)$, $(i = 1, \ldots, n)$.

Coming now to the same problem, but treated at the quantum level, it is possible to show using the methods of quantum optics in the Wigner representation [11], also called the "semiclassical approach [5]," that one can use the classical solution $\bar{E}_i(\boldsymbol{r}, z)$ to determine the quantum fluctuations within the small-quantum-fluctuation limit.

More precisely, let us write the quantum positive frequency field operator $\hat{E}_i(\boldsymbol{r}, z)$ as

$$\hat{E}_i(\boldsymbol{r}, z) = \langle \hat{E}_i(\boldsymbol{r}, z) \rangle + \delta\hat{E}_i(\boldsymbol{r}, z), \tag{9.2}$$

where $\delta\hat{E}_i(\boldsymbol{r}, z)$ is the local quantum fluctuation operator of the field at frequency ω_i, the total electric field Hermitian operator $\hat{E}(\boldsymbol{r}, z)$ being given by the sum $\sum_i \left[\hat{E}_i(\boldsymbol{r}, z) e^{i(k_i z - \omega_i t)} + \hat{E}_i^\dagger(\boldsymbol{r}, z) e^{-i(k_i z - \omega_i t)}\right]$. As long as the field contains a macroscopic number of photons and if the input fields are coherent or nearly coherent states, the quantum mean field is nothing more than the field given by classical nonlinear optics: $\langle \hat{E}_i(\boldsymbol{r}, z) \rangle = \bar{E}_i(\boldsymbol{r}, z)$. Then the fluctuations $\delta\hat{E}_i(\boldsymbol{r}, z)$ remain small compared to the mean fields and obey a simple propagation equation, obtained by linearizing the classical equation of propagation (9.1) around the mean field, namely,

$$2ik_i \frac{\partial}{\partial z} \delta\hat{E}_i(\boldsymbol{r}, z) + \nabla^2 \delta\hat{E}_i(\boldsymbol{r}, z)$$
$$= \sum_j \left(\frac{\partial F_i}{\partial E_j}\right)_{E=\bar{E}} \delta\hat{E}_j(\boldsymbol{r}, z) + \sum_j \left(\frac{\partial F_i}{\partial E_j^*}\right)_{E=\bar{E}} \delta\hat{E}_j^\dagger(\boldsymbol{r}, z). \tag{9.3}$$

In next subsection we will formulate the Green's function method allowing us to write the solution of Eq. (9.3).

9.2.2 Green's Function Approach

A major advantage of the propagation equation (9.3) is its linear character, so that one can find its solution in a given transverse plane $z = z^{\text{out}}$ as a linear combination of the input fluctuations in plane $z = z^{\text{in}}$,

$$\delta\hat{E}_i(\boldsymbol{r}, z^{\text{out}}) = \sum_j \int\int d^2r' G_i^j(\boldsymbol{r}, \boldsymbol{r}') \delta\hat{E}_j(\boldsymbol{r}', z^{\text{in}})$$
$$+ \sum_j \int\int d^2r' H_i^j(\boldsymbol{r}, \boldsymbol{r}') \delta\hat{E}_j^\dagger(\boldsymbol{r}', z^{\text{in}}), \tag{9.4}$$

where $G_i^j(\boldsymbol{r}, \boldsymbol{r}')$ and $H_i^j(\boldsymbol{r}, \boldsymbol{r}')$ are the Green's functions associated with Eq. (9.3). Equation (9.4) is a kind of Huygens–Fresnel integral, which describes the propagation of the quantum fluctuations within the small-fluctuation limit.

The quantum input field fluctuations are those of a vacuum field, or of a coherent field. They obey the following relation [12] in the scalar field and paraxial approximation:

$$\langle \delta\hat{E}_j(\boldsymbol{r}, z^{\text{in}}) \delta\hat{E}_k^\dagger(\boldsymbol{r}', z^{\text{in}}) \rangle = C_j \delta_{jk} \delta^{(2)}(\boldsymbol{r} - \boldsymbol{r}') \tag{9.5}$$

(where $C_j = \hbar\omega_j / 2\varepsilon_0 L$, L being the length along Oz of the quantization box). The input correlation functions not in the antinormal order vanish.

From Eqs. (9.3) and (9.5), one can easily derive the correlation functions between the quantum fluctuations of different operators. One finds, for example,

$$\begin{aligned} &\left\langle \delta\hat{E}_i\left(\boldsymbol{r}, z^{\text{out}}\right) \delta\hat{E}_k^{\dagger}\left(\boldsymbol{r}', z^{\text{out}}\right)\right\rangle \\ &= \sum_j C_j \int\!\!\int d^2 r_1 G_i^j\left(\boldsymbol{r}, \boldsymbol{r}_1\right) G_k^{j*}\left(\boldsymbol{r}', \boldsymbol{r}_1\right). \end{aligned} \tag{9.6}$$

Any correlation function or variance is therefore a given combination of the Green's functions $G_i^j\left(\boldsymbol{r}, \boldsymbol{r}'\right)$ and $H_i^j\left(\boldsymbol{r}, \boldsymbol{r}'\right)$, that we must now calculate. To do so, one uses the propagation equation (9.3): $G_i^j\left(\boldsymbol{r}, \boldsymbol{r}'\right)$, for example, is the solution of this equation when one takes the following initial conditions,

$$\begin{aligned} \delta\hat{E}_k\left(\boldsymbol{r}, z^{\text{in}}\right) &= \delta_{jk}\delta^{(2)}\left(\boldsymbol{r} - \boldsymbol{r}_1\right), \\ \delta\hat{E}_k^{\dagger}\left(\boldsymbol{r}, z^{\text{in}}\right) &= 0. \end{aligned} \tag{9.7}$$

Green's function $G_i^j\left(\boldsymbol{r}, \boldsymbol{r}'\right)$ is approximately evaluated by numerical techniques: one discretizes the transverse plane and one calculates using the split-step method [13] the solution of Eq. (9.3) with a following initial condition: unity on a given pixel and zero on all the other pixels. One must also choose the size of the transverse grid large enough compared to the soliton radius to avoid any spurious effects due to the periodic boundary condition imposed at the limits of this grid by the Fast Fourier Transform (FFT) procedure.

9.2.3 Correlations Between the Photocurrents

In order to measure the local field fluctuations, we use photodetectors assumed to have a quantum efficiency 1 and to be sensitive only to the field component of frequency ω_j. If the photodetector has a very small area dS around the point $\boldsymbol{r}$, photodetection theory implies in the small fluctuation limit that the photocurrent fluctuations in direct photodetection, $\delta\hat{I}_j\left(\boldsymbol{r}\right)$, are given by

$$\delta\hat{I}_j\left(\boldsymbol{r}\right) = \bar{E}_j\left(\boldsymbol{r}, z^{\text{out}}\right) dS \sum_i \int\!\!\int d^2 r_1 \left[A_j^i\left(\boldsymbol{r}, \boldsymbol{r}_1\right) \delta\hat{E}_i\left(\boldsymbol{r}_1, z^{\text{in}}\right) + h.c.\right], \tag{9.8}$$

in the case where the mean field $\bar{E}_j\left(\boldsymbol{r}, z^{\text{out}}\right)$ is real (which is the case for the analytical solitons given in Section 9.3.1), $A_j^i\left(\boldsymbol{r}, \boldsymbol{r}'\right)$ being the following combination of the G and H Green's functions,

$$A_j^i\left(\boldsymbol{r}, \boldsymbol{r}'\right) = G_j^i\left(\boldsymbol{r}, \boldsymbol{r}'\right) + H_j^i\left(\boldsymbol{r}, \boldsymbol{r}'\right)^*. \tag{9.9}$$

One can also use a balanced homodyne scheme, with the help of a local oscillator field of amplitude $E_{loc}\left(\boldsymbol{r}\right) = \left|E_{loc}\left(\boldsymbol{r}\right)\right| e^{i\theta}$, in order to measure the local fluctuations of a given quadrature component. Provided that the local oscillator amplitude is much larger than the one of the field to measure, the photocurrent fluctuations are given in this case by

$$\delta \hat{I}_j(\theta, \boldsymbol{r}) = |E_{loc}(\boldsymbol{r})| \, dS \sum_i C_i \times \int\int d^2 r_1 \left[B_j^i(\theta, \boldsymbol{r}, \boldsymbol{r}_1) \, \delta \hat{E}_i\left(\boldsymbol{r}_1, z^{\text{in}}\right) + h.c. \right], \tag{9.10}$$

with $B_j^i(\theta, r, r')$ being equal to

$$B_j^i(\theta, r, r') = e^{i\theta} G_j^i(r, r') + e^{-i\theta} H_j^i(r, r')^* . \tag{9.11}$$

In particular, when $\theta = 0$ or $\pi/2$, the photocurrent fluctuations give a signal proportional to the quadrature component parallel or orthogonal to the mean amplitude, which are the amplitude and phase quadrature fluctuations.

Knowing these local photocurrent fluctuations as a function of the input fluctuations $\delta \hat{E}_i\left(\boldsymbol{r}, z^{\text{in}}\right)$ from Eqs. (9.8) and (9.10), and the input correlations functions (9.5), one can then derive the covariance between pixels:

$$C(x, x', \theta) = <\delta \hat{I}(\theta, x) \delta \hat{I}(\theta, x')>. \tag{9.12}$$

It is also possible to determine the covariance functions or the local variance in the case of large-area photodetectors by integrating these expressions over the detector surface. For example, the photocurrent variance directly measured by a photodetector of large area S is equal to

$$\langle (\delta \hat{I}_j)^2 \rangle = \sum_i C_i \int\int_S d^2 r \int\int_S d^2 r' \int\int d^2 r_1 \bar{E}_i\left(\boldsymbol{r}, z^{\text{out}}\right) \bar{E}_i\left(\boldsymbol{r}', z^{\text{out}}\right) A_j^i(\boldsymbol{r}, \boldsymbol{r}_1) \, A_j^i(\boldsymbol{r}', \boldsymbol{r}_1)^* . \tag{9.13}$$

The Green's function method can be applied to any configuration of fields propagating in a nonlinear medium. We will consider in this chapter scalar and vector $\chi^{(3)}$ as well as $\chi^{(2)}$ spatial solitons.

9.3 Spatial Solitons: Mean Values

9.3.1 $\chi^{(3)}$ Scalar Spatial Soliton

A scalar soliton can be obtained in a Kerr medium for a single transverse coordinate x by using, for example, a planar waveguide. The classical propagation equation for the complex electric field envelope $U(x, z)$ reads in this case:

$$\frac{\partial E}{\partial z} = \frac{i}{2k} \frac{\partial^2 E}{\partial x^2} + i\gamma |E|^2 E, \tag{9.14}$$

where γ is the usual Kerr coefficient, z the propagation direction, k the longitudinal wave-vector, and x the position in the transverse plane. It can be written in a universal form if one uses a scaling parameter η and the scaled variables $\zeta = \eta z$, $r = x\sqrt{2\eta k}$, and its solution is the well-known ζ invariant hyperbolic secant [14]:

$$u = \frac{\sqrt{2}}{\cosh r}. \tag{9.15}$$

9.3.2 $\chi^{(3)}$ Vector Spatial Soliton

In the last few years, a great number of multicomponent solitons have been demonstrated [15]. Among these, the simplest example is the bimodal vector soliton which is shown in Fig. 9.1a and consists of two opposite circular components that trap each other in a nonbirefringent Kerr medium. Their complex envelopes U and V obey the following coupled nonlinear Schrödinger equations,

$$\begin{aligned} \frac{\partial U}{\partial Z} &= i\frac{1}{2k}\frac{\partial^2 U}{\partial x^2} + i\gamma(|U|^2 U + C|V|^2 U), \\ \frac{\partial V}{\partial Z} &= i\frac{1}{2k}\frac{\partial^2 V}{\partial x^2} + i\gamma(|V|^2 V + C|U|^2 V), \end{aligned} \tag{9.16}$$

where γ is the Kerr coefficient, and C represents the strength of the cross-phase modulation (for instance, $C = 7$ in an isotropic liquid such as CS_2 [16]). As seen from Fig. 9.1b, this soliton bound-state propagates in equilibrium due to incoherent coupling between both circular components. If alone in the Kerr medium, the symmetrical field U tends to self-focus. However, the antisymmetrical field V on the counterrotating polarization tends to diffract and it can be shown that a proper choice of the intensities (in dimensionless units in Fig. 9.1) ensures an exact balance between diffraction and self-focusing. However, this equilibrium is unstable if the cross-phase modulation coefficient is greater than the self-phase modulation coefficient, leading to symmetry-breaking instability, whose experimental demonstration is reported in [17].

9.3.3 $\chi^{(2)}$ Spatial Soliton

In a $\chi^{(2)}$ medium, the complex envelopes of a fundamental field $E_1(\boldsymbol{r}, z)$ $e^{i(k_1 z - \omega t)}$ and a second-harmonic field $E_2(\boldsymbol{r}, z)\, e^{i(k_2 z - 2\omega t)}$ are coupled through the following set of equations [10], without the walk-off,

$$\begin{aligned} 2ik_1\frac{\partial}{\partial z}E_1(\boldsymbol{r}, z) + \nabla^2 E_1(\boldsymbol{r}, z) &= -2\frac{\omega^2}{c^2}\chi^{(2)}E_1^* E_2 e^{-i\Delta k z}, \\ 2ik_2\frac{\partial}{\partial z}E_2(\boldsymbol{r}, z) + \nabla^2 E_2(\boldsymbol{r}, z) &= -4\frac{\omega^2}{c^2}\chi^{(2)}E_1^2 e^{i\Delta k z}, \end{aligned} \tag{9.17}$$

with $\Delta k = 2k_1 - k_2$. Here also, we will limit our analysis to the case of a single transverse dimension x. Equation (9.17) can also be written in a universal form [15],

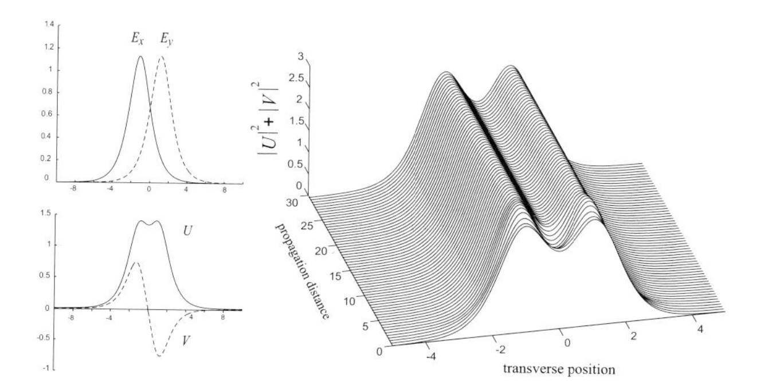

Fig. 9.1. Vector soliton projected either on orthogonal linear polarizations (E_x and E_y) or on counterrotating circular polarizations (U and V). The right plot shows invariant propagation of the soliton bound-state.

$$i\frac{\partial u}{\partial \zeta}+\frac{\partial^2 u}{\partial r^2}-\alpha u+u^* v=0,$$
$$i\sigma\frac{\partial v}{\partial \zeta}+\frac{\partial^2 v}{\partial r^2}-\sigma\left(2\alpha+\beta\right)v+\frac{1}{2}u^2=0, \tag{9.18}$$

when one introduces the scaling parameter η as in the $\chi^{(3)}$ case, and a free longitudinal phase-shift parameter α, writing $\zeta=\eta z, r=x\sqrt{2\eta k_1}, \sigma=k_2/k_1$, $\beta=\Delta k/\eta, u=2(\omega^2\chi^{(2)}/k_1c^2)\frac{E_1}{\eta}e^{-i\alpha\zeta}, v=((\omega^2\chi^{(2)}/(k_1c^2E_2)/\eta))e^{-i(2\alpha+\beta)\zeta}$.

Equations (9.18) have an infinite set of ζ-invariant solutions, each one giving a precise value to the ratio between the powers of the fundamental and the second-harmonic fields. Among these solutions, one is interesting for computational reasons : it is the only one known so far that has an analytical expression. It corresponds to the particular case $\alpha=1$, $\sigma\left(2\alpha+\beta\right)=1$, and is given by

$$u=\frac{3\sqrt{2}}{2\cosh^2 r/2},$$
$$v=\frac{3}{2\cosh^2 r/2}. \tag{9.19}$$

For this analytical soliton, the ratio between the second-harmonic power and the fundamental power is 2 ("equiphotonic" case), and the scaling parameter η has a well-defined value $\eta=k_2((2k_1-k_2)/(k_1-2k_2))\approx-2\Delta k/3$. When the relation between the scaling parameter and the phase mismatch is not fulfilled, the soliton exists, but is no longer analytical. Its shape remains almost the same as in Eq. (9.19). We will therefore use below the analytical soliton (9.19) as a typical example, keeping in mind that the results obtained using a soliton corresponding to the neighboring parameters should not differ significantly from those derived with the help of the analytical soliton.

Let us make precise the orders of magnitude using the experiment described in [18], even though it is not made in the one-transverse-dimensional case. The spatial soliton had a width of 20 μm for an input irradiance of roughly 50 GW/cm^2. In our reduced units, this corresponds to a propagation parameter $\zeta = 1$ for a propagation in the KTP nonlinear crystal over a length of 0.3 mm.

9.4 Squeezing on the Total Beam

Applying the methods of Section 9.2, we are now able to determine the spatial quantum properties of the solitons defined in Section 9.3. We give in this section the properties of global squeezing for the three types of solitons.

9.4.1 $\chi^{(3)}$ Scalar Spatial Soliton

As is well known [19], the Kerr effect produces in the plane-wave case a significant amount of squeezing, which increases monotone-wise with the propagation distance. The squeezing is optimized for a given quadrature component (best squeezing) which is neither the amplitude nor the phase quadrature. However, the amplitude quadrature of the field remains at the shot-noise level. Figure 9.2 gives the results of the Green's function method for the scalar soliton, and shows that the results are quite similar to the plane-wave case, in the case of a photodetector having an area much larger than the soliton spot. Neglecting diffraction, or using single-mode propagation in an optical fiber [1,20] seems to be almost equivalent to compensating diffraction by self-focusing, for a given nonlinear phase ζ. Hence, a spatial soliton appears as a practical means to obtain strong squeezing, with the restriction that it must be detected by homodyne techniques.

9.4.2 $\chi^{(3)}$ Vector Soliton: Total Beam Squeezing and Correlation Between Polarizations

Because vector solitons are not analytically integrable, the impulse response to a Dirac-like field modification of the perfect vector soliton is more complicated to determine, and is calculated as follows. The perturbation consists in a unity single-pixel step, multiplied by a small coefficient before addition to the field, in order to ensure a near-perfect linearity of propagation equations for the perturbation. Both the unmodified and the modified fields are first numerically propagated using Eq. (9.16), and then subtracted from each other. Finally, the subtraction result is divided by the initial multiplication coefficient, in order to retrieve the output corresponding to the input

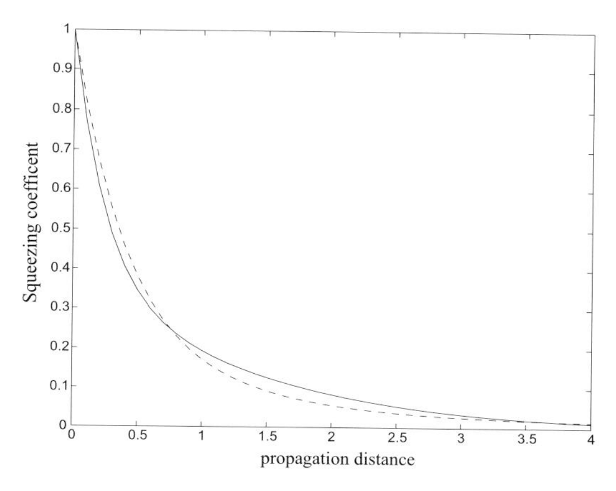

Fig. 9.2. Best total squeezing in shot-noise units, observed on the Kerr scalar soliton (solid line), versus the normalized propagation distance. The dashed line corresponds to the plane-wave case. In both cases, the normalized propagation distance corresponds to the nonlinear phase ζ.

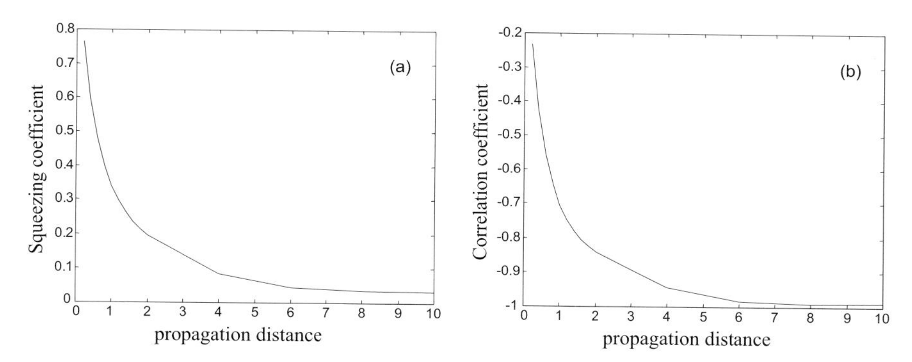

Fig. 9.3. Best total squeezing in shot-noise units (a) and correlation coefficient between circular polarizations (b) versus the normalized propagation distance.

unity single-pixel step. Figure 9.3 shows the global squeezing coefficient of the vector soliton for the best quadrature. This figure is very similar to that obtained for a scalar soliton (Fig. 9.2): strong squeezing appears, although somewhat smaller than in the scalar case for the same propagation distance. This squeezing is due to a strong anticorrelation between circular components, as shown in Fig. 9.3b. It can be verified in Fig. 9.4, which displays the best squeezing of each circular polarization component as a function of propagation, that the best squeezing on a single circular polarization is much weaker. We can conclude that a vector soliton as a whole exhibits quantum properties that are similar to a scalar soliton, and the field on each polarization does not itself constitute a soliton. Figure 9.5 shows that squeezing on

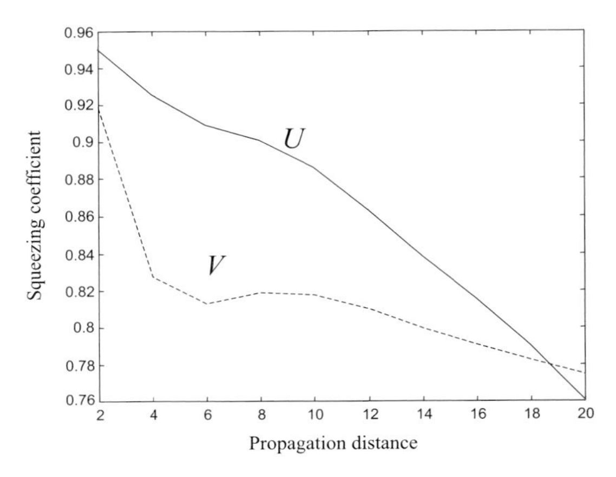

Fig. 9.4. Best squeezing in shot-noise units for each circular polarization versus the normalized propagation distance.

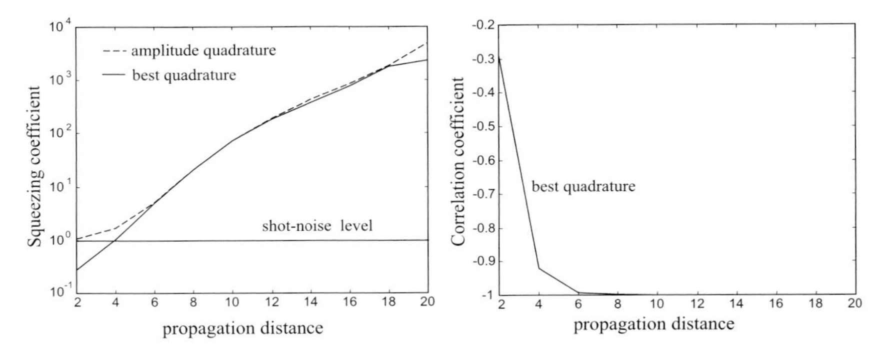

Fig. 9.5. Squeezing coefficient (left) and correlation coefficient (right) of linear polarizations versus the propagation distance.

linear polarizations disappears after a relatively short propagation distance, even for the best quadrature. Moreover, fluctuations grow exponentially. On the other hand, fluctuations on both linear polarizations are perfectly anticorrelated after some distance, as was expected because of the strong squeezing of the global beam. We have verified that the results on this global beam can also be retrieved from a description of the field by projection on linear polarizations.

9.4.3 $\chi^{(2)}$ Spatial Solitons

In $\chi^{(2)}$ media, and in the plane-wave case, when one starts at $\zeta = 0$ from the fundamental field only, second-harmonic generation leads to a gradually increasing intensity squeezing on the total fundamental field [21], which becomes almost perfect at long propagation lengths [22]. The situation is different when one starts from a $\chi^{(2)}$ spatial soliton. In this case, one can show that the squeezing oscillates with the propagation length, and reaches a maximum value of 70%. Figure 9.6 presents the global squeezing of intensity on the second harmonic [23]. The noise on this quadrature oscillates but remains below the shot-noise value, whereas there is no squeezing on the other quadratures (fundamental intensity and phase, second-harmonic phase).

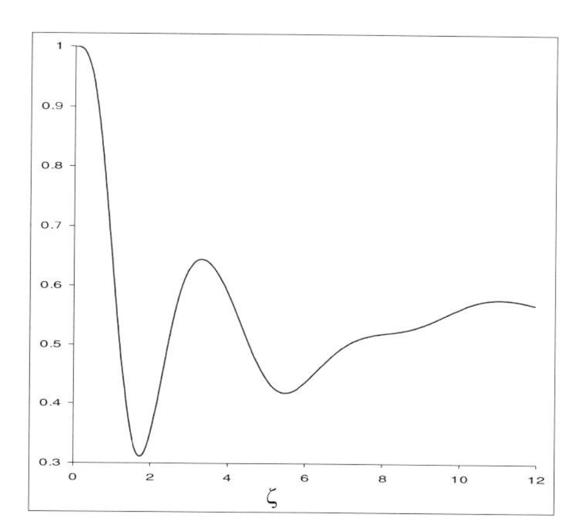

Fig. 9.6. Squeezing coefficient of the amplitude quadrature of the second-harmonic versus the propagation distance.

9.5 Local Quantum Fluctuations

Formula (9.13) and the analogous formula derived from Eq. (9.10) for the case of balanced homodyne detection allow us to determine the photocurrent fluctuations measured with photodetectors having a surface S of any area and shape. The minimum area for which our approximate calculation gives a physical result is the pixel area, having a size equal to the transverse discretization length used to calculate the Green's functions G and H. We present the results for $\chi^{(3)}$ and vector solitons. The results for $\chi^{(2)}$ solitons are very similar to those of $\chi^{(3)}$ solitons with, however, an important difference that all squeezing effects are effective on the amplitude quadrature of the second-harmonic: there is no need of homodyne detection.

9.5.1 $\chi^{(3)}$ Scalar Spatial Soliton

Because of the absence of squeezing on the amplitude quadrature, we simulate a homodyne detector, made of photodetectors of very small area and quantum efficiency equal to unity, and a local oscillator at frequency ω, placed at the output of the nonlinear medium, which is able to monitor the quantum noise on any quadrature component at a point (x, z_{out}). By varying the local oscillator phase to reach the minimum noise level, we obtain a quantity that we call "best squeezing." Figures 9.7a and 9.7b give, for two normalized propagation distances ($\zeta = 0.3$ and $\zeta = 3$), such a quantity as a function of x, together with the phase angle of the local oscillator enabling us to reach the best squeezed quadrature. One observes that the central pixel is the most squeezed. The noise reduction remains small because detecting a small part of the beam is equivalent to introducing losses if the beam is monomode, as is almost the case here [24]. For a longer propagation length, the noise level appears to be above the shot noise for any quadrature. Indeed, Fig. 9.8,

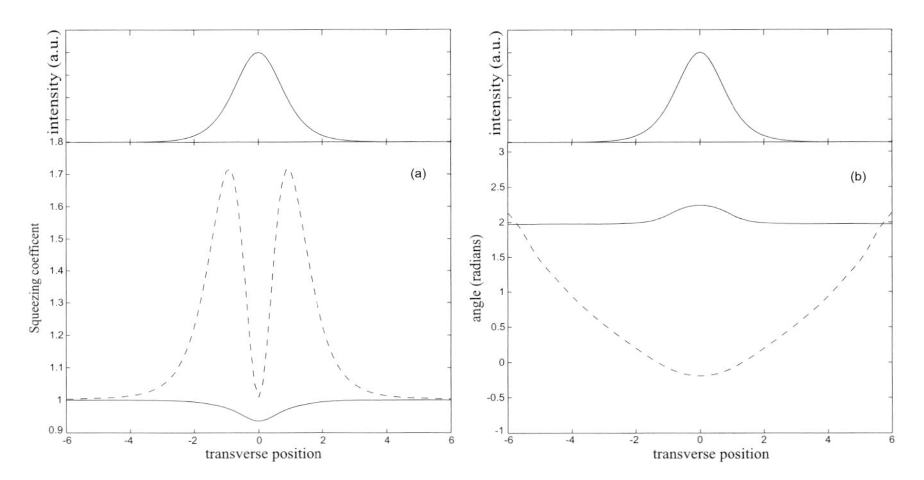

Fig. 9.7. Bottom: (a) best total squeezing in shot-noise units, and (b) corresponding angles, observed on a single pixel, versus its transverse position. Solid line: for the propagation distance equal to 0.3. Dashed line: for the propagation distance equal to 3. Top: corresponding transverse intensity profile of the scalar soliton.

which displays the variation of the best squeezing on the central pixel as a function of the propagation length, shows that the fluctuations on this pixel decrease until an optimal propagation length ($\zeta = 0.3$), then increase and overpass the shot noise for a sufficiently long propagation length. Hence, long propagation lengths produce at the same time local excess noise and squeezing on the total beam (see Fig. 9.2). These results are not contradictory because, as we will see, there is a gradual build-up of anticorrelations between different transverse parts of the soliton due to diffraction. Figure 9.9 shows the best squeezing for a photodetector of variable size, centered on the beam axis. This photodetector can be seen as an iris that allows the detection of only the central part of the beam, with a minimum size corresponding to the central pixel and a maximum size corresponding to the detection of

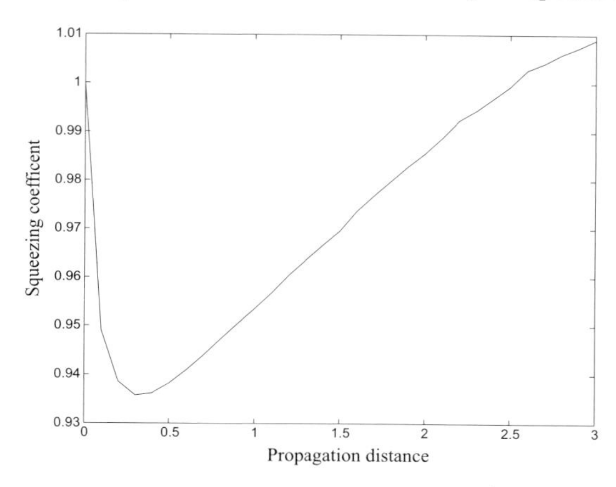

Fig. 9.8. Best squeezing in shot-noise units, observed on the central single pixel, versus the propagation distance.

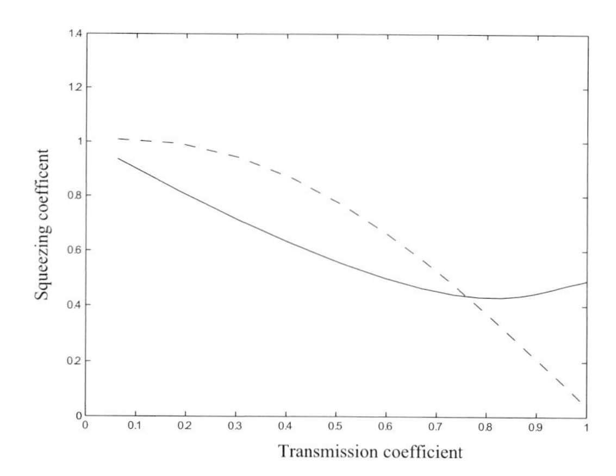

Fig. 9.9. Best squeezing for a photodetector of variable size, versus the transmission coefficient defined as the ratio between the detected and the total intensity. Solid line: for the propagation distance equal to 0.3. Dashed line: for the propagation distance equal to 3.

the entire beam. As expected, the noise on a very small area is close to the shot noise. The squeezing on a small central area is bigger at a small distance, and the squeezing for the entire beam is better at a great distance. For $\zeta = 0.3$ the squeezing is maximum for a finite size of the photodetector (transmission coefficient of 0.8, corresponding to a normalized radius of 1), whereas detecting the entire beam ensures the maximum squeezing for $\zeta = 3$. The explanation is again in building anticorrelations between the two sides of the soliton for long propagation distances (see Section 9.6).

In many circumstances, a spatial soliton behaves as a single-mode object. The present analysis allows us to check such a single-mode character at the quantum level: let us assume that the system is in a single-mode quantum state. This means that it is described by the state vector $|\Psi > \otimes |0 > \otimes \cdots \otimes |0 > \cdots$, with $|\Psi >$ being a quantum state of the mode having the exact spatial variation of the soliton field, and all the other modes being in the vacuum state. It has been shown in [25] that if a light beam is described by such a vector, then the noise recorded on a large photodetector with an iris of variable size in front of it has a linear variation with the iris transmission. We see in Fig. 9.9 that this is not the case for a spatial scalar soliton: as far as its quantum noise distribution is concerned, it is not a single-mode object. Let us also notice that the departure from the linear variation is on the order of 10%, so that the single-mode approach of the soliton remains a good approximation.

9.5.2 Intensity Squeezing by Spatial Filtering

In the results presented in the preceding subsection, squeezing can be measured only by homodyne detection, because no squeezing is present for the amplitude quadrature, which is the easiest to measure as it requires only direct detection and no interferometer. Mecozzi and Kumar [3] have shown

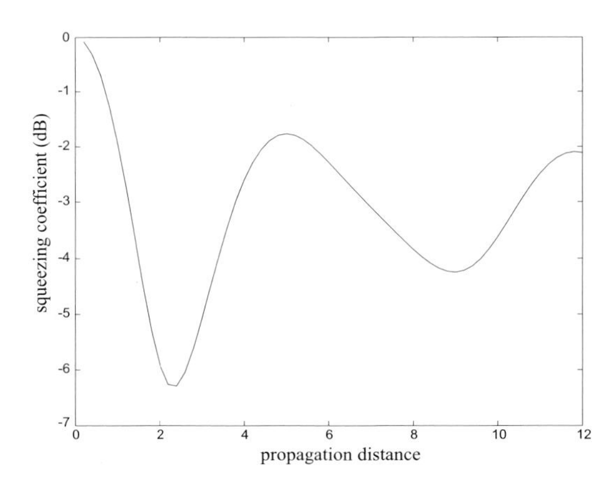

Fig. 9.10. Intensity squeezing (dB) for a photodetector placed at the center of the Fourier plane, with a total width of 0.25 in units of spatial frequencies corresponding to the normalized units of the direct space.

that a simple stop on the central part of the soliton beam ensures intensity squeezing on the remaining light. They have proposed the following intuitive explanation: if the intensity fluctuation is positive for the entire beam, the soliton becomes narrower (its width is inversely proportional to its power) and the stop will produce larger losses which compensate for the extra power due to the fluctuation. In the Fourier plane, a square diaphragm (lowpass filter) will produce even better squeezing. Figure 9.10 shows the intensity squeezing that we have obtained with the Green's function method by considering an aperture placed in the Fourier plane. These results are in full agreement with Fig. 1 of [3], that uses another linearization method. The residual differences come probably from a small difference in the aperture width. We find, as did Mecozzi and Kumar, that sub-Poissonian light can be very easily produced from a spatial soliton by placing a simple aperture in the Fourier plane. These results are the transposition into the spatial domain of a method that has been used to experimentally produce squeezing of temporal fiber solitons [26].

9.5.3 $\chi^{(3)}$ Vector Soliton

Figure 9.11 shows the squeezing coefficient on one pixel versus its transverse position, for the propagation length where this local squeezing is the most intense. The degree of squeezing is similar to the scalar case (see Fig. 9.7) with, however, two peaks of squeezing located on the intensity peaks of the multimode vector soliton. As in the scalar case, local squeezing is maximum for a

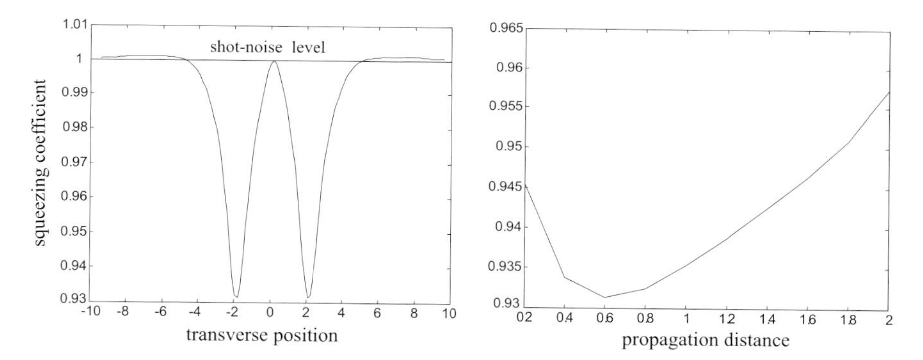

Fig. 9.11. Best squeezing on one pixel, (left): versus the transverse position for $\zeta = 0.6$, (right): versus the propagation distance for the transverse position corresponding to the strongest squeezing

relatively small propagation length (Fig. 9.11b). Figure 9.12 shows the best squeezing for a photodetector of variable size, centered on one of the intensity peaks. In contrast to Fig. 9.9, the variation is far from being linear with the transmission, meaning that the system is not single-mode, as expected. One observes that for long propagation distances, the global squeezing reaches a maximum for a photodetector size corresponding roughly to the peak width. This global squeezing implies the existence of spatial anticorrelations. Indeed, the local squeezing on one pixel disappears for such long distances (see Figs. 9.11 and 9.12).

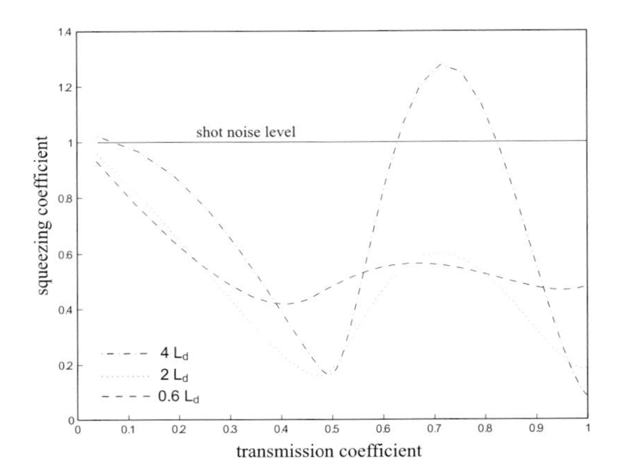

Fig. 9.12. Best squeezing for a photodetector centered on one intensity peak and of variable size, for three different propagation distances, versus the transmission coefficient defined as the ratio between the detected and the total intensity.

9.6 Quantum Correlations Between Field Quadratures at Different Points

Applying the methods of Section 9.5, and using formulas analogous to (9.13), we can also determine the spatial quantum correlations between the different field quadrature components measured in two spatially separated areas.

9.6.1 $\chi^{(3)}$ Scalar Spatial Soliton

Figure 9.13 shows that the covariance between pixels on the best squeezed quadrature (calculated with the help of Eq. (9.12), and where the best squeezing is defined for the total beam) increases and spreads out when passing from $\zeta = 0.3$ to $\zeta = 3$. Indeed, only pixels close to each other are anticorrelated for $\zeta = 0.3$ and the correlation between the adjacent pixels becomes positive for $\zeta = 3$. For this latter distance, squeezing is due to the anticorrelation between the left and the right sides of the soliton. Note that this covariance evolves with the size of the pixel [4].

9.6.2 Vector Solitons

Figure 9.14 shows the spatial correlations, in terms of the covariance functions $C(x, x', \theta)$ of the two polarizations, and the covariance function between these two modes, on the best squeezed quadrature and for two propagation lengths. One sees that for $\zeta = 0.6$, anticorrelations appear between pixels when one measures different circular polarizations. It means that the effect of cross-phase modulation is already noticeable, whereas effects of diffraction and

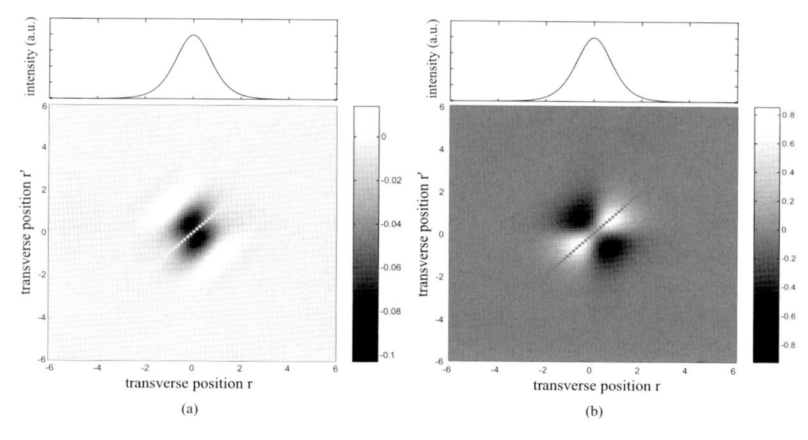

Fig. 9.13. Bottom: the covariance $C(x, x', \theta)$ between pixels on the best squeezed quadrature for (a) the propagation distance equal to 0.3 and (b) the propagation distance equal to 3. The values on the main diagonal (variance) have been removed. Top: corresponding transverse intensity profile of the scalar soliton.

self-phase modulation are still weak. For $\zeta = 2$, anticorrelations appear also between pixels on the same polarization, as for the scalar soliton.

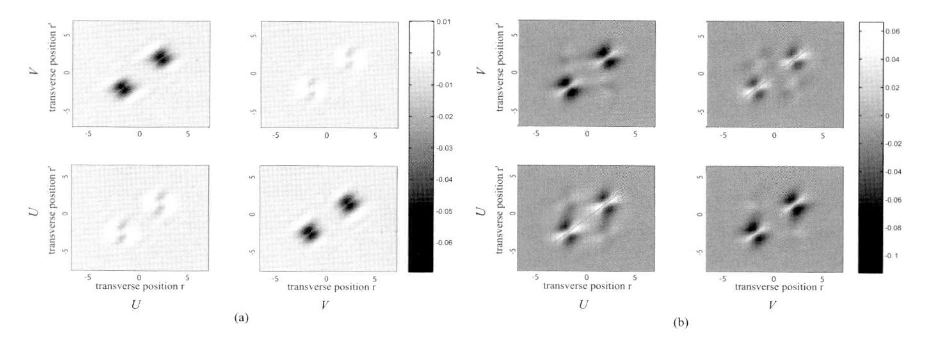

Fig. 9.14. Covariance $C(x, x', \theta)$, in shot-noise units, between pixels on the best squeezed quadrature for two propagation distances (a) 0.6 and (b) 2. The values on the main diagonal (variance) have been removed.

9.6.3 $\chi^{(2)}$Spatial Solitons

The covariance between the fundamental and second-harmonic amplitude quadratures appears to be very similar to that of the scalar $\chi^{(3)}$ soliton for the best quadrature (see Fig. 9.13). It means that anticorrelations build between the left fundamental side and the right second-harmonic side of the soliton (and vice versa). Here again, quantum properties are nonlocal.

9.7 Conclusion

Let us summarize the main results about the spatial distribution of quantum fluctuations in 1D solitons that we have presented in this chapter: as far as measurements on the whole soliton are considered, we have shown that scalar Kerr solitons, as well as vector Kerr solitons, exhibit squeezing properties similar to those of a plane wave propagating in the same medium and for the same type of propagation distance. Because of diffraction, as the soliton propagates, strong anticorrelations develop between symmetrical points inside the scalar soliton spot, and between the polarization components for the vector soliton or frequency components for the $\chi^{(2)}$ soliton. Because its formation is not due to phase effects, squeezing in a $\chi^{(2)}$ soliton can be observed on the amplitude quadrature, without homodyne detection.

In view of applications to quantum information, the next step of these studies is the exploration of the quantum properties of soliton arrays, and of the possible quantum entanglement between the different components of a soliton array. From a classical point of view, Trillo et al. have shown [27] that such an array can propagate without any deformation or recurrence. In such a case, corresponding to 180° out-of-phase adjacent solitons, preliminary quantum results [28] show that solitons behave as independent objects, with squeezing properties for each soliton of the array similar to those of a single scalar spatial soliton. For in-phase solitons on the other hand, modulation instability (i.e., exponential amplification of noise) prevents any squeezing after some propagation distance.

To our knowledge, only one experiment has demonstrated an effect related to the quantum fluctuations in spatial solitons: quantum spatial noise leads, for arrays of Kerr solitons in a planar waveguide to a shot-to-shot jitter that has been experimentally characterized [**?**, 30]. Detailed experimental studies of local quantum noise and correlations, which would be of high interest, are difficult to perform because of the large amount of classical fluctuations of the high-power laser needed to produce the solitons.

References

1. S. Spalter, N. Korolkova, F. Konig, A. Sizmann, and G. Leuchs, Phys. Rev. Lett. **81**, 786 (1998).
2. E. M. Nagasako, R. W. Boyd, and G. S. Agrawal, Opt. Express **3**, 171 (1998).
3. A. Mecozzi and P. Kumar, Quantum and Semiclass. Optics **10**, L21 (1998).
4. N. Treps and C. Fabre, Phys. Rev. A **62** 033816 (2000); Europhys. Lett. **38**, 335 (1997).

5. C. Fabre, *Quantum fluctuations in light beams*, Les Houches Session 63, eds. S. Reynaud, E. Giacobino, and J. Zinn-Justin (North-Holland, Amsterdam, 1997), p. 181.
6. P. D. Drummond and S. J. Carter, J. Opt. Soc. Am. B **4**, 1465 (1987).
7. M. Rosenbluh and R. M. Shelby, Phys. Rev. Lett. **66**, 153 (1991); P. D. Drummond, R. M. Shelby, S. R. Friberg, and Y. Yamamoto, Nature **365**, 307 (1993).
8. E. Lantz, T. Sylvestre, H. Maillotte, N. Treps, and C. Fabre, J. Opt. B: Quantum Semiclass. Opt. **6**, S295 (2004).
9. E. Lantz, N. Treps, C. Fabre, and E. Brambilla, Eur. Phys. J. D **29**, 437 (2004).
10. Y. Shen, *The principles of nonlinear optics* (Wiley, New York, 1984).
11. A. Gatti, H. Wiedemann, L. A. Lugiato, I. Marzoli, G. L. Oppo, and S. Barnett, Phys. Rev. A **56**, 877 (1997).
12. L. A. Lugiato, A. Gatti, and H. Wiedemann, *Quantum fluctuations and nonlinear optical patterns*, Les Houches Session 63, eds. S. Reynaud, E. Giacobino, and J. Zinn-Justin, (North-Holland, Amsterdam, 1997), p. 431.
13. W. H. Press, S. A. Teukolsky, W. T. Vetterling, and B. P. Flamery *Numerical Recipes* (Cambridge University Press, Cambridge, 1992).
14. R. Y. Chiao, E. Garmire, and C. H. Townes, Phys. Rev. Lett. **13**, 479 (1964).
15. Y. S. Kivshar and G. P. Agrawal, *Optical solitons: from fibers to photonic crystals* (Academic, San Diego, 2003).
16. R. W. Boyd, *Nonlinear optics* (Academic, San Diego, 1992).
17. C. Cambournac, T. Sylvestre, H. Maillotte, B. Vanderlinden, P. Kockaert, Ph. Emplit, and M. Haelterman, Phys. Rev. Lett. **89** 083901 (2002).
18. W. E. Torruelas, Z. Wang, D. J. Hagan, E. W. VanStryland, G. I. Stegeman, L. Torner, and C. R. Menyuk, Phys. Rev. Lett. **74**, 5036 (1995).
19. M. Kitagawa and Y. Yamamoto, Phys. Rev. A **34**, 3974 (1986).
20. R. Shelby, M. Levenson, S. Perlmutter, R. De Voe, and D. Walls, Phys. Rev. Lett. **57**, 681 (1986).
21. Z. Ou, Phys. Rev. A**49**, 2106 (1994).
22. M. Olsen, R. Horowicz, L. Plimak, N. Treps, and C. Fabre, Phys. Rev. A **61**, 021803 (2000).
23. G. Keller, *Etude du bruit quantique dans un soliton spatial*, internal report (unpublished).
24. C. Fabre, J. B. Fouet, A. Maître, Opt. Lett. **25**, 76 (2000).
25. N. Treps, V. Delaubert, A. Maître, J.-M. Courty, and C. Fabre, Phys. Rev. A **71**, 013820 (2005).
26. S. Spalter , M. Burk, U. Strößner, M. Böhm, A. Sizmann, and G. Leuchs, Europhys. Lett., **38**, 335 (1997).
27. S. Trillo, S. Wabnitz, and T. A. B. Kennedy, Phys. Rev. A **50**, 1732 (1994).
28. E. Lantz, *Correlations and squeezing in a soliton array in Kerr media*, internal report (unpublished).
29. E. Lantz, C. Cambournac, and H. Maillotte, Optics Express **10**, 942 (2002).
30. G. Fanjoux, E. Lantz, F. Devaux, and H. Maillotte, *Influence of the spatio-temporal coherence of the light on the jitter of spatial soliton arrays propagating in a non instantaneous medium waveguide*, submitted (2005).

10 Quantum Fluctuations in Cavity Solitons

Gian-Luca Oppo and John Jeffers

Department of Physics and Applied Physics, University of Strathclyde,
107 Rottenrow, Glasgow G4 ONG, Scotland, UK
e-mail: gianluca@phys.strath.ac.uk

10.1 Introduction

Spatial structures in extended nonlinear optical devices can display important quantum features. Quantum images in Degenerate Optical Parametric Oscillators (DOPO) show quadrature squeezing in the near field [1, 2], and Einstein–Podolsky–Rosen (EPR) correlations in the far field [3]. These effects are due to the generation of entangled photons in the parametric down-conversion process within the optical cavity. For a review of these effects we refer the reader to [4, 5].

Although deterministic spatial structures are commonplace in many branches of science such as hydrodynamics, morphogenesis, biological populations, extended chemical reactions, and so on, the coupling of quantum fluctuations and nonlinear spatial structures can be seen in optics even at room temperature. Quantum images are noise-driven precursors of the spatial patterns observed above threshold, but are induced by quantum fluctuations in photonic devices such as the DOPO, the OPO [6], Kerr cavities [7], and in intracavity second-harmonic generation [8].

Recent interest in spatiotemporal structures of cavity-based photonic devices has focused on localized states, also known as optical bullet holes [9] or cavity solitons (CS) [10]. One of the main advantages of these structures over spatially extended features such as patterns, is their possible use as elements for information processing and optical memories [10]. Recent experimental observations of CS in semiconductor-based devices [11] have further increased interest in the fundamental aspects of these structures. It is the aim of this chapter to review some features of CS in the presence of quantum fluctuations in models of OPO. Quantum fluctuations can be responsible for the growth of arrays of CS [12, 13] and for the appearance of remarkable quantum correlations in the near and far field [14] of localized states. Many of the results discussed below can be found in other nonlinear optical cavity devices such as Kerr, saturable absorber, and SHG cavities. We restrict the description to the OPO, however, and in particular to the DOPO because in this case quantum correlations are enhanced by the presence of twin photons without loss of generality.

The chapter is divided into four sections. In Sections 10.2 and 10.3 we describe CS in DOPO and the treatment of quantum fluctuations in DOPO models, separately. Section 10.4 is devoted to the growth of arrays of CS induced by quantum fluctuations. Quantum quadratures and correlations in the near and far field of CS are presented in Section 10.5, and conclusions and future developments are discussed in Section 10.6.

10.2 Cavity Solitons in Degenerate Optical Parametric Oscillators

The classical mean-field equations for a phase-matched DOPO, where both pump and signal fields are resonated (see Fig. 10.1) are [15]

$$\begin{aligned} \partial_t A_0 &= \Gamma\left[-A_0 + E - A_1^2\right] + \frac{ia}{2}\nabla^2 A_0, \\ \partial_t A_1 &= -A_1 - i\Delta_1 A_1 + A_0 A_1^* + ia\nabla^2 A_1 \,. \end{aligned} \tag{10.1}$$

The slowly varying amplitudes of the pump and signal fields are denoted by

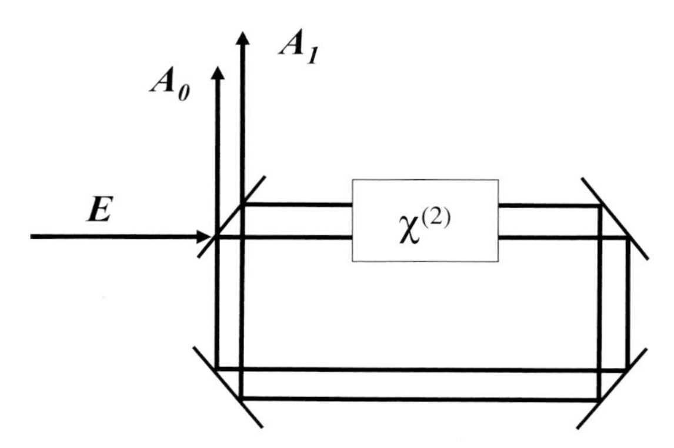

Fig. 10.1. Schematic diagram of the doubly resonant DOPO.

A_0 and A_1, respectively, the time has been normalized by the photon life time in the signal cavity, $\Gamma = \gamma_0/\gamma_1$ is the ratio between the pump and signal cavity decay rates, E is the amplitude of the external pump field (here assumed to be real), Δ_1 is the signal detuning, and $a = c/\gamma_1 k_z$ is the diffraction parameter with c the speed of light and k_z the longitudinal wave-vector of the pump field. We assume that the pump detuning is zero for convenience. Because in the following we will change the ratio between the cavity decay rates, we write the equations without the usual normalizations of the diffraction coefficients with γ_0 and γ_1 [15]. The Laplacian $\nabla^2 = \partial^2/\partial x^2 + \partial^2/\partial y^2$ reduces to $\partial^2/\partial x^2$ in one transverse dimension (1D).

10.2.1 Spatial Equations and Domain Walls with Oscillatory Tails

In 1D all steady states (stable and unstable) have to satisfy ordinary differential equations containing spatial derivatives. For the DOPO the equations are

$$\frac{\partial^2}{\partial x^2} A_0 = \left(\frac{2i\Gamma}{a}\right)\left(-A_0 + E - A_1^2\right),$$
$$\frac{\partial^2}{\partial x^2} A_1 = \left(\frac{i}{a}\right)\left(-A_1 - i\Delta_1 A_1 + A_0 A_1^*\right). \tag{10.2}$$

Above the threshold for signal generation and for positive and zero detunings the DOPO equations admit two steady-state homogeneous solutions given by [16]

$$A_0^s = E - (A_1^s)^2, \qquad A_1^s = \pm\left(\frac{EI_s}{1 + I_s + i\Delta_1}\right)^{1/2} \equiv \sigma e^{(i\theta)},$$
$$I_s = |A_1^s|^2 = \sigma^2 = \sqrt{E^2 - \Delta_1^2} - 1\,, \tag{10.3}$$

where σ and θ are, respectively, the modulus and the phase of A_1^s and

$$\sin(\beta - 2\theta) = \left(\frac{\sigma^2}{2E}\right)\sin\beta, \qquad \beta = arg(1 - i\Delta_1). \tag{10.4}$$

In the phase space of Eqs. (10.2) the homogeneous states correspond to two fixed points. For the cases of interest here, Trillo et al. [16] found numerically that a Domain Wall (DW) solution connecting the two homogeneous states (10.3) is stable in 1D. Such solutions are solitonic in nature, belong to the broader class of CS, and correspond to heteroclinic connections between the two fixed points in the phase space of the fields and their spatial derivatives. An example of the projection of these heteroclinic connections is shown in Fig. 10.2. Solutions of the steady-state Eqs. (10.2) can be found numerically with any required accuracy [17]. We note here that solutions of systems (10.2) almost always diverge to infinity (i.e., with unphysical field intensities) as is easily ascertained by studying the Jacobian of the system in the neighborhood of a generic phase point. It is thus quite remarkable that these equations admit a set of finite intensity solutions starting and ending on the fixed points: namely the heteroclinic and homoclinic orbits described in this chapter.

10.2.2 Cavity Solitons Formed by Locked Domain Walls

An adjacent pair of domain walls, if stationary, represents a homoclinic trajectory in the phase space of Eqs. (10.2): the fields start close to one homogeneous solution for large negative x and end up back at the same homogeneous solution for large positive x. For sufficiently large separations, each DW is

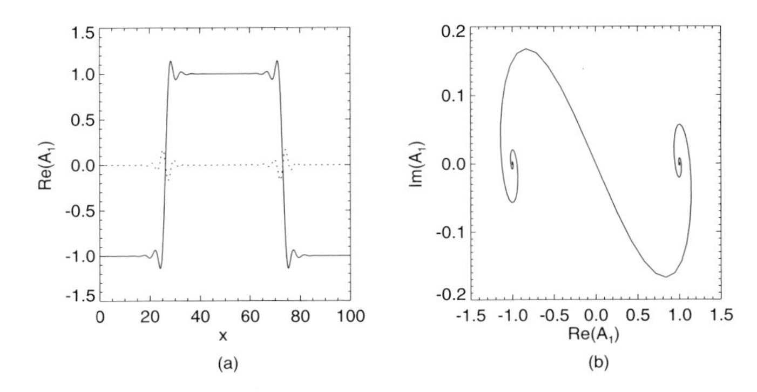

Fig. 10.2. A heteroclinic solution consisting of a pair of domain walls for $\Delta_1 = 0.0$, $a = 0.5$, $E = 2.0$, and $\Gamma = 1.0$; (a) real part (solid line) and imaginary part (dotted line) of the signal field as a function of the transverse coordinate x in the DOPO; (b) the DW pair in (a) plotted in the complex A_1-plane.

essentially independent: the distance between them can be increased or decreased with no apparent constraint. We evaluate the Jacobian of Eqs. (10.1) at a stationary solution consisting of a pair of widely separated DW. Such a Jacobian has two zero eigenvalues: one corresponding to an overall translational invariance of the solution, and the other, to a relative motion of the two DW. The presence of two zero eigenvalues remains true until the DW are close enough for their oscillating tails to interact. When this happens, a locking phenomenon occurs that permits only a discrete set of stationary DW separations. Some examples of the resulting structures together with the largest of the nonzero eigenvalues of the equations linearized about each solution are shown in Fig. 10.3a–f. For any initial condition with a given separation of DW with oscillatory tails, the distance between the DW relaxes to the closest of these equilibrium values. No annihilation process of contiguous DW is observed.

A generic stationary solution A_1^s consists of a trajectory orbiting around the stable homogeneous states and whose real part vanishes at an even number of points $[x_1, \dots, x_{2n}]$ because we are using periodic boundary conditions. The defect cores are located at positions $x_n = x_{n-1} + s_j$, with n, j integers, giving rise to a huge number of possible stable distributions of defects. Note that defect distributions presenting a significant degree of periodicity (where the possible periods are twice the distances s_j) are a very small fraction of the total number. Different final states with arbitrary numbers of defects can be reached by starting from the unstable (zero signal) homogeneous solution with an added random perturbation. The resulting stable 1D structures contain, on average, a wide range of spatial wavelengths giving rise to a continuum background in Fourier space, enhanced by the fact that separations s_j are incommensurate with each other. There is an analogy between this

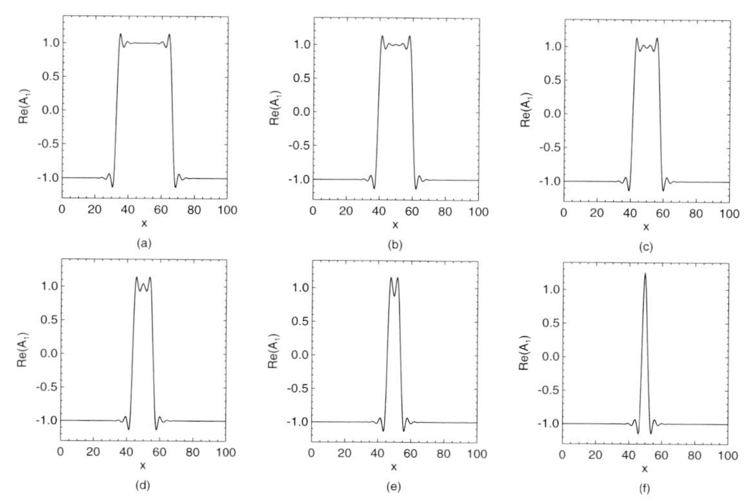

Fig. 10.3. Stationary DW pairs separated by various distances. All solutions shown are both stationary and stable. Parameters as in Fig. 2. The largest nonzero eigenvalues of the equations linearized around these solutions are: (a) -0.0000003685, (b) -0.0001319, (c) -0.0009383, (d) -0.006748, (e) -0.0511, (f) -0.4131.

behavior and temporal chaos, because when $\partial_t = 0$, Eq. (10.2) can be considered as a dynamical system with the variable x assuming the role of time. For this reason these aperiodic (disordered) stable structures have been labeled "spatial chaos" by previous authors [18, 19]. Figure 10.4 shows a typical example of a deterministic, stable, and disordered solution and its spatial power spectrum. Solutions of this kind are generic in the parameter space (E, Γ). The isolated DW and localized structures of locked DW displayed in Fig. 10.3

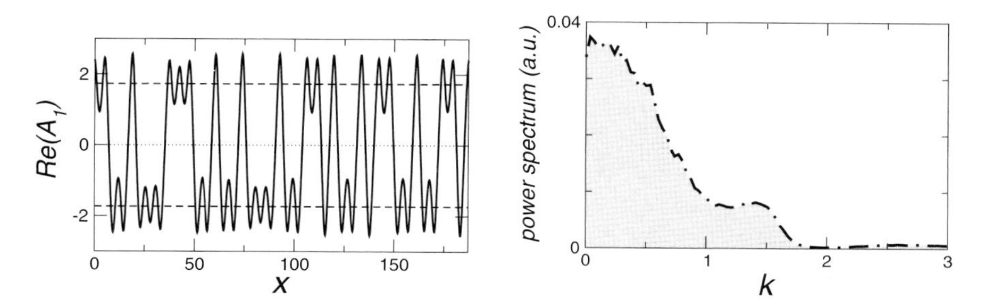

Fig. 10.4. Left panel: typical asymptotic distribution of the real part of the signal field in the absence of fluctuations for $E = 4$ and $\Gamma = 0.2$. Right panel: the spatial power spectrum averaged over many realizations starting from the unstable zero state.

are the cavity solitons of interest for DOPO. They survive in two transverse dimensions although the thresholds for their observation can change due to local curvature effects [17].

10.3 Quantum Fluctuations in DOPO

Open quantum optical systems are often simulated using quantum Langevin equations. These are first-order temporal differential equations for the field, and include stochastic terms to take account of the noise due to the interaction of the system with its environment. The Heisenberg uncertainty principle restricts the minimum size of the noise. In this section we give an overview of the derivation of Langevin equations containing such stochastic noise terms, and show two examples for the DOPO with transverse effects.

Langevin equations are derived from the master equation for the density operator of the open system by taking coherent state matrix elements [20]. This converts the master equation into a Fokker–Planck (FP) equation for a quasi-probability, a distribution analogous to the probability distribution in classical physics. The FP equation can then be mapped onto a stochastic Langevin equation with quantum-limited noise terms. Various FP equations can be derived for the same system for quasi-probabilities associated with different operator orderings. The two most typical are the equations for the Wigner function, which is associated with symmetrical ordering, and the Q- or Husimi function, which is associated with antinormal ordering [21]. The derivation of Langevin equations can be problematic, however, as the FP equations can contain nonlinear terms, third-order derivatives, or have negative diffusion coefficients. In practice higher-order terms are often neglected, leading to Langevin equations associated with what is known as "stochastic electrodynamics".

10.3.1 Wigner Representation

In optical cavities with finite transverse extent a similar procedure can be used to derive Langevin equations from Fokker–Planck equations for quasi-probability functionals, in which the field is associated with each point in the transverse plane. The set of equations for the degenerate optical parametric oscillator in the Wigner representation are as follows [2,22],

$$\begin{aligned} \dot{\alpha_0}(x) &= \Gamma\left[-\alpha_0 + E\right] - \frac{1}{2}\alpha_1^2 + \frac{i}{2}\frac{\partial^2}{\partial x^2}\alpha_0 + \sqrt{\frac{2\Gamma}{n_{th}}}\zeta_0, \\ \dot{\alpha_1}(x) &= -\alpha_1 - i\Delta_1\alpha_1 + \alpha_0\alpha_1^* + i\frac{\partial^2}{\partial x^2}\alpha_1 + \sqrt{\frac{2}{n_{th}}}\zeta_1, \end{aligned} \tag{10.5}$$

where $\alpha_{0(1)}(x)$ is the pump (signal) field (now referred to with Greek letters to reflect their fluctuating nature), n_{th} is the number of photons at threshold, a convenient parameter for describing the nonlinearity of the system, and $\zeta_{0(1)}$ is the pump (signal) Langevin noise term, which satisfies $\langle\zeta_i(x,t)\zeta_j^*(x',t')\rangle = \frac{1}{2}\delta_{ij}\delta(x-x')\delta(t-t')$. In the above equations the pump detuning has been set to zero, as before.

The Wigner functional provides Langevin equations that can be used to evaluate symmetrically ordered expectation values. These are particularly useful when discussing squeezing, as four-port homodyne detection of squeezed light is a symmetrical detection scheme. There are problems associated with the use of the Wigner function(al) in nonlinear optics, however. In the OPO there is a threshold pump power below which no signal field is generated, but the light is squeezed. As this threshold is approached from below, higher-order terms in the FP equation become significantly large so that neglecting them is not a good approximation.

10.3.2 Q-Representation

A way around this difficulty has been proposed using Q-function-based Langevin equations for the signal and pump fields. Details are omitted, and the reader is directed to reference [14] where a detailed exposition is given. The equations are for $\Gamma = 1$

$$
\begin{aligned}
\dot{\alpha_0}(x) &= -\left[1 - i\frac{\partial^2}{\partial x^2}\right]\alpha_0(x) + E - \frac{1}{2}\alpha_1^2(x) + \sqrt{\frac{2}{n_{th}}}\xi_0, \\
\dot{\alpha_1}(x) &= -\left[1 - 2i\frac{\partial^2}{\partial x^2}\right]\alpha_1(x) + \alpha_0(x)\alpha_1^*(x) + \sqrt{\frac{2}{n_{th}}}\xi_1,
\end{aligned}
\tag{10.6}
$$

where a is the ratio of the diffraction coefficients for the pump and signal fields, and ξ_0 is a Langevin noise source that satisfies $\langle\xi_0(x,t)\xi_0^*(x',t')\rangle = \delta(x-x')\delta(t-t')$. The noise source for the signal field is phase sensitive, and also depends on the pump field value:

$$
\xi_1(x,t) = \left[\frac{-\alpha_{0I}}{2\sqrt{2+\alpha_{0R}}} + \frac{i}{2}\sqrt{2+\alpha_{0R}}\right]\Phi(x,t) + \sqrt{\frac{1-|\alpha_0|^2/4}{2+\alpha_{0R}}}\Psi(x,t), \tag{10.7}
$$

where $\alpha_0 = \alpha_{0R} + i\alpha_{0I}$ and Φ and Ψ are noise sources with expectation values similar to that of the pump fluctuations ξ_0. The Q-function treatment is valid provided that the pump field magnitude remains below twice the threshold for signal generation, so the theory is valid in the nonlinear regime. It has been applied successfully to derive quantum correlations both below and above threshold, where patterns emerge in the signal field [14].

10.4 Arrays of CS Induced by Quantum Fluctuations

In this section we study the effect of quantum fluctuations on the distributions of DW and defects in the signal field. In particular, we identify a clear change of behavior in the spatial spectrum decreasing the ratio Γ between the cavity finesse of the pump and the signal fields, respectively. When considering Eqs. (10.5) in the case of small Γ we see that the fluctuations of the pump field become progressively irrelevant. In this case, the quantum behavior of the DOPO is described by a "classical" equation of the pump field such as the first equation of (10.1) and a Langevin-like equation for α_1 as in Eq. (10.5) [13, 23].

Figure 10.5a shows the 1D evolution of the signal field under the action of quantum fluctuations for $\Gamma = 0.02$, $E = 1.5$, and $n_{th} = 1000$ obtained from these equations in the Wigner representation. In the plot, local fluctuations have been filtered out by introducing a threshold at $Re(\alpha_1) = 0$ so that the dark (white) regions in Fig. 10.5a represent positive (negative) values of $Re(\alpha_1)$. Moreover, we eliminate those defect pairs whose separation is less than a definite critical distance because they are doomed to disappear [12]. After a transient whose duration increases exponentially with increasing n_{th}, we reach a stationary equilibrium regime where the average number of defects remains constant. This corresponds to a balance between the rates of appearance and disappearance of pairs of DW. We stress that all quantities considered below are evaluated at equilibrium.

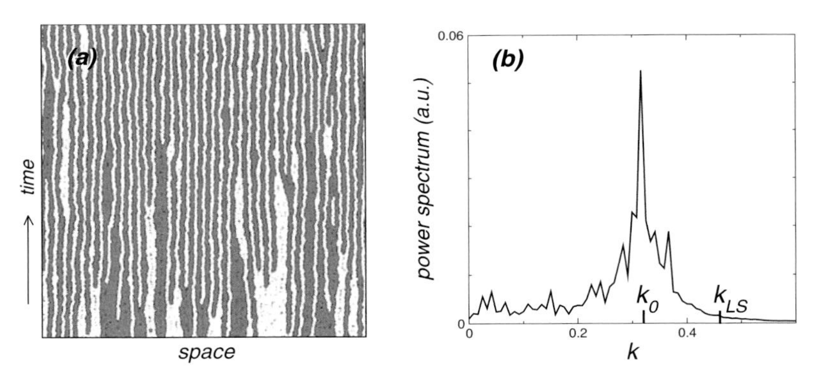

Fig. 10.5. (a) Temporal evolution driven by quantum fluctuations of the real part of the output signal field for $\Delta_1 = 0$, $E = 1.5$, $\Gamma = 0.02$, and $n_{th} = 1000$. Arrays of cavity solitons are clearly visible after the transient has ended. (b) The Fourier spectrum of the spatial autocorrelation function $C[Re(\alpha_1)]$, averaged over time after the equilibrium state has been reached; k_0 corresponds to the wave-vector of arrays of cavity solitons and k_{LS} to the largest eigenvalue of the linear stability analysis of the stable homogeneous solution, away from $k = 0$.

For low Γ, a large number of locked cavity solitons are clearly visible in the near field. They form arrays that jitter under the action of the fluctuations. The lengths of the arrays are an arbitrary multiple of the soliton size s_0. The average length of the arrays increases with increasing ratio of pump-to-signal finesse. For example, in Fig. 10.5a this average length is larger than the transverse simulation size. The presence of arrays of cavity solitons is reflected in the far field where a huge peak at k_0 appears. The power spectrum shown in Fig. 10.5b is nothing more than the Fourier transform of the spatial autocorrelation function $C[g(x,t)] = \int_{-\infty}^{+\infty} h(x+x',t)h^*(x',t)dx'$ averaged over time. In our case $h = Re(\alpha_1)$. The average size of the arrays of cavity solitons is given by the inverse of the decay rate of the averaged spatial correlation function. A large peak at k_0 in the power spectrum signals the presence of large spatial correlations at distances of the order of s_0.

These arrays should not be confused with patterns above a modulational instability because they are formed by progressive locking of localized structures and not by the instability of a given wave vector. The phenomenon described here is analogous to the noise-induced suppression of spatial chaos presented in [12] but is now entirely due to quantum fluctuations. Arrays of cavity solitons in the signal intensity in the limit of small Γ and the corresponding off-axis peak in the far field are a new "quantum structure" in that they are induced by quantum noise after long transients. Note that without the quantum fluctuations the far field is broad-band (see Fig. 10.4) and displays no correlations at any particular wave vector. We will see in Section 10.5 that photons at the k_0 peak display large important quantum correlations.

Realistic quantum fluctuations can induce arrays of cavity solitons in a 1D configuration of a DOPO. Figure 10.6 shows the duration of the transient before reaching the equilibrium regime against $n_{th}\Gamma$, a parameter that measures the inverse of the fluctuation strength and depends on the pump wavelength, the diffraction in the cavity, the material nonlinearity, and the finesse of both cavities. It is important to note that Γ cannot be pushed below, say, 10^{-3} to maintain the validity of the mean-field limit [24], and large values of n_{th} can lead to undesirably long transients. Staying above but close to the threshold of signal generation also helps to achieve optimal balance among the parameters and for this reason we have chosen to work within the range of $1.5 \leq E \leq 3$. It is important to show that the quantum structure described here is fundamentally different from the quantum images observed around a modulational instability as described for DOPO in [1, 2], for OPO in [6], for Kerr cavities in [7], and for intracavity second-harmonic generation in [8]. In all these cases, the quantum image is associated with a noisy precursor of the pattern that forms above a modulational instability. This means that if we switch the noise off after the formation of the quantum image, the far-field peak at the critical wave vector k_c of pattern formation above the modulational threshold disappears. In the present case of arrays of cavity solitons, instead, if we remove the noise after the arrays have been induced by the quantum fluctuations, the arrays will survive indefinitely because they are one of many stable stationary solutions of the system. In this respect it is important to note that our arrays of cavity solitons are induced but not sustained by quantum fluctuations.

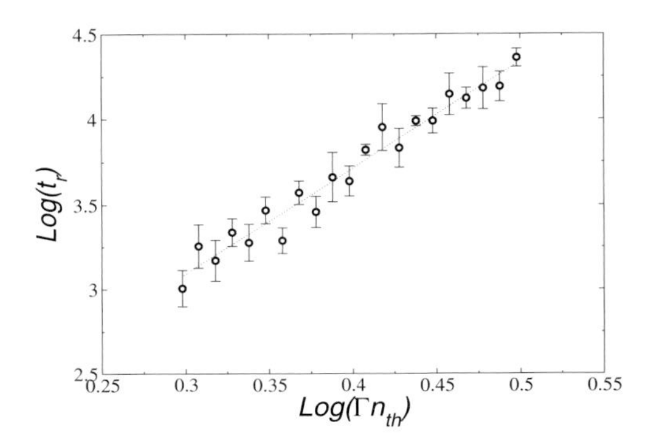

Fig. 10.6. Duration of the transient t_r before the equilibrium regime is reached versus Γn_{th} on a logarithmic scale for $E = 2.6$.

To explain why quantum fluctuations select regular arrays of cavity solitons for small Γ because stable homogeneous solutions exist also for large Γ we present a simple argument suggested in [12]. In the limit of small Γ, local oscillations at the tails of the locked DW have large amplitude [17]. Once a single-peak cavity soliton is excited by the fluctuations, the probability of exciting another soliton peak in the vicinity of the first one is spatially inhomogeneous due to the presence of the oscillatory tails. In particular, fluctuations much smaller than those necessary to excite a single-peak soliton from a homogeneous background can excite a new peak in the vicinity of the large amplitude oscillations of the soliton tail. This was determined by finding the unstable single-soliton solution that provides the critical magnitude of such excitations. In the limit of small Γ the critical amplitudes for erasing a soliton peak are larger than those for its excitation, as shown in [12], and so the equilibrium density of defects is large. Therefore, the average separation distance between defects is small and, because this cannot be smaller than s_0 (the characteristic size of a soliton), arrays of solitons form.

The heuristic argument provided above explains the critical role played by the parameter Γ in the stochastic selection of the final solutions. In particular, the condition of small Γ has a twofold relevance: it increases both the role played by the signal noise and the size of the local oscillations of the soliton tails. In agreement with this argument, we observe for increasing Γ that the average size of the arrays of cavity solitons tends to decrease and larger patches of homogeneous solutions progressively appear. This means that the peak in the far field gradually decreases and eventually disappears on increasing Γ. We note, however, that by increasing Γ the validity of the quantum model based on a classical pump equation becomes questionable because the fluctuations of the pump field cannot be neglected any longer [23]. If we use Eqs. (10.5) for increasing Γ, we observe arrays of solitons progressively decreasing in size and finally disappearing well before $\Gamma = 1$, leaving the dynamics to be dominated by domain walls performing random walks [12].

10.5 Quantum Features in the Near and the Far Field of CS

We now turn our attention to the behavior of quantum fluctuations inside DW and CS in DOPO. We separate our presentation into the near- and the far-field measurements.

10.5.1 Quantum Correlations of CS in the Near Field

We consider the output of a broad area one-dimensional DOPO above threshold ($E = 1.2$) and compare the quadrature components of the fluctuations coming from the center of the DW with those of the homogeneous part in the transverse plane. Figure 10.7 shows the averaged distribution of the real part of the signal field $A_1 =< \alpha_1 >$ at resonance for $E = 1.2$, the size and the two positions of the near-field detectors. The first position is straight at the center of the DW whereas the second one is in the homogeneous region. We have evaluated the normally ordered correlation function

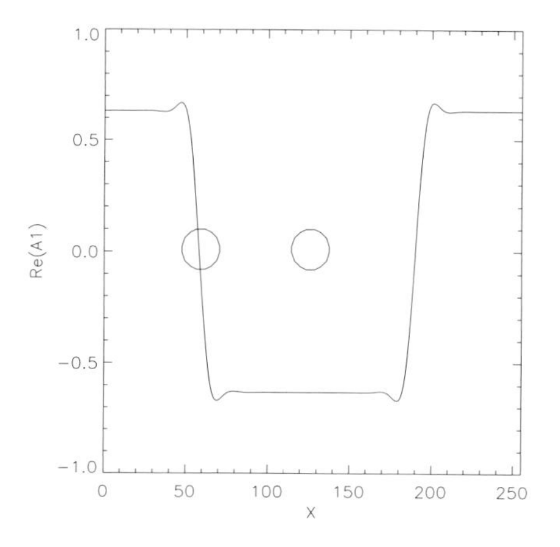

Fig. 10.7. Averaged transverse distribution of $Re(A_1) =< Re(\alpha_1) >$ for $E = 1.2$. The two circles correspond to the positions of the near-field quadrature detectors.

$$\Gamma_\phi =<: \Lambda_\phi \Lambda_\phi :> \qquad \Lambda_\phi = (\alpha_1 - < \alpha_1 >)\, e^{-i\phi} + c.c. \qquad (10.8)$$

where $< . >$ denotes a temporal average after transients have been discarded and α_1 is obtained via the simulation of the Langevin equations (10.5). Numerical simulations in the Q-representation (10.6) will be presented elsewhere [25]. The results of the evaluation of Γ_ϕ are presented in Fig. 10.8.

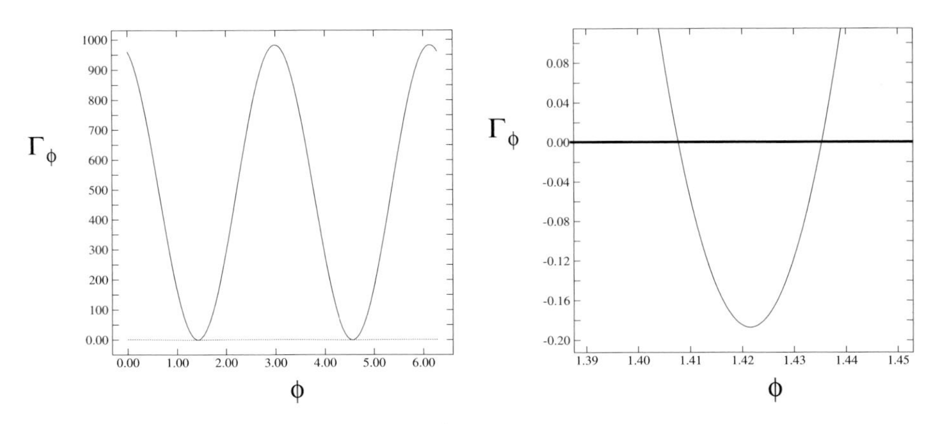

Fig. 10.8. Left panel: the near-field quadrature correlation Γ_ϕ for the detector on the DW (solid line) and on the homogeneous solution (dotted line) for $E = 1.2$. Right panel: magnification of Γ_ϕ for the DW. The solid line at zero represents the shot noise.

The quadrature correlations of the noise in the CS clearly show an amplification [26] of the noise amplitude for certain angles ϕ of the quadrature phase when compared with those of the homogeneous solution. The amplification is huge (almost three orders of magnitude for the simulations presented in Fig. 10.8) and leads to squeezing ellipses that are hugely asymmetric. A considerable amount of squeezing is present in the near-field detection of photons coming from the DW (see the right panel of Fig. 10.8). Such squeezing is much larger than that described in [2] for a DOPO below threshold which progressively vanishes approaching the threshold.

A comparison between the squeezing ellipses for the DW and the homogeneous solution is presented in Fig. 10.9. The amplification of the amplitude of the fluctuations in a precise direction of the quadrature phase ϕ is so large that the squeezing ellipse of the homogeneous solution is invisible in the left panel of Fig. 10.9. Such amplification is due to the breaking of the transverse symmetry due to the presence of the DW. It is well known that the Goldstone mode associated with such symmetry breaking has a zero eigenvalue and consequently a marginal stability. Although the homogeneous solution maintains transverse symmetry and its fluctuations are damped for every value of ϕ,

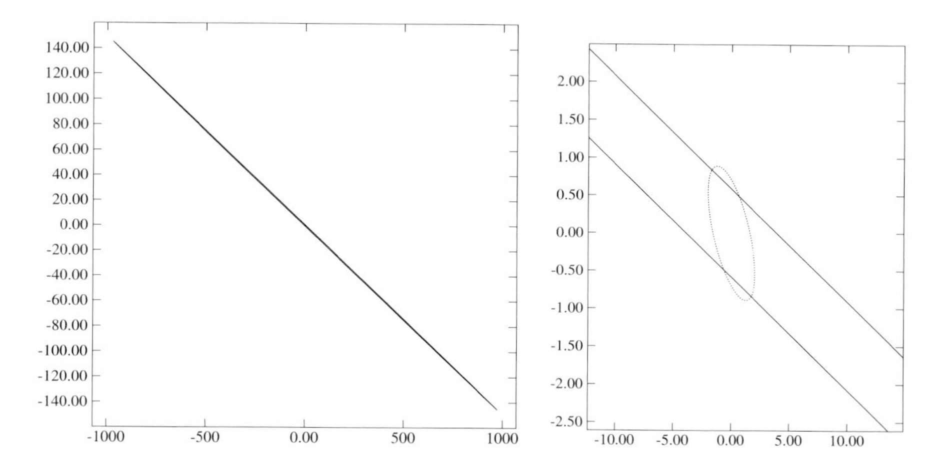

Fig. 10.9. Left panel: the near-field quadrature correlation Γ_ϕ in a polar representation for the detector on the DW. Right panel: magnification of the central part of the squeezing ellipses. Parameters are the same as in Fig. 10.8.

DW and CS break the translational invariance and their fluctuations appear amplified in the direction of marginal stability when compared with those of the homogeneous solution. The zero eigenvalue of the Goldstone mode is not capable of damping fluctuations leading to jitter (and diffusion) in the position of the DW.

Finally, from the magnification of the central part of the squeezing ellipses presented in the right panel of Fig. 10.9, we see that the levels of squeezing in the DW and in the homogeneous part of the signal field are comparable in magnitude. This suggests that far-field detection can be more appropriate for an easy detection of nonclassical features in these structures.

10.5.2 Quantum Correlations of CS in the Far Field

Quantum features of photons generated in CS are more easily observed in far-field detection. Here photons emitted in homogeneous regions of the transverse space accumulate in the $k = 0$ mode (on-axis emission) and photons generated by spatial structures such as CS and patterns are scattered off-axis with a spatial wave-vector $k \neq 0$. We consider here the quantum fluctuations of two configurations of CS in DOPO: two well-separated DW (similar to those described in Fig. 10.7) and an irregular sequence of locked DW similar to those described in Sections 10.2 and 10.4. Without loss of generality, the chosen DOPO parameters correspond to those selected in [14]; that is, $E = 1.5$, $n_{th} = 10,000$, and $\Delta_1 = -0.18$. In Fig. 10.10 we show the near- and far-field distributions of the averaged signal and the pump fields for these two configurations. We note that in [14] numerical simulations are performed in the Q-representation with Eqs. (10.6).

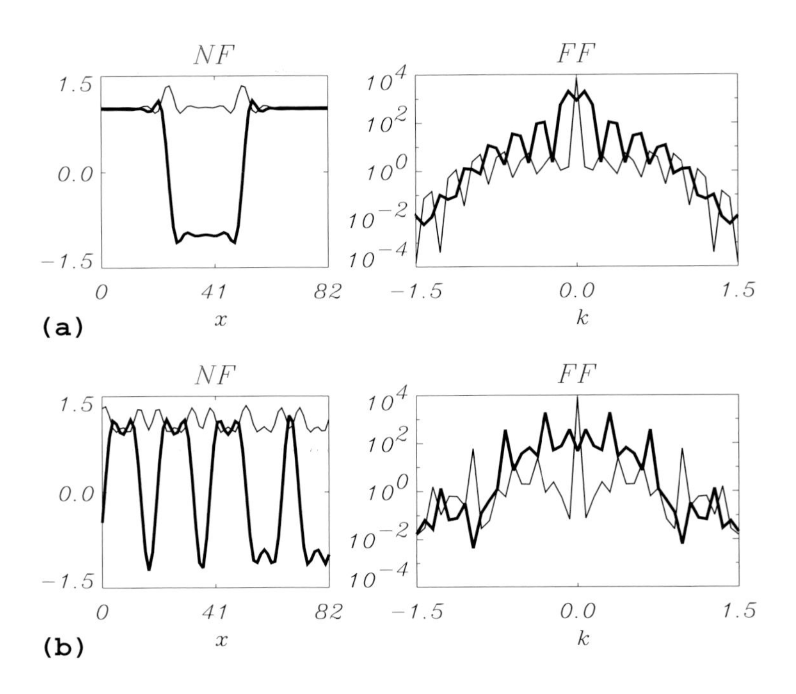

Fig. 10.10. Averaged transverse distributions of the real part of the near-field and intensity (log scale) of the far-field for the pump (thin line) and signal (thick line) for $E = 1.5$, $n_{th} = 10,000$, and $\Delta_1 = -0.18$; (a) two DW, (b) spatial chaos. Reproduced from [14].

Nonclassical features can be detected in the intensities of twin beams in the far-field by evaluating the normal-ordered variance in the difference of the two signal intensities at opposite k wave-vectors [14]:

$$\mathcal{V}(k) = \frac{\left\langle [\delta I(k) - \delta I(-k)]^2 \right\rangle}{\mathcal{N}(k)} \qquad \delta I(k) = I(k) - < I(k) > \tag{10.9}$$

where I is the signal intensity, $< . >$ denotes time averages, and $\mathcal{N}(k)$ is the shot noise proportional to the sum of the averaged signal intensities at $\pm k$. Negative values of $\mathcal{V}(k)$ indicate sub-Poissonian statistics in the intensity difference of the two signal beams at $\pm k$ [27]. For standard quantum images below the threshold, $\mathcal{V} = -0.5$ for any pump intensity and far-field wave vector [2, 27]. This is an obvious artifact due to the normalization $\mathcal{N}(k)$ which instead strongly depends on the pump intensity and far-field wave-vector. To clarify this issue, Zambrini et al. [14] plotted $\mathcal{V}(k)$ below the signal generation threshold ($E = 0.99$) as displayed in the left panel of Fig. 10.11. Small deviations of the numerical simulations from the analytical value $-1/2$ are due to the smallness of the shot noise used in the normalization. The shot noise is in fact proportional to the averaged signal intensity in the farfield shown in the insert of the left panel of Fig. 10.11. The left panel of Fig. 10.11 shows the

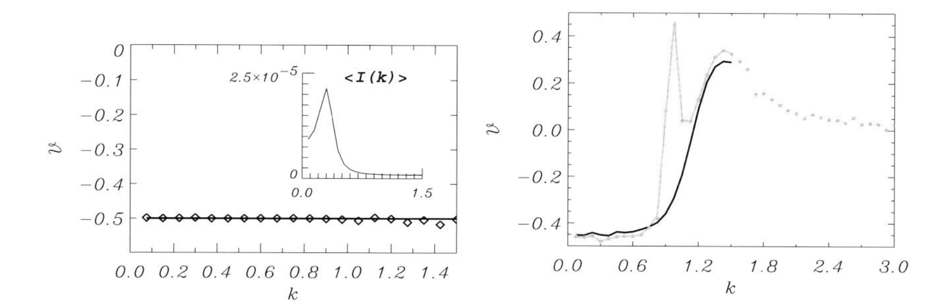

Fig. 10.11. Left panel: the twin-beam variance $\mathcal{V}(k)$ below the threshold for signal generation ($E = 0.99$, $n_{th} = 10,000$, and $\Delta_1 = -0.18$). The insert shows the averaged signal intensity. Right panel: $\mathcal{V}(k)$ above threshold ($E = 1.5$). The black line corresponds to the CS configuration of Fig. 10.10a; the grey line to that of Fig. 10.10b. Reproduced from [14].

spatial spectrum of the far-field variance $\mathcal{V}(k)$ for the two CS configurations of interest. As in the case below threshold, quantum correlated beams are also found in the presence of CS. Note that the scattered photons (i.e., $k \neq 0$) are now entirely due to the presence of the localized CS structures. A difference with the case below the threshold, however, is that a cut-off for the negative variance appears at large k ($k \sim 1$). This cut-off is much higher than all the main spatial components in the CS structures as shown by the far-field spectra in Fig. 10.10. Correlations become classical for large wave-vectors where the signal is depleted more than the pump field and reaches asymptotically the level of a coherent state (see the dotted gray line in Fig. 10.11) [14].

The interpretation of a band of quantum-correlated photons in the far field of the signal of CS in the DOPO is simple. Any photon scattered by the spatial structure of CS appears in the far field with its twin at the opposite wave-number. Negative variances $\mathcal{V}(k)$ reflect the quantum nature of the twin-beam photon generation in the $\chi^{(2)}$ crystal and are then independent of the photon output direction. It is obvious, however, that not all k-modes are equally populated in the presence of CS. For example, in the case of arrays of CS driven by quantum fluctuations described in Section 10.3 (see Fig. 10.5), the largest number of photons in the far field appear at the peak k_s. All these photons display quantum correlations in the far field with negative variance $\mathcal{V}(k)$, and are thus quantum structures.

10.6 Conclusions and Acknowledgments

In this chapter we have explored the relation between cavity solitons and quantum fluctuations in the DOPO. This is the ideal system to explore such a relation, as it readily exhibits both of these properties.

The first effect of quantum fluctuations that we have considered is quantum noise-induced stabilization of locked arrays of cavity solitons. In the neighborhood of a domain wall quantum noise can be large enough to induce the creation of another pair of such walls, which locks to the first domain wall at a particular distance. This phenomenon builds up to form large arrays of domain walls or equivalently, cavity solitons.

We have also considered quadrature correlations in the near and far field of domain walls. In the near field the unusual nature of the domain wall center, where the pump is above the threshold and the signal field vanishes, suggests that there should be extreme features associated with squeezing around such points. In the domain wall there is large amplification of the quantum fluctuations with a particular phase when we compare them with those of the homogeneous solution. The resultant fluctuations are many times larger than the fluctuations at a position away from the domain wall, a consequence of the breaking of the transverse symmetry imposed by the domain wall.

Quantum correlations have also been found in the far-field intensity difference of twin beams emitted from the cavity soliton structures. These correlations are a consequence of the twin-photon generation process in the $\chi^{(2)}$ crystal. The spatial wave-vectors at which the correlations appear are determined by the underlying transverse structure of the signal field.

Our final conclusion from the work presented here is that arrays of CS in DOPO for small values of the decay ratio Γ are the *Quantum CS Structure* "par excellence" because they are generated by the quantum fluctuations and display remarkable quantum features in both the near and the far field.

We would like to thank the following people for useful discussions, and/or for the permission to use some of their results in this chapter: Roberta Zambrini, Andrew Scroggie, Ivan Rabbiosi, Graeme McCartney, Steve Barnett, and Pere Colet. We would also like to thank the European Commission (via the Quantim and FunFACS networks), SGI, the Royal Society—Leverhulme Trust, and the University of Strathclyde (SRIF-II) for funding in support of this research.

References

1. A. Gatti and L. A. Lugiato, Phys. Rev. A **52**, 1675 (1995).
2. A. Gatti, H. Wiedemann, L. A. Lugiato, I. Marzoli, G.-L. Oppo and S. M. Barnett, Phys. Rev. A **56**, 877 (1997).
3. I. Marzoli, A. Gatti, and L. A. Lugiato, Phys. Rev. Lett. **78**, 2092 (1997).
4. L. A. Lugiato, M. Brambilla, and A. Gatti, Adv. Atom. Mol. and Opt. Phys. **40**, 229 (1999).

5. M. I. Kolobov, Rev. Mod. Phys. **71**, 1539 (1999).
6. C. Szwaj, G.-L. Oppo, A. Gatti, and L. A. Lugiato, Eur. Phys. J. D **10**, 433 (2000).
7. R. Zambrini, M. Hoyuelos, A. Gatti, P. Colet, L. A. Lugiato, and M. San Miguel, Phys. Rev. A **62**, 063801 (2000).
8. M. Bache, P. Scotto, R. Zambrini, M. San Miguel, and M. Saffman, Phys. Rev. A **66**, 013809 (2002).
9. W. J. Firth and A. J. Scroggie, Phys. Rev. Lett. **76**, 1623 (1996).
10. W. J. Firth and C. O. Weiss, Optics and Photonics News **13**, 54 (2002).
11. S. Barland et al., Nature **419**, 699 (2002).
12. I. Rabbiosi, A. J. Scroggie, and G.-L. Oppo, Phys. Rev. Lett. **89**, 254102 (2002).
13. I. Rabbiosi, A. J. Scroggie, and G.-L. Oppo, Eur. Phys. J. D **22**, 453 (2003).
14. R. Zambrini et al., Eur. Phys. J. D **22**, 460 (2003).
15. G.-L. Oppo, M. Brambilla, and L. A. Lugiato, Phys. Rev. A **49**, 2028 (1994); G.-L. Oppo, M. Brambilla, D. Camesasca, A. Gatti, and L. A. Lugiato, J. Mod. Opt. **41**, 1151 (1994);
16. S. Trillo, M. Haelterman, and A. Sheppard, Opt. Lett. **22**, 970 (1997).
17. G.-L. Oppo, A. J. Scroggie, and W. J. Firth, Phys. Rev. E **63**, 66209 (2001).
18. P. Coullet, C. Elphick, and D. Repaux, Phys. Rev. Lett. **58**, 431 (1987).
19. M. C. Cross and P. C. Hohenberg, Rev. Mod. Phys., **65**, 851 (1993).
20. H. Carmichael, *An Open Systems Approach to Quantum Optics* (Springer, Berlin, 1993).
21. There are other useful functions, such as the positive P, but we do not consider these here.
22. A. Gatti et al., Opt. Express **1**, 21 (1997).
23. R. Zambrini, S. M. Barnett, P. Colet, and M. San Miguel, Phys. Rev. A **65**, 023813 (2002); see also erratum by the same authors and San Maxi Miguel, ibid **65**, 049901(2002).
24. The mean field limit has been obtained as a first-order expansion in the transmittivity of the cavity mirrors. By reducing Γ below say 10^{-3} a new perturbation expansion is necessary to include higher-order terms. Note also that the limit of $\Gamma \to 0$ is singular even in the absence of noise and diffraction.
25. J. Jeffers, R. Zambrini, A. J. Scroggie, G. McCartney, and G.-L. Oppo, in preparation (2005).
26. We use the word "amplification" when comparing the amplitude of the near-field fluctuations in the DW and homogeneous regions. Because the eigenvalue of the Goldstone mode is zero, the DW fluctuations are neither damped or amplified when considered separately.
27. R. Zambrini and M. San Miguel, Phys. Rev. A **66**, 023807 (2002).

11 Quantum Holographic Teleportation and Dense Coding of Optical Images

Ivan V. Sokolov

V. A. Fock Physics Institute, St. Petersburg State University, 198504
St. Petersburg, Russia, sokolov@is2968.spb.edu

11.1 Introduction

Quantum information has emerged as an actively developing field in the past decade [1, 2]. The goal of this novel area of research in theory and experiment is to apply the laws of the quantum world to information processing and communication. One can mention here quantum cryptography, quantum computing, quantum teleportation, quantum dense coding, and other applications. It seems natural to extend the concepts and approaches developed in quantum imaging to the phenomena of quantum information, thus introducing parallelism and all-optical methods to the latter field of research.

In this chapter we discuss the extension of two continuous-variable protocols of quantum information: quantum teleportation and quantum dense coding, onto the optical images. Similar to most quantum information phenomena, the basic resource here is provided by quantum entanglement. The key notions for our consideration are the continuous-variable spatially multimode squeezing and entanglement, introduced in previous chapters and discussed below.

It is well known that, by following the principles of quantum physics, it is possible to transport an arbitrary quantum state of the electromagnetic field or another object from one place to another using a classical information exchange in combination with a quantum channel that exploits quantum entangled states. This operation, named quantum teleportation, was initially proposed for discrete variables [3] and later extended to continuous-variable schemes [4–6]. Experimental demonstrations for discrete variables were achieved in [7] for single-photon polarization states, and in [8, 9] for continuous variables. Apart from its intrinsic fundamental relevance, the interest of quantum teleportation arises also from its potential applications in the fields of quantum error correction [10], quantum dense coding [11], and quantum cryptography [12].

The first proposals of the teleportation schemes considered the case of single-mode fields or, at most, broadband teleportation of time-dependent signals [13]. However, the spatial degrees of freedom offer an opportunity for substantially increasing the number of channels in which teleportation can be realized in parallel. A protocol that allows us to teleport a spatially multimode state of the field was proposed recently in [14, 15]. Teleportation with these features has been called *quantum holographic teleportation*. Such a teleportation scheme has far greater potential as compared to a single-mode case because it allows for a simultaneous teleportation of two-dimensional optical images or other two-dimensional data sets. This generalized protocol opens new potential applications of teleportation as quantum interface in two-dimensional parallel quantum computing, in parallel quantum communication, quantum memory, error correction and so on.

Quantum holographic teleportation of optical images is considered in Section 11.3 of this chapter.

Quantum dense coding is a quantum-entanglement-based scheme for communication channels. The basic feature of the quantum dense coding protocol is the use of two channels in the quantum entangled state. The signal is created by the sender (Alice) in the first channel. Due to effective quantum entanglement, the second channel plays the role of perfect reference system for the first one. The receiver (Bob) detects the signal by means of a Bell-type measurement, performed simultaneously with two channels. Quantum entanglement allows for signal detection with sensitivity beyond the standard quantum limit for a single channel.

Quantum dense coding has been first proposed [16] and experimentally realized [17] for discrete variables, *qubits*, and later elaborated [11] and experimentally realized [18] for continuous variables. In Section 11.4 we discuss the continuous-variables quantum dense coding protocol for optical images [19]. This scheme extends the protocol [11] to the essentially multimode in space and time optical communication channels. The spatially multimode generalization exploits the inherent parallelism of optical communications and allows for simultaneous parallel dense coding of an input image with many elements. The capacity of the parallel dense coding scheme greatly exceeds that of its single-mode version.

11.2 Continuous-Variable Squeezing and Entanglement for Spatially Multimode Light Fields

11.2.1 Spatial Scales of Quantum Correlations in Squeezed Light

Light fields in the squeezed state are typically produced by nonlinear parametric interactions. Since the first experimental demonstration [20], the phenomenon of squeezing was observed by three- and four-wave mixing in nonlinear crystals, resonant media, and optical fibers, both in continuous-wave and pulsed regimes [21].

The problem of generation of multivariate squeezing is crucial for quantum imaging. The theoretical proposals and experimental efforts aiming at effective spatially multimode squeezing both in traveling-wave and cavity-based configurations are reviewed in other chapters of this book.

To be specific, we consider in this chapter the sources of squeezed and entangled light fields based on traveling-wave type-I OPAs (Optical Parametric Amplifiers). The intensive nondepleting classical pump field is taken as a monochromatic plane wave with frequency ω_p and wave vector $\boldsymbol{k}_p$ along the z-direction, (see Fig. 11.1). The nonlinear crystal of length l has a shape of a plane slab with large dimensions along the x- and y-directions. By the type-I parametric down-conversion, a pump photon produces a pair of photons of equal polarizations with frequencies $\omega \pm \Omega$, where $\omega = \omega_p/2$, and transverse components of wave-vectors $\pm\boldsymbol{q}$. These conditions follow from the frequency and the transverse momentum conservation. The wave with frequency $\omega + \Omega$ and transverse component $\boldsymbol{q}$ has the wave-vector $\boldsymbol{k}(\boldsymbol{q}, \Omega)$. As

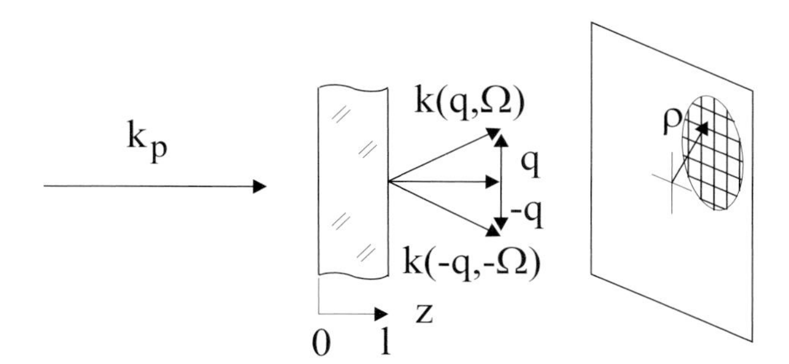

Fig. 11.1. Schematic of traveling wave OPA.

will be shown below, in the case of collinear and frequency-degenerate phase matching one needs two independent OPAs, OPA$_1$ and OPA$_2$, in order to produce two squeezed and entangled light beams. We shall use the space- and time-dependent photon annihilation and creation operators $\hat{S}_n(\boldsymbol{\rho}, t)$ and $\hat{S}_n^\dagger(\boldsymbol{\rho}, t)$, $n = 1, 2$, of the squeezed fields. Here $\boldsymbol{\rho}$ is the 2D transverse coordinate. These operators obey [22, 23] the free-field commutation relations,

$$[\hat{S}_n(\boldsymbol{\rho}, t), \hat{S}_{n'}^\dagger(\boldsymbol{\rho}', t')] = \delta_{n,n'}\delta(\boldsymbol{\rho} - \boldsymbol{\rho}')\delta(t - t'), \tag{11.1}$$
$$\left[\hat{S}_n(\boldsymbol{\rho}, t), \hat{S}_{n'}(\boldsymbol{\rho}', t')\right] = 0,$$

and are normalized so that $\langle \hat{S}_n^\dagger(\boldsymbol{\rho}, t)\hat{S}_n(\boldsymbol{\rho}, t)\rangle$ gives the mean value of the irradiance, expressed in photons per cm^2 per second.

The transformation of the input fields $\hat{A}_n(\boldsymbol{\rho}, t)$ of the OPAs in the vacuum state into the output fields $\hat{S}_n(\boldsymbol{\rho}, t)$ in the broadband multimode squeezed state is described in terms of the Fourier components of these operators in the frequency and spatial-frequency domain,

$$\hat{s}_n(\boldsymbol{q}, \Omega) = \int d\boldsymbol{\rho}\, dt \exp[i(\Omega t - \boldsymbol{q} \cdot \boldsymbol{\rho})]\hat{S}_n(\boldsymbol{\rho}, t). \tag{11.2}$$

In what follows we shall use similar notation for $\hat{a}_n(\boldsymbol{q}, \Omega)$ and other field operators. The squeezing transformation, performed by the OPAs, can be written as follows,

$$\hat{s}_n(\boldsymbol{q}, \Omega) = U_n(\boldsymbol{q}, \Omega)\hat{a}_n(\boldsymbol{q}, \Omega) + V_n(\boldsymbol{q}, \Omega)\hat{a}_n^\dagger(-\boldsymbol{q}, -\Omega), \qquad (11.3)$$

where the coefficients $U_n(\boldsymbol{q}, \Omega)$ and $V_n(\boldsymbol{q}, \Omega)$ depend on the pump-field amplitudes of the OPAs, their nonlinear susceptibilities, and the phase-matching conditions. For the type-I phase-matched traveling-wave OPAs, the coefficients $U_n(\boldsymbol{q}, \Omega)$ and $V_n(\boldsymbol{q}, \Omega)$ are given in Appendix A.

In what follows we assume the frequency-degenerate and collinear phase-matching condition in both OPAs with equal coupling constant $|g_1| = |g_2| = g$ and equal degree of squeezing (see Appendix A), where

$$r_1(\boldsymbol{q}, \Omega) = r_2(\boldsymbol{q}, \Omega) \equiv r(\boldsymbol{q}, \Omega), \qquad (11.4)$$

and $r(0,0) = g$. A simple physical interpretation of wideband squeezing can be given in terms of squeezing ellipses, introduced in the (x, y) plane of complex field amplitude for any pair of conjugate waves with the Fourier amplitudes $\hat{s}_n(\boldsymbol{q}, \Omega)$ and $\hat{s}_n(-\boldsymbol{q}, -\Omega)$. The $(\boldsymbol{q}, \Omega)$ ellipse, representing the quantum uncertainty area for the given pair of conjugate waves, is stretched and squeezed by the factor $\exp[\pm r_n(\boldsymbol{q}, \Omega)]$. The major axis of the ellipse is oriented at the angle $\psi(\boldsymbol{q}, \Omega)$ in the plane of complex field amplitude.

The spatiotemporal scales of squeezing and entanglement are sensitive to the rotation of the squeezing ellipses with frequencies $\boldsymbol{q}, \Omega$, that is, to the frequency dispersion of squeezing. As a result of the rotation, the noise suppression in the given field quadrature goes over to the noise amplification at higher frequencies, as shown in Fig. 11.2a. The rotation in dependence on

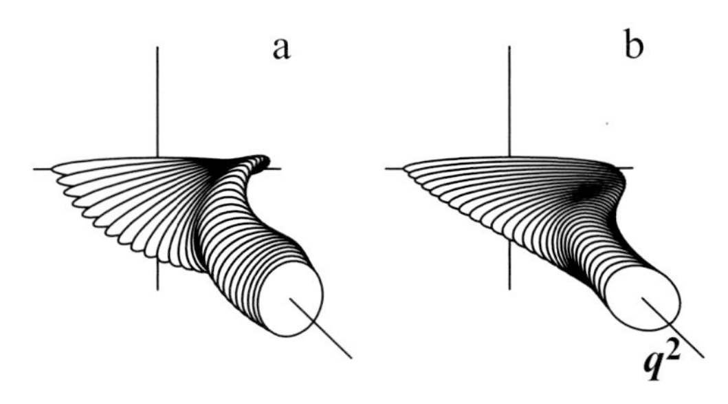

Fig. 11.2. The squeezing ellipses (a) for the broadband in space–time field $\hat{s}_1(\boldsymbol{q}, \Omega)$ in dependence on the mismatch $\delta(\boldsymbol{q}, \Omega)$ (arbitrary units). The type-I collinear degenerate phase matching, $\exp[r(0,0)] = 3$. (b) The same squeezing ellipses for the imaging system with a properly inserted lens.

the spatial frequency is due to diffraction. This rotation can be effectively eliminated by a properly inserted lens imaging system, as shown in [22, 23] and illustrated in Fig. 11.2b.

11.2.2 Spatially Multimode Entanglement

The EPR (Einstein–Podolsky–Rosen) beams $\hat{E}_n(\boldsymbol{\rho}, t)$, n = 1,2, can be created by the interference mixing at the 50:50 beamsplitter of two broadband multimode squeezed beams $S_m(\boldsymbol{\rho}, t)$, created by two degenerate traveling-wave OPAs:

$$\hat{E}_n(\boldsymbol{\rho}, t) = \sum_{m=1,2} R_{nm} \hat{S}_m(\boldsymbol{\rho}, t). \tag{11.5}$$

Here

$$\{R_{nm}\} = \frac{1}{\sqrt{2}} \begin{pmatrix} 1 & 1 \\ -1 & 1 \end{pmatrix}, \tag{11.6}$$

is the scattering matrix of the beamsplitter. This generation scheme of the spatially multimode EPR entanglement is a generalization of the scheme previously used [8] for the generation of entanglement, which is broadband in time, but single-mode in space.

Alternatively, the two multimode EPR beams can be generated by a single traveling-wave OPA, degenerate in frequency, with type-II phase matching [24, 25] (in the latter case the two beams have orthogonal polarizations), or by a single-frequency or momentum nondegenerate OPA.

In analogy to the single-mode EPR beams, the multimode EPR beams are created if squeezing in both channels is effective, and the squeezing ellipses are oriented in the orthogonal directions. For simplicity we shall assume that OPA_1 and OPA_2 have such properties that

$$\begin{aligned} U_1(\boldsymbol{q}, \Omega) &= U_2(\boldsymbol{q}, \Omega) \equiv U(\boldsymbol{q}, \Omega), \\ V_1(\boldsymbol{q}, \Omega) &= -V_2(\boldsymbol{q}, \Omega) \equiv V(\boldsymbol{q}, \Omega). \end{aligned} \tag{11.7}$$

These assumptions provide,

$$\begin{aligned} \psi_1(\boldsymbol{q}, \Omega) &= \psi_2(\boldsymbol{q}, \Omega) \pm \pi/2 \equiv \psi(\boldsymbol{q}, \Omega), \\ \phi_1(\boldsymbol{q}, \Omega) &= \phi_2(\boldsymbol{q}, \Omega) \pm \pi/2 \equiv \phi(\boldsymbol{q}, \Omega). \end{aligned} \tag{11.8}$$

The physical properties of the generated entanglement are illustrated in Fig. 11.3. Consider the corresponding coherence volumes $V_c = cT_cS_c$ in two incident squeezed beams. Here $T_c = 2\pi/\Omega_c$ and $S_c = (2\pi/q_c)^2$ are, respectively, the coherence time and the coherence area, related to the frequency- and spatial-frequency bandwidths Ω_c and q_c of effective noise suppression in the low-noise quadratures.

The left and right lower ellipses in Fig. 11.3 and vectors inside represent the effective local values of the broadband field fluctuations in these coherence volumes. The vectors represent only the stretched (amplified) quadrature amplitudes of the fields $\hat{S}_1$ and $\hat{S}_2$. The squeezed quadrature amplitudes are negligible for $\exp[r(0,0)] \gg 1$ and are not shown. After the scattering, the outgoing fields $\hat{E}_1$ and $\hat{E}_2$ in the corresponding coherence volumes are composed of the same amplified quadrature amplitudes (see left and right upper plots). This means (in the limit of effective squeezing) the effective correlation and entanglement between the scattered fields: the quadrature amplitudes of the fields $\hat{E}_1$ and $\hat{E}_2$ coincide (up to the sign, introduced by the unitary transform (11.6)). In the spatiotemporal domain the entanglement between the broadband fields is local, "volume to volume." In the frequency

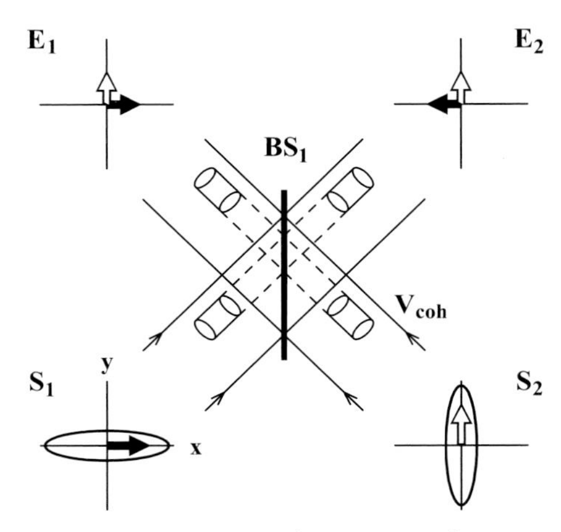

Fig. 11.3. Generation of the EPR fields $\hat{E}_1(\boldsymbol{\rho}, t)$ and $\hat{E}_2(\boldsymbol{\rho}, t)$, locally entangled in space–time, via interference of the illuminating squeezed fields $\hat{S}_1(\boldsymbol{\rho}, t)$ and $\hat{S}_2(\boldsymbol{\rho}, t)$.

the domain the EPR fields are entangled for the frequencies Ω and spatial frequencies $\boldsymbol{q}$ within the phase matching of the OPA.

11.3 Quantum Holographic Teleportation of Optical Images

In this section we describe the optical scheme of quantum holographic teleportation and formulate criteria for achieving high fidelity of the quantum state teleportation.

Quantum holographic teleportation has physical features that are not present in the single-mode teleportation scheme. One of these features is the possibility of controlling the performance of holographic teleportation by optical elements properly inserted into the light beams propagated in the scheme. This possibility of optical control is related to the phenomenon of diffraction which is less important for the single-mode light fields.

Another important difference between the spatially multimode scheme and its single-mode counterpart follows from the fact that in the case of quantum holographic teleportation one has an input signal with a large number of degrees of freedom. The usually discussed measure of the quality of teleportation, the fidelity of teleportation of the quantum state of a global system, does not apply here because it tends to be very close to zero. One has to define, therefore, a reduced fidelity of teleportation related to the pertinent degrees of freedom to be teleported (see Section 11.3.4). Such a reduced fidelity can be made close to unity by a proper choice of the spatial bandwidth of the EPR beams used in the protocol as well as by optimization of the scheme with optical devices (lenses) properly inserted in the light beams.

A peculiar characteristic of the multimode teleportation schemes, based on multipixel light detection, is the need for using a coarse-grained description of the input and output signals in order to characterize the quality of the multimode teleportation.

In an important case of the input image field with quantum Gaussian statistics (e.g., input field in spatially multimode squeezed or coherent state) we find the fidelity of teleportation in an explicit form. It turns out that the fidelity not only depends on the features of the quantum noise in each mode of the input state, and on the degree of entanglement in the quantum channel (which would be true also in the case of single-mode teleportation), but it also depends on the number and choice of image elements or *pixels* that one wants to teleport in parallel.

Quantum holographic teleportation can be viewed as an extension to the quantum domain of conventional nonstationary holography (see Section 11.3.5). In fact, one can recognize in the quantum holographic teleportation protocol all basic elements present in holography. The novel feature, which converts holography into quantum holographic teleportation and allows for suppression of quantum fluctuations in the teleported (reconstructed) image field beyond the standard quantum limit, is the spatially multimode quantum entanglement.

11.3.1 Basics of Quantum Teleportation

The basic components of quantum teleportation are: (i) the so-called quantum channel (a pair of quantum objects in EPR state), and (ii) the EPR measurement. For some important physical situations, an EPR state can be interpreted as the state with precisely defined values of physical variables, associated with the relative motion of the objects of the EPR pair. By contrast, the knowledge of individual variables for the objects constituting the EPR pair, is minimal.

A general scheme of quantum teleportation is illustrated in Fig. 11.4. An input object 1 is in an unknown quantum state $|\psi^{(\mathrm{in})}>_1$. This state has to be teleported by Alice and Bob onto the output object 3. The stages of quantum teleportation are:

(a) Alice and Bob prepare the objects 2 and 3 in certain EPR state $|\psi_{\mathrm{in}}^{EPR}>_{2,3}$ (quantum channel). This means that the variables of the relative motion of the objects 2 and 3 are precisely defined and known both to Alice and Bob.

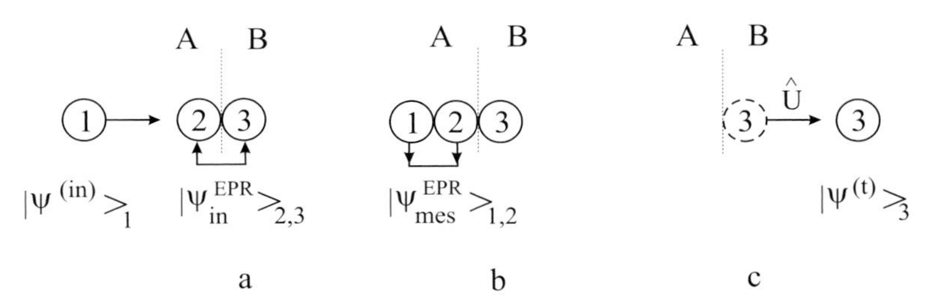

Fig. 11.4. General scheme of quantum teleportation.

(b) Alice performs the measurement of the EPR quantum state of the input object 1 and object 2. As a result of the measurement, the state of the objects 1 and 2 is reduced to some EPR state $|\psi_{mes}^{EPR} >_{1,2}$. This gives to Alice and Bob the definite value of variables of the relative motion of the objects 1 and 2.

(c) During the preparation and measurement of EPR states and after the information exchange over a classical channel, Alice and Bob gain the precise knowledge of variables of the relative motion of the objects 1 and 3. This allows to Bob to perform a physical action, which transforms (by means of the unitary transformation $\hat{U}$) the quantum state of the object 3 to the input quantum state of the object 1, $|\psi^{(t)} >_3 = |\psi^{(\mathrm{in})} >_{1\to 3}$, thus achieving the teleportation of quantum state.

11.3.2 Optical Scheme for Quantum Teleportation of Images

The scheme is an extension onto the spatially multimode light fields of that proposed and realized in [5, 8, 9], and is shown in Fig. 11.5.

The EPR pair of light beams is prepared by optical mixing at the beamsplitter BS_1 of two fields in the spatially multimode squeezed state from two OPAs, as described in Section 11.2.2. The input light field to be teleported from Alice to Bob is described by the field operator $\hat{A}_{\mathrm{in}}(\boldsymbol{\rho}, t)$, where $\boldsymbol{\rho}$ is the 2D transverse coordinate in a cross-section of the beam.

In order to detect two quadrature components of the input field $\hat{A}_{\mathrm{in}}(\boldsymbol{\rho}, t)$, this field is split at the beamsplitter BS_2. Another input port of the beamsplitter is illuminated by the EPR beam $\hat{E}_1(\boldsymbol{\rho}, t)$. In the absence of the EPR beam, which is an essential part of the teleportation scheme, this input port would be illuminated by a broadband in space–time flux of vacuum fluctuations.

Two quadrature components of the scattered by BS_2 light fields $\hat{B}_x(\boldsymbol{\rho}, t)$ and $\hat{B}_y(\boldsymbol{\rho}, t)$ are detected "point-by-point" by two homodyne detectors D_x and D_y formed by high-efficiency multipixel photodetection matrices (CCDs). The spatiotemporal quantum fluctuations of the quadrature components $\hat{B}_x(\boldsymbol{\rho}, t) + \hat{B}_x^\dagger(\boldsymbol{\rho}, t)$ and $\hat{B}_y(\boldsymbol{\rho}, t) - \hat{B}_y^\dagger(\boldsymbol{\rho}, t)$ are locally imprinted into the photocurrents $\hat{I}_x(\boldsymbol{\rho}, t)$ and $\hat{I}_y(\boldsymbol{\rho}, t)$ on the output of individual pixels of the CCD cameras. The photocurrents are sent from Alice to Bob via two multichannel parallel classical communication lines. This part of the protocol corresponds to the EPR measurement by Alice (see Section 11.3.1).

Bob uses the photocurrents $\hat{I}_x(\boldsymbol{\rho}, t)$ and $\hat{I}_y(\boldsymbol{\rho}, t)$ for the reconstruction of the field $\hat{A}_{\text{out}}(\boldsymbol{\rho}, t)$ via two multichannel modulators M_x and M_y which modulate in space and time the relevant quadrature components of an incoming plane coherent light wave. Due to the multimode nature of entanglement our

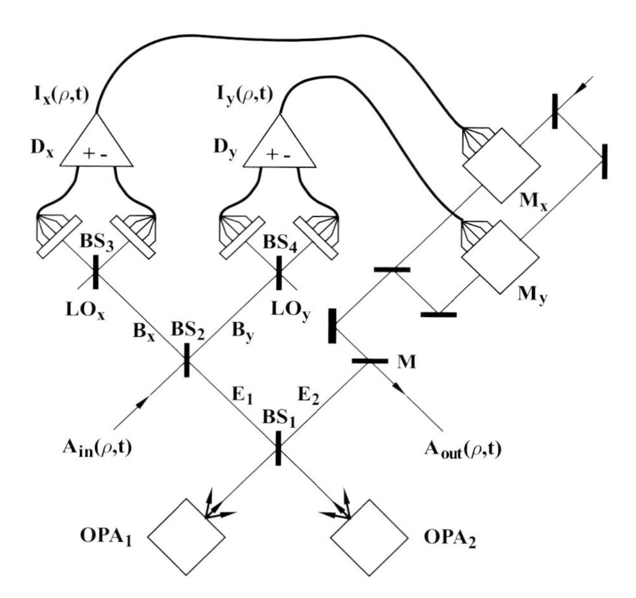

Fig. 11.5. Scheme of holographic teleportation.

scheme allows for parallel teleportation of N elements of the input wavefront, preserving their space–time correlations. This number is estimated [14, 15] as the ratio of the beam cross-section to the coherence area of the light created by the OPAs. In the previous teleportation schemes [5, 8, 9] $N = 1$.

11.3.3 Quantum Statistics of the Teleported Field

The input fields of the balanced homodyne detectors D_x and D_y, used for detection of the x and y field quadratures, read

$$\hat{B}_{x,y}(\boldsymbol{\rho}, t) = \frac{1}{\sqrt{2}}\big(\pm \hat{A}_{\text{in}}(\boldsymbol{\rho}, t) + \hat{E}_1(\boldsymbol{\rho}, t)\big), \tag{11.9}$$

with the $+(-)$ sign corresponding to $x(y)$. We assume that the scattering matrix of the beamsplitter BS_2 is given by Eq. (11.6). These fields in turn are mixed with the local oscillator fields LO_x and LO_y having complex amplitudes $B_x^{(H)} = B_0$ and $B_y^{(H)} = iB_0$, where B_0 is real. For parallel teleportation of the local spatiotemporal quantum correlations of the input field, we have first to measure those correlations with spatiotemporal resolution. Temporal resolution can be achieved by choosing properly the frequency bandwidth of a photodetector. On the contrary, to resolve spatially the quantum fluctuations we have to use multipixel arrays of photodetectors (such as CCD cameras) with pixel size much smaller than the typical spatial scale of quantum correlations. For the case of homodyne detection of spatially single-mode quantum fields, the time-dependent difference photocurrent operators were considered in [26] and, in the more general case, in [27,28]. In Appendix B we shall demonstrate (see Eq. (11.90)) that the difference photocurrent density operators in the balanced homodyne measurement with spatial resolution, performed by D_x and D_y, are expressed through the field operators at the surface of the detectors in analogy to the earlier investigated observables for detection without spatial resolution:

$$\begin{aligned} \hat{I}_x(\boldsymbol{\rho}, t) &= B_0\big[\hat{B}_x(\boldsymbol{\rho}, t) + \hat{B}_x^\dagger(\boldsymbol{\rho}, t)\big], \\ \hat{I}_y(\boldsymbol{\rho}, t) &= B_0\frac{1}{i}\big[\hat{B}_y(\boldsymbol{\rho}, t) - \hat{B}_y^\dagger(\boldsymbol{\rho}, t)\big]. \end{aligned} \tag{11.10}$$

In particular, these equations provide a correct expression for the spatiotemporal shot noise in the balanced homodyne detection scheme.

The photocurrent densities $\hat{I}_x(\boldsymbol{\rho}, t)$ and $\hat{I}_y(\boldsymbol{\rho}, t)$ are sent from Alice to Bob via two multichannel classical communication lines. These signals are used by Bob for the independent local modulation of two quadrature components of an external coherent wave, both phase matched with the relevant quadratures of the EPR fields. In the modulated beam the field component $\propto \hat{I}_x(\boldsymbol{\rho}, t) - i\hat{I}_y(\boldsymbol{\rho}, t)$ is created. The teleported field $\hat{A}_{\text{out}}(\boldsymbol{\rho}, t)$ is obtained by interference mixing at the mirror M with high reflectivity of the modulated field with the second EPR beam $\hat{E}_2$ (see Fig. 11.5),

$$\hat{A}_{\text{out}}(\boldsymbol{\rho}, t) = \hat{E}_2(\boldsymbol{\rho}, t) + g_c\big(\hat{I}_x(\boldsymbol{\rho}, t) - i\hat{I}_y(\boldsymbol{\rho}, t)\big). \tag{11.11}$$

Here g_c is the coupling constant that takes into account the efficiency of modulation and the transmission of the mirror M. The teleportation takes place when $g_c B_0\sqrt{2} = 1$. For a perfect balancing of the mirror reflectivity and modulation gain, as described in [5], the teleported field $\hat{A}_{\text{out}}(\boldsymbol{\rho}, t)$ takes the form:

$$\hat{A}_{\text{out}}(\boldsymbol{\rho}, t) = \hat{A}_{\text{in}}(\boldsymbol{\rho}, t) + \hat{F}(\boldsymbol{\rho}, t), \tag{11.12}$$

where

$$\hat{F}(\boldsymbol{\rho}, t) = \hat{E}_2(\boldsymbol{\rho}, t) + \hat{E}_1^\dagger(\boldsymbol{\rho}, t), \tag{11.13}$$

is the noise field added by the teleportation process. In the ideal case of perfect entanglement of two EPR beams at all frequencies Ω and spatial frequencies

$\boldsymbol{q}$ the noise terms $\hat{E}_2(\boldsymbol{\rho}, t)$ and $\hat{E}_1^\dagger(\boldsymbol{\rho}, t)$ are perfectly anticorrelated and their quantum fluctuations cancel each other.

One can easily illustrate this noise cancellation by constructing a superposition (11.13) of the complex field amplitudes at the left and right upper plots in Fig. 11.3. Here the field $\hat{E}_1(\boldsymbol{\rho}, t)$ is taken complex conjugate, thus accounting for the sign inversion by the term $\propto i\hat{I}_y(\boldsymbol{\rho}, t)$ in the reconstructed field (11.11). The perfect noise cancellation would correspond to the perfect point-to-point in space and instantaneous in time teleportation of the quantum state of the input field with an arbitrary distribution in space and time, $\hat{A}_{\rm out}(\boldsymbol{\rho}, t) = \hat{A}_{\rm in}(\boldsymbol{\rho}, t)$. However such teleportation would require infinitely large energy of EPR beams. Indeed, first as in the single-mode case, one would have to achieve an infinite squeezing per single coherence volume of an EPR beam. In addition because now we have broadband multimode entanglement, one would need an infinite number of elementary coherence volumes in the EPR beams. In practice teleportation will never be point-to-point in space and instantaneous in time but always "on average" within some spatial area and within some finite time interval.

By using Eqs. (11.5) and (11.3), we obtain for the noise amplitude $\hat{f}(\boldsymbol{q}, \Omega)$ given by Eq. (11.13) in the Fourier domain,

$$\hat{f}(\boldsymbol{q}, \Omega) = \xi^*(-\boldsymbol{q}, -\Omega)\hat{c}_1^\dagger(-\boldsymbol{q}, -\Omega) + \xi(\boldsymbol{q}, \Omega)\hat{c}_2(\boldsymbol{q}, \Omega), \tag{11.14}$$

where

$$\xi(\boldsymbol{q}, \Omega) = U(\boldsymbol{q}, \Omega) - V^*(-\boldsymbol{q}, -\Omega), \tag{11.15}$$

and the field operators

$$\hat{c}_{1,2}(\boldsymbol{q}, \Omega) = \frac{1}{\sqrt{2}}\big(\pm \hat{a}_1(\boldsymbol{q}, \Omega) + \hat{a}_2(\boldsymbol{q}, \Omega)\big), \tag{11.16}$$

with the $+(-)$ sign corresponding to 1(2), are unitary superpositions of two independent vacuum fields on the inputs of the OPAs. The fields $\hat{c}_{1,2}(\boldsymbol{q}, \Omega)$ are also in the vacuum state. An immediate consequence of Eqs. (11.82) and (11.14) is that the noise operators have the commutation relations of a classical field:

$$\begin{aligned}
[\hat{f}(\boldsymbol{q}, \Omega), \hat{f}^\dagger(\boldsymbol{q}', \Omega')] = (2\pi)^3\delta^2(\boldsymbol{q} - \boldsymbol{q}')\delta(\Omega - \Omega')\times \\
\big(|\xi(\boldsymbol{q}, \Omega)|^2 - |\xi(-\boldsymbol{q}, -\Omega)|^2\big) = 0, \\
[\hat{f}(\boldsymbol{q}, \Omega), \hat{f}(\boldsymbol{q}', \Omega')] = 0,
\end{aligned} \tag{11.17}$$

and thus can be considered as classical noise forces. Actually, as a consequence of Eq. (11.86), one has $|V(\boldsymbol{q}, \Omega)| = |V(-\boldsymbol{q}, -\Omega)|$, and the function $\xi(\boldsymbol{q}, \Omega)$ in (11.14) is found in the form

$$\xi(\boldsymbol{q}, \Omega) = e^{-i\phi(\boldsymbol{q},\Omega)}\left\{e^{-r(\boldsymbol{q},\Omega)}\cos\psi(\boldsymbol{q}, \Omega) + ie^{r(\boldsymbol{q},\Omega)}\sin\psi(\boldsymbol{q}, \Omega)\right\}. \tag{11.18}$$

In the spatiotemporal domain the noise field $\hat{F}(\boldsymbol{\rho}, t)$ is given by

$$\hat{F}(\boldsymbol{\rho}, t) = \frac{1}{(2\pi)^3} \int d\boldsymbol{\rho}_0 dt_0 \Big\{ \xi^*(\boldsymbol{\rho} - \boldsymbol{\rho}_0, t - t_0) \hat{C}_1^\dagger(\boldsymbol{\rho}_0, t_0)$$

$$+ \xi(\boldsymbol{\rho} - \boldsymbol{\rho}_0, t - t_0) \hat{C}_2(\boldsymbol{\rho}_0, t_0) \Big\}, \tag{11.19}$$

and the second-order correlation functions of the noise field are found in the form

$$\langle \hat{F}(\boldsymbol{\rho}, t) \hat{F}^\dagger(\boldsymbol{\rho}', t') \rangle = G(\boldsymbol{\rho} - \boldsymbol{\rho}', t - t'), \tag{11.20}$$

$$\langle \hat{F}(\boldsymbol{\rho}, t) \hat{F}(\boldsymbol{\rho}', t') \rangle = 0 \,. \tag{11.21}$$

The Fourier transform of the Greens function $G(\boldsymbol{\rho}, t)$ reads,

$$G(\boldsymbol{q}, \Omega) = |\xi(\boldsymbol{q}, \Omega)|^2 =$$

$$e^{-2r(\boldsymbol{q}, \Omega)} \cos^2 \psi(\boldsymbol{q}, \Omega) + e^{2r(\boldsymbol{q}, \Omega)} \sin^2 \psi(\boldsymbol{q}, \Omega). \tag{11.22}$$

The statistics of a light field are determined in the most general form by the characteristic functional, which is Gaussian [14,15] for the noise field (11.19). Incidentally, a similar Greens function describes the photocurrent correlations in space–time by the homodyne detection of multimode squeezed light [29].

When squeezing and entanglement are not present, $r(\boldsymbol{q}, \Omega) = 0$, the Green's function is δ-correlated in space–time,

$$G(\boldsymbol{\rho}, t) = \delta(\boldsymbol{\rho}) \delta(t). \tag{11.23}$$

In the presence of an effective entanglement with the scales S_c, T_c and by the optimum phase matching in the teleportation scheme, $\psi(0,0) = 0$, the positive δ-correlated term is accompanied by a negative term due to spatiotemporal anticorrelations on the scales S_c, T_c. This spatiotemporal anticorrelation of the noise field allows for suppression of the averaged noise field, as we shall discuss in detail below.

The spatial scales of the noise suppression are sensitive to the rotation of squeezing ellipses in dependence of the spatial frequency, shown in Fig. 11.2a. In the spatial frequency domain this is evident from the behavior of the Green's function (11.22), where the amplified quadrature amplitudes of noise are present with the weight $\propto \sin^2 \psi(\boldsymbol{q}, \Omega) \neq \sin^2 \psi(0,0) = 0$. The misalignment of squeezing ellipses increases with propagation along the crystal and in free space and leads to the diffraction spread of the coherence area. A properly inserted lens imaging system compensates [22, 23] this misalignment, as shown in Fig. 11.2b, and brings the size of the coherence area S_c to its optimum value.

In order to compensate diffraction, one can insert the lenses directly into the EPR beams $\hat{E}_n$, $n = 1, 2$. In the general case this effect is described by quadratic in q phase shifts $\theta_n(q) = \gamma_n q^2$:

$$\hat{E}_n(\boldsymbol{q},\Omega) \to \hat{\tilde{E}}_n(\boldsymbol{q},\Omega) = \hat{E}_n(\boldsymbol{q},\Omega)e^{i\theta_n(q)}. \tag{11.24}$$

By accounting for this phase correction in Eqs. (11.12) and (11.13) we find the corrected orientation angle $\psi(\boldsymbol{q},\Omega) \to \tilde{\psi}(\boldsymbol{q},\Omega) = \psi(\boldsymbol{q},\Omega) + \theta(q)$, where $\theta(q) = (\theta_1(q) + \theta_2(q))/2$. This angle should be substituted into the Green's function (11.22), $G(\boldsymbol{q},\Omega) \to \tilde{G}(\boldsymbol{q},\Omega)$. The best result that can be obtained with lenses is to set

$$\theta(q) = -[\mathrm{d}^2\psi/\mathrm{d}q^2]_{q=0}\, q^2, \tag{11.25}$$

as in Fig. 11.2b. With this choice $\tilde{\psi}(\boldsymbol{q},0) \approx 0$ and $\tilde{G}(q,0) \approx e^{-2r(q,0)}$ over a broad range of q. Physically speaking, lenses compensate the effect of diffraction on the spatial scale of entanglement.

The multimode teleportation will always be "on average" within some finite spatial area and some finite time interval. Therefore, for characterizing quantitatively the performance of teleportation we have to introduce a coarse-grained description of the input and output variables. We consider averaging of the field variables over a pixel S_j of area $S = \Delta^2$ and over a time window T_i of duration T:

$$\hat{A}_{\text{out}}(j,i) = \frac{1}{\sqrt{ST}} \int_{S_j} d\boldsymbol{\rho} \int_{T_i} dt\, \hat{A}_{\text{out}}(\boldsymbol{\rho},t), \tag{11.26}$$

with analogous definitions for the input field. The averaged field operators obey standard commutation relations

$$[\hat{A}_{\text{out}}(j,i), \hat{A}^{\dagger}_{\text{out}}(j',i')] = \delta_{j,j'}\delta_{i,i'}\ , \tag{11.27}$$

and hence correspond to a discrete subset of field oscillators.

Next, we consider generic field quadrature operators of the output and input field

$$\hat{X}^{\varphi}_{\text{out/in}}(j,i) = \hat{A}_{\text{out/in}}(j,i)e^{-i\varphi} + \hat{A}^{\dagger}_{\text{out/in}}(j,i)e^{i\varphi}, \tag{11.28}$$

$$\hat{Y}^{\varphi}_{\text{out/in}}(j,i) = -i\hat{A}_{\text{out/in}}(j,i)e^{-i\varphi} + i\hat{A}^{\dagger}_{\text{out/in}}(j,i)e^{i\varphi}\ . \tag{11.29}$$

These are observables that can be measured by means of homodyne detection with a high-efficiency CCD camera. By using Eq. (11.12) we obtain,

$$\hat{X}^{\varphi}_{\text{out}}(j,i) = \hat{X}^{\varphi}_{\text{in}}(j,i) + \hat{\mathcal{X}}_{\varphi}(j,i)\ , \tag{11.30}$$

$$\hat{Y}^{\varphi}_{\text{out}}(j,i) = \hat{Y}^{\varphi}_{\text{in}}(j,i) + \hat{\mathcal{Y}}_{\varphi}(j,i)\ , \tag{11.31}$$

where the excess noise added by the teleportation process on the measured field quadrature is given by

$$\hat{\mathcal{X}}_{\varphi}(j,i) = \frac{1}{\sqrt{ST}} \int_{S_j} d\boldsymbol{\rho} \int_{T_i} dt \left[\hat{F}(\boldsymbol{\rho},t)e^{-i\varphi} + \hat{F}^{\dagger}(\boldsymbol{\rho},t)e^{i\varphi}\right]\ , \tag{11.32}$$

$$\hat{\mathcal{Y}}_{\varphi}(j,i) = \frac{1}{\sqrt{ST}} \int_{S_j} d\boldsymbol{\rho} \int_{T_i} dt \left\{-i\left[\hat{F}(\boldsymbol{\rho},t)e^{-i\varphi} - \hat{F}^{\dagger}(\boldsymbol{\rho},t)e^{i\varphi}\right]\right\}\ . \tag{11.33}$$

These operators are a linear combination of Gaussian stochastic variables, independent of the input. Hence the set

$$\left\{\hat{\mathcal{X}}_\varphi(j,i)\,,\,\hat{\mathcal{Y}}_\varphi(j,i)\right\}_{j,i}\,, \tag{11.34}$$

represents a set of classical Gaussian stochastic variables. Here $j = 1,\ldots,N$, $i = 1,\ldots,K$, is a finite set of indices labeling the pixels and the time intervals of interest. These variables are independent of the input field and have zero mean values. Their statistical properties are completely described in terms of a covariance matrix

$$\begin{aligned}\mathcal{C}\,(j,j'\,;\,i,i') &= \langle\hat{\mathcal{X}}_\varphi(j,i)\,\hat{\mathcal{X}}_\varphi(j',i')\rangle \\ &= \langle\hat{\mathcal{Y}}_\varphi(j,i)\,\hat{\mathcal{Y}}_\varphi(j',i')\rangle\,.\end{aligned} \tag{11.35}$$

The covariance matrix elements can be expressed in terms of the Green's function (11.22) as

$$\mathcal{C}\,(j,j'\,;\,i,i') = 2\int \mathrm{d}\boldsymbol{q}\,B_\Delta(\boldsymbol{q})B_T(\Omega)\cos[\boldsymbol{q}\cdot(\boldsymbol{\rho}_j-\boldsymbol{\rho}_{j'})-\Omega(t_i-t_{i'})]\tilde{G}(\boldsymbol{q},\Omega), \tag{11.36}$$

where $\boldsymbol{\rho}_j$ is the center of the pixel j, and t_i is the center of the ith time interval. Here $\tilde{G}(\boldsymbol{q},\Omega)$ is the Green's function (11.22) with the corrected value of the orientation angle $\tilde{\psi}(\boldsymbol{q},\Omega) = \psi(\boldsymbol{q},\Omega) + \theta(\boldsymbol{q})$; see (11.25). In Eq. (11.36) functions $B_\Delta(\boldsymbol{q})$ $B_T(\Omega)$ arise from the coarse-graining operation. For example, a square pixel of side $\Delta = \sqrt{S}$ they read,

$$B_\Delta(\boldsymbol{q}) = \frac{\Delta^2}{4\pi^2}\,\mathrm{sinc}^2\!\left(\frac{q_x\Delta}{2}\right)\mathrm{sinc}^2\left(\frac{q_y\Delta}{2}\right) \stackrel{\Delta\to\infty}{\longrightarrow} \delta(\boldsymbol{q}), \tag{11.37}$$

$$B_T(\Omega) = \frac{T}{2\pi}\,\mathrm{sinc}^2\!\left(\frac{\Omega T}{2}\right) \stackrel{T\to\infty}{\longrightarrow} \delta(\Omega). \tag{11.38}$$

The covariance matrix (11.36) does not depend on the phase φ of the local oscillator, used for the homodyne detection, so that the added noise is the same for any quadrature component.

Ideal teleportation takes place when the Gaussian distribution of noise has a vanishing small width in all directions of phase space, so that it approximates a multivariate Dirac δ-function. This can be realized if both the time window T and the pixel sizeΔ are large enough, so that the functions (11.37) and (11.38) in the integral of Eq. (11.36) filter a band of temporal and spatial frequencies well inside the squeezing bandwidths, where $G(\tilde{\boldsymbol{q}},\Omega) \ll 1$. When the pixel size Δ and the time window T are much larger than the OPA coherence length l_c, and the OPA coherence time T_c, respectively, we obtain

$$\lim_{\Delta\to\infty\,,T\to\infty} \mathcal{C}\,(j,j'\,;\,i,i') = 2\delta_{j,j'}\delta_{i,i'}\exp(-2r(0,0))\,. \tag{11.39}$$

In addition, as in the single-mode case, a large degree of EPR correlation (large squeezing parameter r) is required in order to achieve a good quality teleportation.

For a broadband OPA, where the parametric crystal length is on the order of millimeters, the coherence time is of the order of femtoseconds to picoseconds, so that the usual detection time windows overcome by several order of magnitudes the coherence time. Hence, in Eq. (11.36) it is reasonable to assume the limit $T \gg T_c$, under which the noise added by the teleportation scheme is uncorrelated in time,

$$\mathcal{C}\,(j, j'\,;\, i, i') = \langle \hat{\mathcal{Y}}_\varphi(j, i)\, \hat{\mathcal{Y}}_\varphi(j', i')\rangle = \delta_{i,i'} \mathcal{C}(j, j')\,. \tag{11.40}$$

However, the same is not true for the spatial domain. For example, for an OPA with a 3 mm long crystal, at $\lambda = 0.712\,\mu\text{m}$, taking as a rough estimate for the coherence length l_c the diffraction spread l_d at the crystal exit, we arrive at

$$l_c \sim l_d = \sqrt{l/2k} = 13\,\mu\text{m}\,. \tag{11.41}$$

Provided that the spatial extent in the transverse plane where EPR correlations do exist is limited by the pump spot size, choosing $\Delta \gg l_c$ would amount to integrating over the whole beam cross-section and losing all spatial information.

In Fig. 11.6 we illustrate the role of the pixel size for the noise added by the teleportation process. Precisely, we plot our numerical calculations for some elements of the covariance matrix in the limit $T \gg T_c$ as a function of the ratio of the pixel size to the diffraction length $D = \Delta/l_d$, where l_d is defined by Eq. (11.41). As our system has an overall translational symmetry, the covariance matrix elements $\mathcal{C}(j, j')$ depend only on the relative distance and on the orientation of pixels with respect to the difference position vector $(\boldsymbol{\rho}_j - \boldsymbol{\rho}_{j'})$. In particular, all the diagonal elements $\mathcal{C}(j, j)$ have the same value.

Figure 11.6a shows the diagonal element $\mathcal{C}(j, j)$ of the covariance matrix. The wide and narrow solid lines correspond to the observation with (bold lines) and without (thin lines) diffraction phase-shift compensation with the use of a lens arrangement. In both cases the plot for the diagonal element $\mathcal{C}(j, j)$ shows the classical limit $\mathcal{C}(j, j) \to 2$ for small pixel size, when the contribution of the high-frequency Fourier components of the noise field, $q \gg q_c$, remaining in the vacuum state, dominates. On the other side, when the pixel size is of the same order of magnitude as the coherence length, both the wide and narrow lines rapidly approach the $\Delta \to \infty$ limit of Eq. (11.39), $\mathcal{C}_(j, j) \to 2\exp[-2r(0, 0)]$ (dashed line), which corresponds to the single-mode quantum teleportation limit. This behavior should be compared with the covariance matrix of the noise in a classical teleportation scheme, that is, in the absence of the EPR correlations. Then $r(\boldsymbol{q}, \Omega) = 0$, $\mathcal{C}_{\text{cl}}(j, j') = 2\delta_{j,j'}$, and $\mathcal{C}_{\text{cl}}(j, j) = 2$. In this limit two units of vacuum noise are added at each pixel, just as in the case of the single-mode teleportation [8]. Figure 11.6b shows some off-diagonal elements of the covariance matrix as a function of

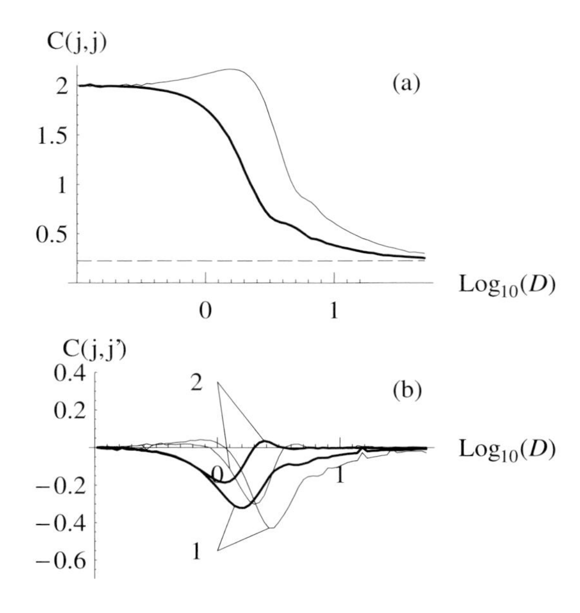

Fig. 11.6. Covariance matrix elements of the noise added by the teleportation scheme as a function of $D = \Delta/l_d$. The squeezing parameter is $\exp[r(0,0)] = 3$. Figure 11.6a shows the diagonal element $\mathcal{C}(j,j)$ of the covariance matrix for observation without (thin line) and with (bold line) a correcting lens. Figure 11.6b shows off-diagonal elements $\mathcal{C}(j,j')$ with $|\boldsymbol{\rho}_j - \boldsymbol{\rho}_{j'}| = \Delta$ (lines 1), and with $|\boldsymbol{\rho}_j - \boldsymbol{\rho}_{j'}| = \sqrt{2}\Delta$ (lines 2). Bold and thin lines have the same meaning as in Fig. 11.6a.

Δ/l_d. They show correlations between the nearest-neighbor pixels in a row or a column, $|\boldsymbol{\rho}_j - \boldsymbol{\rho}_{j'}| = \Delta$ (lines 1), and between the pixels on a diagonal $|\boldsymbol{\rho}_j - \boldsymbol{\rho}_{j'}| = \sqrt{2}\Delta$ (lines 2), of the detector matrix. As can be seen from these plots, when the pixel size is small compared to the coherence length, our teleportation scheme not only adds noise on each pixel (as in the single-mode scheme) but also introduces correlations between pixels. The existence of spatial correlations over distances on the order of the coherence length is typical of the multimode squeezed light, and has been investigated in detail, for example, in [29, 30].

11.3.4 Global and Reduced Fidelity of Holographic Teleportation

Using Eq. (11.30) we can obtain the explicit relation between the spatio-temporal correlation functions of the field quadratures in the output and the input,

$$\langle \delta\hat{X}^{\varphi}_{\rm out}(j,i)\,\delta\hat{X}^{\varphi}_{\rm out}(j',i')\rangle = \langle \delta\hat{X}^{\varphi}_{\rm in}(j,i)\,\delta\hat{X}^{\varphi}_{\rm in}(j',i')\rangle + \mathcal{C}(j,j';i,i'). \tag{11.42}$$

We see that because all elements $\mathcal{C}(j,j')$ are small provided $\Delta \geq l_c$, the teleportation preserves spatiotemporal pixel correlations. The same holds also for the higher-order correlation functions because the added noise is Gaussian and independent of the input.

This conclusion is based on the added noise power. Some other criteria for the quantum teleportation were suggested in the literature; see, for example, [6].

The quality of reconstruction of the quantum state $|\Psi_{\rm in}\rangle$ of the input field in the teleportation process is usually quantified via the fidelity parameter F. For simplicity we consider here the fidelity of teleportation of a pure quantum state, which is defined as

$$F = |\ \langle \Psi_{\rm in} | \Psi_{\rm out} \rangle\ |^2\ . \tag{11.43}$$

This definition works well for teleportation of a single degree of freedom of the quantized field. But in our case of quantum teleportation of a multimode light field, distributed in space and time, the definition (11.43) meets some obvious difficulties that stem from the multimode nature of the field. Let us, for instance, assume that the input state to be teleported has no correlations in space and time (as, e.g., for a coherent image), and that our teleportation protocol does not add any correlations between spatial and/or temporal modes (as it happens for a large enough detection time window and pixel size). In this case the global fidelity of the teleportation protocol factors in the product of the single-mode fidelities. Even if each mode is teleported with almost perfect fidelity, close to, but slightly less than, unity, the global fidelity in the limit of a large number of modes will be close to zero, and will always reduce to zero for an infinite number of modes. For this reason, in the case of quantum teleportation of a multimode field it is important to identify the relevant set of degrees of freedom and to introduce the notion of the reduced fidelity for this set of degrees of freedom, and of the average fidelity per mode.

An alternative definition of fidelity can be given in terms of the superposition of the Wigner functions describing the state in the input and in the output (see, e.g., [5]). A discussion of the teleportation fidelity for multivariate quantum objects (such as quantum images) can be found in [15]. One can explicitly calculate the fidelity for Gaussian input states (e.g., squeezed states, coherent states, EPR beams), for which one can assume without loss of generality that

$$\langle \hat{X}_{\rm in}(j,i) \rangle = \langle \hat{Y}_{\rm in}(j,i) \rangle = 0\ , \tag{11.44}$$

and

$$\begin{aligned} \langle \hat{X}_{\rm in}(j,i)\, \hat{X}_{\rm in}((j',i') &= V^X(j,j'\,;\, i,i') \rangle\ , \\ \langle \hat{Y}_{\rm in}(j,i)\, \hat{Y}_{\rm in}((j',i') &= V^Y(j,j'\,;\, i,i') \rangle\ , \end{aligned} \tag{11.45}$$

are the input covariance matrices. The fidelity is found in the form

$$F = \frac{1}{\det\left[V^X(j,j';i,i') + \frac{1}{2}\mathcal{C}(j,j';i,i')\right]^{1/2}} \times \frac{1}{\det\left[V^Y(j,j';i,i') + \frac{1}{2}\mathcal{C}(j,j';i,i')\right]^{1/2}}. \quad (11.46)$$

In particular for an input multimode coherent state

$$V^X(j,j';i,i') = V^Y(j,j';i,i') = \delta_{j,j'}\delta_{i,i'},$$

and

$$F = \frac{1}{\det\left[\delta_{j,j'}\delta_{i,i'} + \frac{1}{2}\mathcal{C}(j,j';i,i')\right]}. \quad (11.47)$$

These results have to be compared with the results of a classical teleportation protocol, that is, in the absence of EPR correlations, where $\mathcal{C}_{cl}(j,j';i,i') = 2\delta_{j,j'}\delta_{i,i'}$. For a coherent input, in the classical case we have

$$F_{cl} = \frac{1}{2^{N\times K}}. \quad (11.48)$$

From the above formulas the claim that we made at the beginning of this section should be clear; that is, the global fidelity may rapidly approach zero for a large number of degrees of freedom, and hence lose any quantitative meaning. A good strategy is to identify the relevant degrees of freedom of the system. If, for instance, the state to be teleported is the quantum state of the coherent image, restricted to an array of N_A pixels, no one will probably be interested in the quality of teleportation of the vacuum state of the region of space outside the image.

Because we assumed a plane-wave pump, our model is translationally space and time invariant, and the number of available pixels and time intervals is in principle infinite. This kind of model describes a realistic system well, provided that the pump spot size is much larger than the amplifier coherence area and that the pump pulse duration is much longer than the amplifier coherence time [25]. Obviously, one has also to require that the beam whose state is to be teleported is well confined inside the region where significant gain is available, both in space and time.

Let us assume we divide our system in two subsystems, say A and B, where subsystem A corresponds to a subset $\{j,i\}_A$ of pixels and of time intervals of interest for a given measurement, and subsystem B consists of the remaining $\{j,i\}_B$ pixels and time intervals. By tracing out over the degrees of freedom of subsystem B, one can demonstrate [15] that the formulas (11.46) and (11.47) hold true for the fidelity F_A of the reduced set of degrees of freedom, provided that the covariance matrices in those formulas are the covariance matrices of the second-order moments of operators on the pixels and time intervals of subsystem A.

In Fig. 11.7 we show the reduced fidelity dependence on the pixel size and on the number of pixels for some simple patterns of pixels, in the case of a coherent image in the input. The observation is assumed to be performed within the same time window, so that the dimension of the covariance matrix in our case is given by the number of pixels in the pattern. This is very reasonable for a traveling-wave OPA, because in a realistic configuration in order to obtain large gain a pulsed operation is required. Therefore, the detection time window will be probably longer or of the same order as the duration of the pump pulse. Notice that the same is not necessarily true for a cavity configuration, because the CW operation in this case permits us to resolve the temporal degrees of freedom.

The shape of the patterns for 1, 2, and 4 pixels is shown at the top of Fig. 11.7. We plot our numerical calculations of F_A given by Eq. (11.46) for the degree of squeezing $\exp[r(0,0)] = 3$, as a function of the ratio of the pixel size to the diffraction length $D = \Delta/l_d$. For these patterns only the diagonal covariance matrix elements $\mathcal{C}(j,j)$, and those describing nearest-neighbor correlations in a row, in a column, and in a diagonal are of importance. The wide and narrow solid lines correspond to the observation with (wide lines) and without (narrow lines) diffraction phase-shift compensation with the use of a lens arrangement. For small pixel size, $D \ll 1$, all curves attain the classical limit $F_N \to 0.5^N$. For large pixel size, $D \gg 1$, the curves tend to the limit, imposed by the degree of squeezing: $F_N \to (1+\exp[-2r(0,0)])^{-N} = 0.9^N$ (in

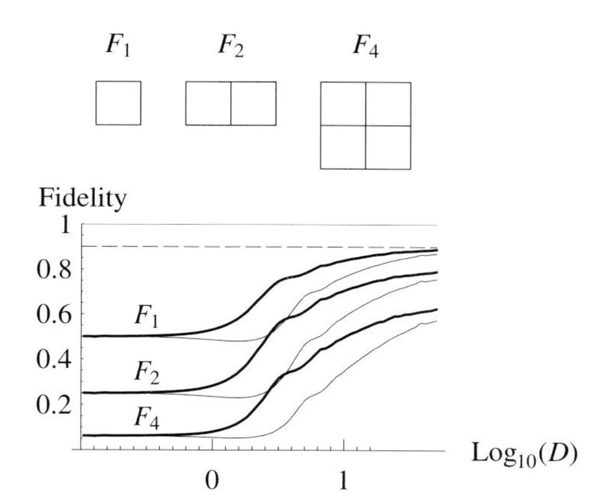

Fig. 11.7. Reduced fidelity of quantum holographic teleportation for patterns of 1, 2, and 4 pixels, shown at the top of the figure (plots F_1, F_2, and F_4, respectively), as a function of $D = \Delta/l_d$. The squeezing parameter is $\exp[r(0,0)] = 3$.

our example). As seen from Fig. 11.7, the fidelity of teleportation decreases with the number of pixels. This qualitatively means that the multipixel observables, dependent on the correlations between pixels, are more sensitive to the absence of entanglement in the quantum channel than the observables for a single pixel.

The effect of optimization of the spatial scales in our teleportation protocol with the use of a lens arrangement is significant for a pixel width $\Delta \approx l_c$. As shown in Fig. 11.7, the optimization allows us to achieve the same value of fidelity for the pixel size, smaller by factor $(2 \ldots 3)$ than in the absence of such optimization.

As is clear from inspection of the curves in Fig. 11.7, the fidelity for N degrees of freedom scales with the N-th power. A useful quantity for the estimation of reduced fidelity F_N for a large array of pixels, $N \gg 1$, is hence the average fidelity per pixel, defined as

$$F_{\mathrm{av}} = (F_N)^{1/N} . \tag{11.49}$$

Consider a single temporal degree of freedom $K = 1$. For a large $N = M \times M$ array of square pixels, the covariance matrix becomes translationally invariant, and it can be diagonalized [15] by means of a discrete Fourier transformation. Figure 11.8 shows the behavior of the average fidelity per pixel for quantum teleportation of a large array of pixels in a coherent state. The

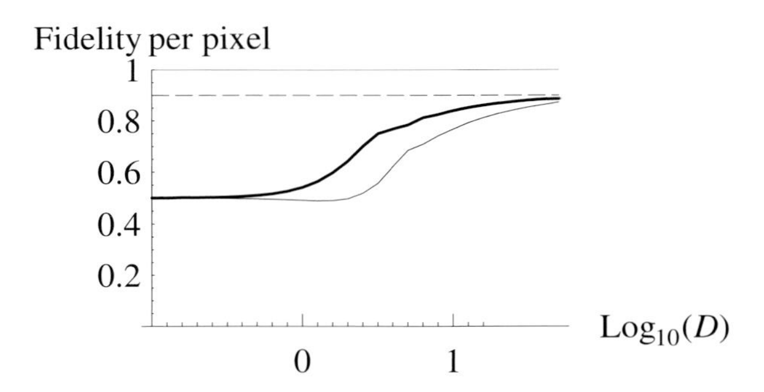

Fig. 11.8. Average fidelity per pixel for the quantum holographic teleportation of a large number of pixels in a coherent state as a function of $D = \Delta/l_d$. The squeezing parameter is $\exp[r(0,0)] = 3$.

reduced fidelity F_1 for one pixel (Fig. 11.7) and the average fidelity per pixel F_{AV} (Fig. 11.8) tend to the same limits for large and small pixel size and are slightly different for medium pixel size $\Delta \propto l_d$, thus reflecting an effect of cross–correlation of the noise field at closely located pixels on F_{AV}.

11.3.5 Quantum Holographic Teleportation and Holography

Quantum teleportation of optical images described in this chapter can be called *quantum holographic teleportation.* In this three-dimensional generalization of the continuous variable teleportation protocol [4, 8] one can recognize an extension to the quantum domain of conventional nonstationary holography.

As in holography, the distributed-in-space–time input field is mixed with local oscillator waves. Classical photocurrent densities $I_x(\boldsymbol{\rho}, t)$ and $I_y(\boldsymbol{\rho}, t)$ are equivalent to nonstationary holograms.

In elementary schemes of holography it is common to perform optical mixing of a signal wave with only one reference wave. One creates and uses a single hologram for field reconstruction. In the general case this results in a loss of information carried by the field quadrature, orthogonal to the reference wave, and in the reconstruction of the signal wave in superposition to its conjugate duplicate. In order to improve this elementary scheme, one could split the input wave by means of a symmetrical beamsplitter and record both quadrature components of the signal field. Both quadrature components of the reconstructed wave could be produced by superposition of outputs of two holograms, illuminated by two plane laser waves shifted by $\lambda/4$. This improved configuration is nothing but the scheme illustrated in Fig. 11.5, but without illumination by two EPR fields.

Although in the classical approach the right input port of beamsplitter BS_2 and the lower input port of M are idle (not illuminated), in the quantum description one should take into account vacuum noise fields incident on these beamsplitters.

The novel feature, which converts holography to the quantum holographic teleportation is a pair of multimode EPR beams shared by Alice and Bob. As discussed in this chapter, only perfect point–by–point and instant–by–instant correlation at the quantum level of these twin fields allows for effective quantum noise suppression in the teleported field. In practical implementation one should provide a strict matching of EPR fields in space and time.

The above-discussed limits on quantum noise and fidelity of quantum holographic teleportation of optical images can be viewed as an improvement of quantum limits on holographic reconstruction of the distributed-in-space light waves, achieved in the presence of Q *quantum channel:* a pair of entangled (correlated at quantum level) spatially multimode light beams.

In the framework of images, the analogy between teleportation and holography mentioned [31] in the context of single-mode fields, becomes complete and precise.

11.4 Quantum Dense Coding of Optical Images

In Sections 11.4.1 and 11.4.2 of this chapter we review the basics of quantum dense coding and present a model of quantum dense coding for optical images, where we assume arbitrarily large transverse dimensions of propagating light beams and the unlimited spatial resolution of a photodetection scheme.

We introduce Shannon mutual information for a stream of classical input images in a coherent state (see Section 11.4.3). An important quantity is the spatiotemporal density of the information stream in bits per $\mathrm{cm}^2 \cdot$ s. This density depends on the degree of squeezing and entanglement in nonclassical illuminating light. Two sets of spatiotemporal parameters play an important role in our protocol: (i) the transverse coherence length and the coherence time of spatially multimode squeezing and entanglement, and (ii) the spatiotemporal parameters of a stream of input images. In our consideration we assume that the sender (Alice) produces a uniform ensemble of images with the Gaussian statistics, characterized by a certain resolution in space and time (the Alice grain).

The information capacity of the dense coding scheme for optical images is discussed in Section 11.4.4. In the case of optical images, one of the bounds on the density of the information stream is imposed by diffraction. We demonstrate how the effect of diffraction can be partially compensated by lenses properly inserted in the scheme. It is shown that the information capacity per properly introduced field degree of freedom is in agreement with general estimates for quantum dense coding.

An important difference between the ideal classical communication channel (i.e., with vacuum fluctuations at the input of the scheme instead of spatially multimode entangled light) and its quantum counterpart is that in the quantum case there exists an optimum spatial density of the signal image elements, which should be matched with the spatial frequency band of entanglement.

11.4.1 Basics of Quantum Dense Coding

In Fig. 11.9 we illustrate two possible schemes of a communication line between Alice (the sender) and Bob (the receiver). Both schemes are based on two parallel channels, but in the conventional coding scheme (a) the channels are independent, and, by contrast, the dense coding scheme (b) essentially exploits quantum entangled states of two objects 1 and 2 propagating in the channels.

In the case of conventional coding, the input quantum state of two channels is a product over channels, $|\psi^{(\mathrm{in})}\rangle_{1,2} = |\psi^{(\mathrm{in})}\rangle_1 |\psi^{(\mathrm{in})}\rangle_2$. Alice independently prepares the objects 1 and 2 in any of N orthogonal quantum states. Bob detects the resulting state $|\psi_n\rangle_1 |\psi_m\rangle_2$, $n, m = 1, \ldots, N$, by means of independent measurement in both channels. Evidently, this allows sending a letter of the alphabet, which is composed of N^2 letters, per cycle.

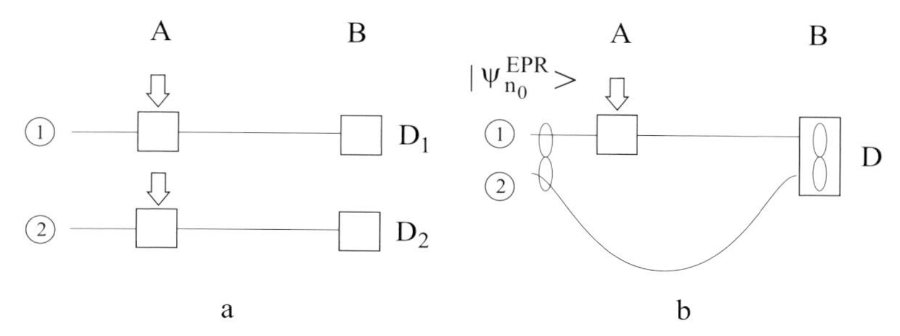

Fig. 11.9. Conventional and dense coding.

In the dense coding scheme, the input state of two channels is one of quantum entangled states $|\psi^{(\mathrm{in})}\rangle_{1,2} = |\psi^{EPR}_{n_0}\rangle_{1,2}$. A complete set of orthogonal EPR states for two channels is composed of N^2 states $|\psi^{EPR}_{n}\rangle_{1,2}$, $n = 1, \ldots, N^2$. An important feature of the EPR basis is that Alice can prepare any of N^2 states by physically operating only in one of two channels, leaving another one untouched. In the dense coding scheme Bob detects the EPR quantum state of the received signal by means of an EPR detector. This provides equal capacity of two schemes: N^2 letters per cycle, without any physical action in the idle channel 2 in the case of dense coding.

Physically speaking, in the presence of entanglement an idle channel provides an optimal reference system for detecting minor physical actions performed by Alice in channel 1.

11.4.2 Optical Scheme for Quantum Dense Coding of Images

The optical scheme implementing the continuous-variable dense coding protocol for optical images is shown in Fig. 11.10. Compared to the generic continuous-variable dense coding scheme [11], here the light fields are assumed to be spatially multimode. At the input, the spatially multimode squeezed light beams with the slow field amplitudes $\hat{S}_1(\boldsymbol{\rho}, t)$ and $\hat{S}_2(\boldsymbol{\rho}, t)$ (we use the Heisenberg representation), are mixed at the symmetrical beamsplitter BS_1. For a properly chosen orientation of the squeezing ellipses of the input fields the scattered fields $\hat{E}_1(\boldsymbol{\rho}, t)$ and $\hat{E}_2(\boldsymbol{\rho}, t)$ are in the entangled state with correlated field quadrature components, as illustrated in Figs. 11.3 and 11.10.

The classical signal image field $\hat{A}(\boldsymbol{\rho}, t)$ is created by Alice in the first beam by means, for example, of the controlled (with a given resolution in spacetime) mixing device Mod with almost perfect transmission for the nonclassical field $\hat{E}_1(\boldsymbol{\rho}, t)$. The receiver (Bob) detects the entangled state of two beams by means of an optical mixing at the symmetrical output beamsplitter BS_2 and the homodyne detection of quadrature components of the output fields $\hat{B}_1(\boldsymbol{\rho}, t)$ and $\hat{B}_2(\boldsymbol{\rho}, t)$. This allows for the measurement of both quadrature components of the image field with effective quantum noise reduction.

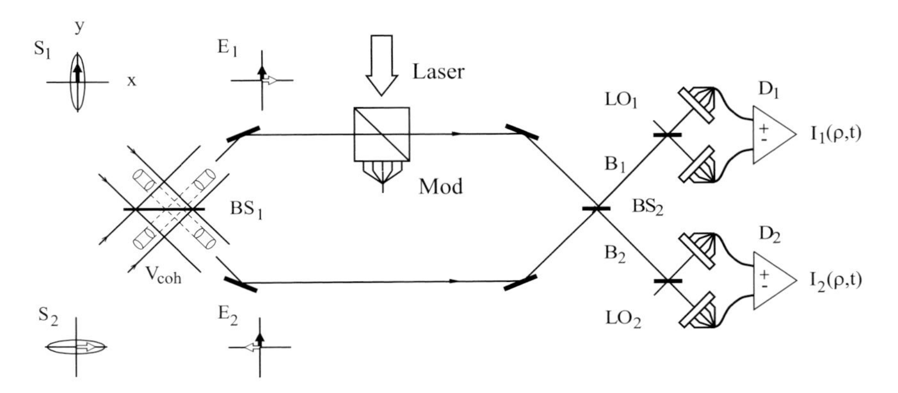

Fig. 11.10. Optical scheme for spatially multimode dense coding.

One can give a more straightforward explanation of the sub-shot-noise detection of the signal in the scheme shown in Fig. 11.10. For the symmetrical scattering matrix of the beamsplitters

$$\{R_{nm}\} = \frac{1}{\sqrt{2}}\begin{pmatrix} 1 & 1 \\ 1 & -1 \end{pmatrix}, \tag{11.50}$$

and equal optical paths of two beams, the effective Mach–Zehnder interferometer directs the input squeezed field $\hat{S}_1(\boldsymbol{\rho}, t)$ onto the detector D_1, and similarly for $\hat{S}_2(\boldsymbol{\rho}, t)$, thus allowing for sub-shot-noise detection of the squeezed quadrature components in both beams.

The fields at the inputs of the homodyne detectors D_1 and D_2 are

$$\hat{B}_n(\boldsymbol{\rho}, t) = \hat{S}_n(\boldsymbol{\rho}, t) + \frac{1}{\sqrt{2}}\hat{A}(\boldsymbol{\rho}, t), \tag{11.51}$$

where $n = 1, 2$. In the paraxial approximation, the slow amplitude of light field $\hat{B}_n(\boldsymbol{\rho}, t)$ is related to the creation and annihilation operators $\hat{b}_n^\dagger(\boldsymbol{q}, \Omega)$ and $\hat{b}_n(\boldsymbol{q}, \Omega)$ for the plane waves with the transverse component of the wave vector $\boldsymbol{q}$ and frequency Ω by

$$\hat{B}_n(\boldsymbol{\rho}, t) = \frac{1}{\sqrt{L^2 T}} \sum_{\boldsymbol{q}, \Omega} \hat{b}_n(\boldsymbol{q}, \Omega) e^{i(\boldsymbol{q}\cdot\boldsymbol{\rho} - \Omega t)}. \tag{11.52}$$

In this section we shall characterize the information capacity of our scheme by means of the density of the information stream in bits per cm$^2\cdot$ s. In order to introduce this relative quantity, it is convenient to consider a large quantization volume with the transverse and longitudinal dimensions L and cT. The summation is performed over the following values of $\boldsymbol{q}$ and Ω: $\boldsymbol{q} = (q_x, q_y)$, $q_x = 2\pi/L n_x, q_y = 2\pi/L n_y$ and $\Omega = 2\pi/T n$ with n_x, n_y and n taking the values $0, \pm 1, \pm 2, \ldots$.

The free-field commutation relations are given by

$$\begin{aligned}\left[\hat{B}_n(\boldsymbol{\rho},t),\hat{B}^{\dagger}_{n'}(\boldsymbol{\rho}\,',t')\right] &= \delta_{n,n'}\delta(\boldsymbol{\rho}-\boldsymbol{\rho}\,')\,\delta(t-t'),\\ \left[\hat{b}_n(\boldsymbol{q},\Omega),\hat{b}^{\dagger}_{n'}(\boldsymbol{q}\,',\Omega')\right] &= \delta_{n,n'}\,\delta_{\boldsymbol{q},\boldsymbol{q}\,'}\,\delta_{\Omega,\Omega'}.\end{aligned}\tag{11.53}$$

The value of the irradiance (in photons per $\mathrm{cm}^2\cdot\mathrm{s}$) is equal to $\hat{B}^{\dagger}_n(\boldsymbol{\rho},t)\hat{B}_n(\boldsymbol{\rho},t)$, and the number of photons in the field mode $(\boldsymbol{q},\Omega)$, localized in the quantization volume L^2cT, is $\hat{b}^{\dagger}_n(\boldsymbol{q},\Omega)\hat{b}_n(\boldsymbol{q},\Omega)$. The observed photocurrent densities are considered in this section as continuous in space and time variables. That is, we assume an arbitrarily high resolving power of the detectors and do not examine the effect of finite pixel size of the CCD matrices on the information capacity. The photocurrent densities

$$\begin{aligned}\hat{I}_1(\boldsymbol{\rho},t) &= B_0\left[\hat{B}_1(\boldsymbol{\rho},t)+\hat{B}^{\dagger}_1(\boldsymbol{\rho},t)\right],\\ \hat{I}_2(\boldsymbol{\rho},t) &= B_0\frac{1}{i}\left[\hat{B}_2(\boldsymbol{\rho},t)-\hat{B}^{\dagger}_2(\boldsymbol{\rho},t)\right],\end{aligned}\tag{11.54}$$

have the following Fourier amplitudes,

$$\begin{aligned}\hat{\imath}_1(\boldsymbol{q},\Omega) &= B_0\left[\hat{b}_1(\boldsymbol{q},\Omega)+\hat{b}^{\dagger}_1(-\boldsymbol{q},-\Omega)\right],\\ \hat{\imath}_2(\boldsymbol{q},\Omega) &= B_0\frac{1}{i}\left[\hat{b}_2(\boldsymbol{q},\Omega)-\hat{b}^{\dagger}_2(-\boldsymbol{q},-\Omega)\right],\end{aligned}\tag{11.55}$$

where B_0 (taken as real) and iB_0 are the local oscillator amplitudes used in homodyne detection (see the discussion in Appendix B). Here and in what follows we denote the Fourier amplitudes of the fields and the photocurrent densities by lower-case symbols.

The squeezing transformation performed by the optical parametric amplifiers illuminating the inputs of the scheme, can be written (see Section 11.2 and Appendix A for details) as follows,

$$\hat{s}_n(\boldsymbol{q},\Omega) = U_n(\boldsymbol{q},\Omega)\hat{c}_n(\boldsymbol{q},\Omega)+V_n(\boldsymbol{q},\Omega)\hat{c}^{\dagger}_n(-\boldsymbol{q},-\Omega),\tag{11.56}$$

where the coefficients $U_n(\boldsymbol{q},\Omega)$ and $V_n(\boldsymbol{q},\Omega)$ depend on the pump-field amplitudes of the OPAs, their nonlinear susceptibilities, and the phase matching conditions. The input fields $\hat{c}_n(\boldsymbol{q},\Omega)$ of the OPAs are assumed to be in the vacuum state.

After some calculation we obtain for the Fourier amplitudes of the photocurrent densities:

$$\hat{\imath}_n(\boldsymbol{q},\Omega) = B_0\left\{\hat{f}_n(\boldsymbol{q},\Omega)+a_n(\boldsymbol{q},\Omega)\right\},\tag{11.57}$$

where

$$\hat{f}_1(\boldsymbol{q},\Omega) = \left[e^{r_1(\boldsymbol{q},\Omega)}\cos\psi_1(\boldsymbol{q},\Omega)+ie^{-r_1(\boldsymbol{q},\Omega)}\sin\psi_1(\boldsymbol{q},\Omega)\right]e^{-i\phi_1(\boldsymbol{q},\Omega)}\hat{c}_1(\boldsymbol{q},\Omega)$$

$$+ \left[h.c., (\boldsymbol{q}, \Omega) \to (-\boldsymbol{q}, -\Omega)\right], \tag{11.58}$$

and

$$\hat{f}_2(\boldsymbol{q}, \Omega) = \left[e^{-r_2(\boldsymbol{q},\Omega)} \cos \psi_2(\boldsymbol{q}, \Omega) + i e^{r_2(\boldsymbol{q},\Omega)} \sin \psi_2(\boldsymbol{q}, \Omega)\right] e^{-i\phi_2(\boldsymbol{q},\Omega)} \hat{c}_2(\boldsymbol{q}, \Omega) \\ + \left[h.c., (\boldsymbol{q}, \Omega) \to (-\boldsymbol{q}, -\Omega)\right], \tag{11.59}$$

represent the quantum fluctuations of the fields at both photodetectors, and

$$\begin{aligned} a_1(\boldsymbol{q}, \Omega) &= \frac{1}{\sqrt{2}} \left[a(\boldsymbol{q}, \Omega) + a^*(-\boldsymbol{q}, -\Omega)\right], \\ a_2(\boldsymbol{q}, \Omega) &= \frac{1}{i\sqrt{2}} \left[a(\boldsymbol{q}, \Omega) - a^*(-\boldsymbol{q}, -\Omega)\right], \end{aligned} \tag{11.60}$$

are the components detected by Bob of the Alice signal image. Here $a(\boldsymbol{q}, \Omega)$ are the Fourier amplitudes of classical field $A(\boldsymbol{\rho}, t)$, defined in analogy to Eq. (11.52).

11.4.3 Shannon Mutual Information for Images

In order to estimate the channel capacity one has to define the degrees of freedom of the noise and the signal in our spatially multimode scheme.

We shall assume that all elements of the scheme: the OPA's nonlinear crystals, beamsplitters, modulator, and the CCD matrices of detectors, have large transverse dimensions. The squeezed light fields are the stationary in time and uniform in the cross-section of the beams' random variables. That is, all correlation functions of these fields are translationally invariant in the $\boldsymbol{\rho}, t$ space. For the observed photocurrent densities this implies that any pair of the Fourier noise amplitudes (11.58) and (11.59) for given $(\boldsymbol{q}, \Omega)$ and $(-\boldsymbol{q}, -\Omega)$ results from squeezing of the input fields $c(\boldsymbol{q}, \Omega)$ and $c(-\boldsymbol{q}, -\Omega)$ and therefore is independent of any other pair.

On the other hand, because the observed photocurrent densities are real, the Fourier amplitudes $\hat{i}_n(\boldsymbol{q}, \Omega)$ and $\hat{i}_n^\dagger(-\boldsymbol{q}, -\Omega)$ are not independent, where

$$\hat{i}_n(\boldsymbol{q}, \Omega) = \hat{i}_n^\dagger(-\boldsymbol{q}, -\Omega). \tag{11.61}$$

For this reason we consider as independent random variables only the noise terms in Fourier amplitudes $i_n(\boldsymbol{q}, \Omega)$ for $\Omega > 0$. The real and imaginary parts of the complex amplitudes $i_n(\boldsymbol{q}, \Omega)$ for $\Omega > 0$ are related to the amplitudes of the real photocurrent noise harmonics $\sim\cos(\boldsymbol{q}\cdot\boldsymbol{\rho} - \Omega t)$ and $\sim\sin(\boldsymbol{q}\cdot\boldsymbol{\rho} - \Omega t)$, directly recovered by Bob from his measurements.

The Fourier amplitudes of the photocurrent densities (11.57) satisfy the relation (11.61) and therefore it is sufficient to take into account only $\Omega > 0$. A random signal sent by Alice is assumed to be stationary and uniform in the cross-section of the beams. The amplitudes $a_n(\boldsymbol{q}, \Omega)$ for $\Omega > 0$, $n = 1, 2$, are taken as independent complex Gaussian variables with variance $\sigma^A(\boldsymbol{q},\Omega)$ depending on $(\boldsymbol{q},\Omega)$. Because the transformation (11.60) is unitary, this implies that the Fourier classical amplitudes $a(\boldsymbol{q},\Omega)$ for any $(\boldsymbol{q},\Omega)$ are also taken as statistically independent, and the quantity

$$\sigma^A(\boldsymbol{q},\Omega) = \langle |a(\boldsymbol{q},\Omega)|^2\rangle, \tag{11.62}$$

is the mean photon number in the Alice signal wave $(\boldsymbol{q},\Omega)$ in the quantization volume, where $\sigma^A(\boldsymbol{q},\Omega) = \sigma^A(-\boldsymbol{q},-\Omega)$. Here the statistical averaging over the Gaussian ensemble of the Alice signal is performed with the complex weight function

$$\mathcal{P}^A_{\boldsymbol{q},\Omega}(a(\boldsymbol{q},\Omega)) = \frac{1}{\pi\sigma^A(\boldsymbol{q},\Omega)}\exp\left\{-\frac{|a(\boldsymbol{q},\Omega)|^2}{\sigma^A(\boldsymbol{q},\Omega)}\right\}. \tag{11.63}$$

In what follows we assume a Gaussian spectral profile of width q_A for the ensemble of input images in the spatial frequency domain,

$$\sigma^A(\boldsymbol{q},\Omega) = (2\pi)^3\frac{P}{\pi(q_A/2)^2}\exp\left(-\frac{q_x^2+q_y^2}{(q_A/2)^2}\right)\Pi(\Omega),$$

$$\Pi(\Omega) = \begin{cases} 1/\Omega_A & |\Omega| \le \Omega_A/2, \\ 0 & |\Omega| > \Omega_A/2, \end{cases} \tag{11.64}$$

and, for the sake of simplicity, the narrow rectangular spectral profile $\Pi(\Omega)$ of width Ω_A and height $1/\Omega_A$ in the temporal frequency domain. Because

$$\sum_{\boldsymbol{q},\Omega}\sigma_A(\boldsymbol{q},\Omega) = L^2TP, \tag{11.65}$$

the total average density of photon flux in the image field per $\mathrm{cm}^2\cdot\mathrm{s}$ is P. The variances of the observables $i_n(\boldsymbol{q},\Omega)$ are finally found in the form

$$\langle\frac{1}{2}\left\{\hat{i}_n(\boldsymbol{q},\Omega),\hat{i}^\dagger_n(\boldsymbol{q},\Omega)\right\}_+\rangle = B_0^2\left[\sigma_n^{BA}(\boldsymbol{q},\Omega)+\sigma^A(\boldsymbol{q},\Omega)\right], \tag{11.66}$$

where $\{\ ,\ \}_+$ denotes the anticommutator. The quantum noise variances in both detection channels are given by

$$\sigma_n^{BA}(\boldsymbol{q},\Omega) = \langle\frac{1}{2}\left\{\hat{f}_n(\boldsymbol{q},\Omega),\hat{f}^\dagger_n(\boldsymbol{q},\Omega)\right\}_+\rangle, \tag{11.67}$$

$$\sigma_1^{BA}(\boldsymbol{q},\Omega) = e^{2r_1(\boldsymbol{q},\Omega)}\cos^2\psi_1(\boldsymbol{q},\Omega) + e^{-2r_1(\boldsymbol{q},\Omega)}\sin^2\psi_1(\boldsymbol{q},\Omega), \tag{11.68}$$

$$\sigma_2^{BA}(\boldsymbol{q},\Omega) = e^{-2r_2(\boldsymbol{q},\Omega)}\cos^2\psi_2(\boldsymbol{q},\Omega) + e^{2r_2(\boldsymbol{q},\Omega)}\sin^2\psi_2(\boldsymbol{q},\Omega). \tag{11.69}$$

Using these results we can evaluate the Shannon mutual information for our dense coding scheme.

It is well known that in the case of a single-mode squeezed light field the statistics of its quadrature amplitudes are Gaussian and can be characterized, for example, by a Gaussian weight function in the Wigner representation. In the homodyne detection of squeezed light, the statistics of the photocounts are also Gaussian due to the linear relation between the field amplitude and the photocurrent density. The discussion of homodyne detection in terms of the characteristic function can be found in [28]. Some considerations for homodyne detection of spatially multimode fields [15] are given in Appendix B.

Finally, in our quantum dense coding scheme the statistically independent degrees of freedom of the noise and the signal are labeled by the frequencies $(\boldsymbol{q}, \Omega)$ for $\Omega > 0$. One can consider our quantum channel as a collection of statistically independent parallel Gaussian communication channels in the Fourier domain. The mutual information between Alice and Bob for a given detector and frequencies $(\boldsymbol{q}, \Omega)$ is defined as

$$I_n^S(\boldsymbol{q}, \Omega) = H_n^B(\boldsymbol{q}, \Omega) - \overline{H_n^{(B|A)}(\boldsymbol{q}, \Omega)}^A . \tag{11.70}$$

Here $H^B(\boldsymbol{q}, \Omega)$ is the entropy of Bob's observable, and

$$\overline{H_n^{(B|A)}(\boldsymbol{q}, \Omega)}^A$$

is averaged over the ensemble of Alice's signals' entropy of noise, introduced by the channel [32]. For the Gaussian channel the mutual information is given by

$$I_n^S(\boldsymbol{q}, \Omega) = \ln\left(1 + \frac{\sigma^A(\boldsymbol{q}, \Omega)}{\sigma_n^{BA}(\boldsymbol{q}, \Omega}\right) . \tag{11.71}$$

The quantum noise suppression within the frequency range of effective squeezing and entanglement increases the signal-to-noise ratio at the right side of (11.71). The total mutual information I^S, associated with the large area L^2 and the large observation time T, is defined as a sum over all degrees of freedom and is related to the density of the information stream J in bits (more precisely, in nits) per $\mathrm{cm}^2 \cdot$ s:

$$I^S = \sum_{n,\boldsymbol{q},\Omega>0} I_n^S(\boldsymbol{q}, \Omega) = L^2 T\, J, \tag{11.72}$$

where

$$J = \frac{1}{(2\pi)^3} \int d\boldsymbol{q} \int_{\Omega>0} d\Omega \sum_{n=1,2} I_n^S(\boldsymbol{q}, \Omega). \tag{11.73}$$

11.4.4 Channel Capacity

For qualitative and numerical analysis it is natural to associate such quantities as the density of the information stream and of the photon flux with the physical parameters present in our quantum dense coding scheme. Squeezing and entanglement, produced by type-I optical parametric amplifiers, are characterized by the effective spectral widths q_c and Ω_c in the spatial and temporal frequency domain. The coherence area in the cross-section of the beams and the coherence time are introduced as $S_c = (2\pi/q_c)^2$ and $T_c = 2\pi/\Omega_c$. For simplicity, we assume that both OPAs have the same coherence area and coherence time. The correlation area S_A and the correlation time T_A of non-stationary images, sent by Alice, are related to the spectral widths of the signal q_A and Ω_A by $S_A = (2\pi/q_A)^2$ and $T_A = 2\pi/\Omega_A$. We consider the broadband degenerate collinear phase matching in the traveling-wave type-I OPAs. The coherence time T_c of the spontaneous down-conversion will be typically short compared to the time duration T_A of Alice movie frame.

The dimensionless information stream $\mathcal{J}$ and the dimensionless input photon flux $\mathcal{P}$ are defined by $\mathcal{J} = S_c T_A J$, $\mathcal{P} = S_c T_A P$. That is, we relate both quantities to the time duration of Alice's movie frame and the coherence area of squeezing and entanglement.

The optimum entanglement conditions in the OPAs are achieved provided

$$\begin{aligned} r_1(\boldsymbol{q},\Omega) &= r_2(\boldsymbol{q},\Omega) \equiv r(\boldsymbol{q},\Omega), \\ \psi_1(\boldsymbol{q},\Omega) &= \psi_2(\boldsymbol{q},\Omega) \pm \pi/2 \equiv \psi(\boldsymbol{q},\Omega), \\ \psi(0,0) &= \pi/2. \end{aligned} \tag{11.74}$$

By using the above-introduced definitions, we find the dimensionless information stream $\mathcal{J}$ in the following form,

$$\mathcal{J} = \int d\boldsymbol{\kappa} \ln\left\{1 + \mathcal{P}\frac{1}{\sigma^{BA}(\boldsymbol{\kappa},0)}\frac{1}{\pi(d_A/2)^2}\exp\left(-\frac{\kappa_x^2+\kappa_y^2}{(d_A/2)^2}\right)\right\}, \tag{11.75}$$

where

$$\sigma^{BA}(\boldsymbol{\kappa},0) = e^{2r(\boldsymbol{\kappa},0)}\cos^2\psi(\boldsymbol{\kappa},0) + e^{-2r(\boldsymbol{\kappa},0)}\sin^2\psi(\boldsymbol{\kappa},0), \tag{11.76}$$

and the dimensionless spatial frequency is defined as $\boldsymbol{\kappa} = \boldsymbol{q}/q_c$. The relative spectral width of the Alice signal $d_A = q_A/q_c = (S_c/S_A)^{1/2}$ can be interpreted as the number of image elements per coherence length, that is, the relative linear density of image elements. In what follows we assume a simple estimate $q_c/2 = \sqrt{2k/l}$, related to the diffraction spread of parametric down-conversion light inside the OPA crystal, where k is the wave number and l is the crystal length.

Quantum noise in the dense coding scheme is effectively reduced for optimum phase matching of squeezed beams. As discussed in Section 11.2, an important factor is the spatial frequency dispersion of squeezing, that is, the $\boldsymbol{q}$-dependence of the squeezed quadrature component phase. This dependence is due to the diffraction inside the OPA. A thin lens properly inserted into the light beam can effectively correct the $\boldsymbol{q}$-dependent orientation of squeezing ellipses, as illustrated in Fig. 11.2b.

The improvement in the signal-to-noise ratio for different spatial frequencies can be characterized by the inverse noise variance $\sigma^{BA}(\boldsymbol{\kappa}, 0)$ shown in Fig. 11.11. As seen in this figure, the phase correction by means of a lens allows for the low-noise signal transmission within the spatial-frequency band of the effective squeezing.

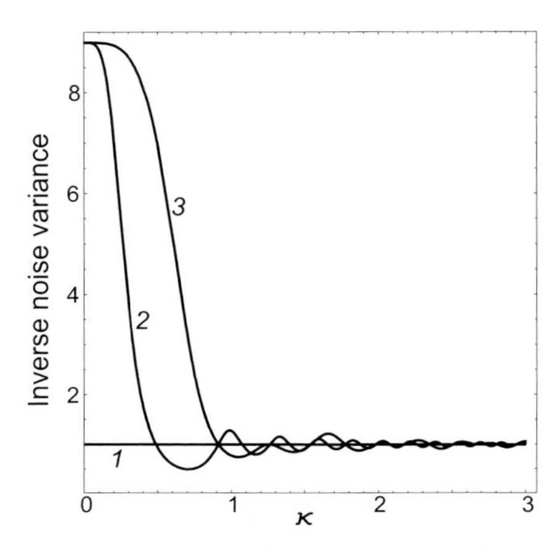

Fig. 11.11. Inverse noise variance in dependence on the spatial frequency κ for vacuum noise at the input (1), and for squeezing with $\exp[r(0,0)] = 3$, without (2) and with (3), phase correction.

In our plots for the mutual information density $\mathcal{J}$ we keep constant the coherence area S_c, the degree of squeezing $r(0,0)$, and the density of signal photon flux $\mathcal{P}$.

The dependence of mutual information density on the relative linear density d_A of the image elements is shown in Fig. 11.12. For $d_A \ll 1$ (large image elements, $S_A \gg S_c$), the mutual information density increases linearly with d_A, because this implies an improvement of spatial resolution in the input signal. In the classical limit (vacuum noise at the input of the scheme), the increase of mutual information density with the density of image elements takes place until the information per Alice's image element becomes of the order of or less than one bit:

$$\ln\left\{1 + \frac{4}{\pi}\frac{\mathcal{P}}{d_A^2}\right\} \sim \frac{\mathcal{P}}{d_A^2} \leq 1. \tag{11.77}$$

The further increase of d_A has no effect because it is completely compensated by the decrease of information per image element. In our plots this corresponds to $d_A \sim \sqrt{\mathcal{P}} \sim 1$ for $\mathcal{P} = 1$, $d_A \sim 1.7$ for $\mathcal{P} = 3$, and $d_A \sim 3$ for $\mathcal{P} = 10$ (see Fig. 11.12a,b,c correspondingly).

It is instructive to estimate the effect of squeezing and entanglement on the information capacity of our dense coding channel. A standard assumption

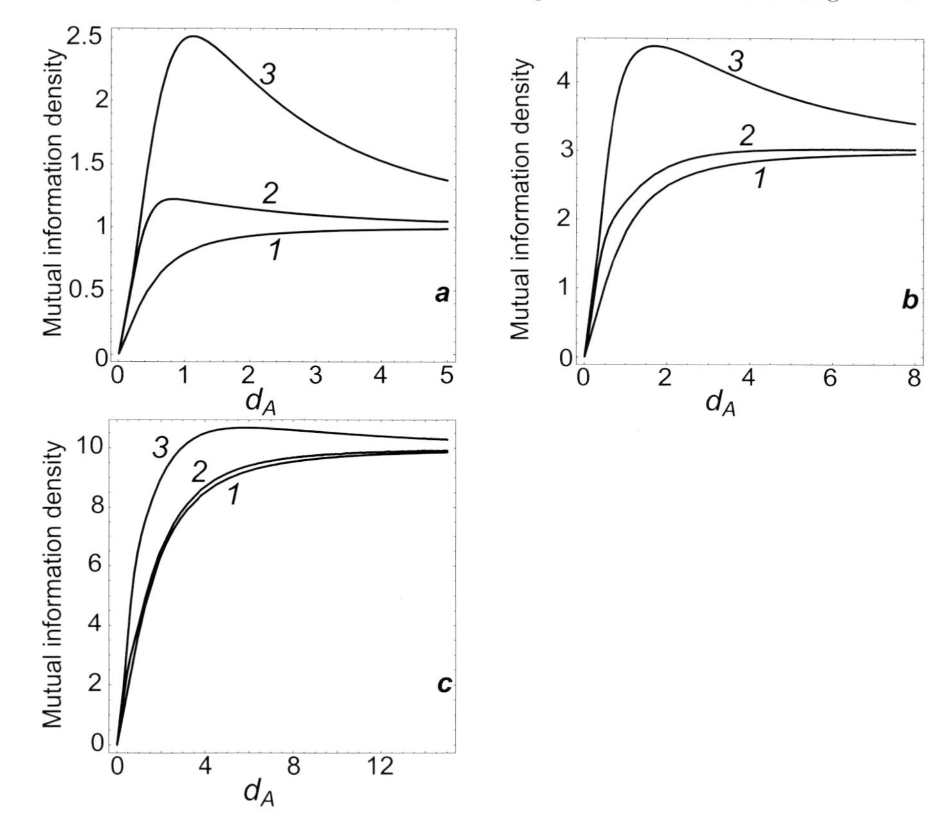

Fig. 11.12. Mutual information density for the vacuum noise at the input of the scheme (1), and for squeezing with $\exp(r(0,0)) = 3$, without (2) and with (3) phase correction. The density of signal photons is $\mathcal{P} = 1$ (a), $\mathcal{P} = 3$ (b), $\mathcal{P} = 10$ (c).

for such an estimate reads

$$\langle n_{squeezed} \rangle \sim \langle n_{signal} \rangle, \tag{11.78}$$

and implies that the energy costs of squeezing and entanglement (the number of photons in squeezed light per mode at a given detector) are of the order of the signal photon number per mode. Here

$$\langle n_{squeezed} \rangle = \sinh^2 r \sim \frac{e^{2r}}{4}. \tag{11.79}$$

Let us take for simplicity $d_A \sim 1$, when the size S_A of the image element is of the order of coherence area S_c of squeezed light. Under this condition one can consider the coherence volume cS_AT_A as a degree of freedom for both the signal and the squeezed field. Then

$$\langle n_{signal} \rangle \sim \mathcal{P}, \tag{11.80}$$

and the assumption (11.78) means,

$$\mathcal{P} \sim \frac{e^{2r}}{4}. \tag{11.81}$$

In our plots $e^{2r} = 9$, and $\mathcal{P} = 1 < e^{2r}/4$, $\mathcal{P} = 3 \sim e^{2r}/4$, $\mathcal{P} = 10 > e^{2r}/4$ in Fig. 11.12a,b,c correspondingly. By inspecting the curves for $d_A \leq 1$ one can observe that for $\langle n_{squeezed} \rangle \sim \langle n_{signal} \rangle$ (the curves 3 and 1 in Fig. 11.12b) the information capacity of the dense coding channel exceeds that of the classical channel by a factor of $\sim$2 .

This result is in agreement with the general properties of quantum dense coding, and with the estimate [11] for the single-mode continuous variables dense coding scheme.

For $\mathcal{P} < e^{2r}/4$ (the curves 3 and 1 in Fig. 11.12a) the superiority of the quantum channel is more significant, but in this case the energy costs of squeezing and entanglement exceed the power of the signal itself. The curves 3 and 1 in Fig. 11.12c illustrate an opposite limit: relatively low energy costs of the quantum channel and a small increase of its information capacity.

For $d_A \gg 1$ (image elements much smaller than the coherence length), the effect of entanglement on the channel capacity is washed out and $\mathcal{J}$ goes down to the classical limit. This is due to the fact that in the limit $S_A \ll S_c$ almost all spatial frequencies of the signal are outside the spatial-frequency band of effective noise suppression, and the channel capacity is finally limited by vacuum noise.

The phase correction of squeezing and entanglement significantly improves the channel capacity, because it brings the spatial frequency band of the effective noise suppression to its optimum value. It eliminates the destructive effect of the amplified (stretched) quadrature of the noise field at the higher spatial frequencies, as seen from Fig. 11.12, curves 3 and 2.

11.5 Conclusions and Outlook

In this chapter we have discussed the model of quantum holographic teleportation of optical images proposed in [14, 15], and have demonstrated the ability of the scheme to transfer a quantum state of an input image from one place to another with high fidelity. We have extended the continuous-variable dense coding protocol onto the optical images [19]. We have shown that such multimode extension of the quantum communication channel provides much higher channel capacity due to its intrinsic parallel nature.

The above-discussed models of quantum holographic teleportation and quantum dense coding of optical images can be viewed as examples of an interplay between quantum information and quantum imaging. Of evident interest is a search for applications of more general kinds of the essentially multivariate quantum entanglement in quantum information phenomena.

One can mention here, for instance, quantum entanglement between light waves of different frequencies, associated with frequency nondegenerate squeezing. Frequency nondegenerate squeezed states of light were observed in some experiments, for example, in [33]. Quantum teleportation of optical images based on this kind of entanglement could allow for the transfer of a quantum state of light to a desirable tunable frequency.

An important problem of quantum information is the problem of exchange of the quantum state between light and the long-lived matter degrees of freedom (quantum memory). An experimental demonstration of quantum memory for light based on the off-resonant interaction of light with a spin-oriented atomic ensemble was achieved in [34]. One could expect that extension of quantum memory schemes, involving essentially multivariate light–matter entanglement, could increase their information capacity.

Of interest could be quantum information schemes, based on the other versions of multivariate entanglement, as, for example, in the domain of orbital angular momentum (OAM) of light (see Chapter 12).

As mentioned in Section 11.4, the superiority of the dense coding scheme over a single communication channel is related to the fact that in the presence of quantum entanglement an idle channel provides an optimal reference system for detecting minor physical perturbation in the signal channel. Such an entanglement-based increase of sensitivity is common for quantum dense coding and some recently discussed schemes of quantum measurements. In analogy to the dense coding of optical images, the use of essentially multimode entanglement in quantum measurements could allow for parallel measurements at improved quantum limits of the distributed-in-space light fields or the multivariate quantum objects. For instance, the measurement at the Heisenberg limit $\delta\varphi \sim 1/n$ of the phase perturbations in a distributed-in-space faint phase object [35] can be achieved in the scheme similar to that of dense coding.

Acknowledgments

The author is indebted to his collaborators who contributed to the works reviewed in this chapter, particularly to A. Gatti, Yu. M. Golubev, T. Yu. Golubeva, M. I. Kolobov, L. A. Lugiato, and D. Vasilyev.

A Properties of Spatially Multimode Squeezing

The basic results for spatially multimode squeezing are summarized in [23]. The coefficients of the squeezing transformation (11.3), (11.56) satisfy the conditions

$$|U_n(\boldsymbol{q},\Omega)|^2 - |V_n(\boldsymbol{q},\Omega)|^2 = 1, \qquad (11.82)$$
$$U_n(\boldsymbol{q},\Omega)V_n(-\boldsymbol{q},-\Omega) = U_n(-\boldsymbol{q},-\Omega)V_n(\boldsymbol{q},\Omega),$$

which are necessary and sufficient for the preservation of the free-field commutation relations (11.1), (11.53). The spatial and temporal parameters of squeezed and entangled light fields essentially depend on the orientation angle $\psi_n(\boldsymbol{q},\Omega)$ of the major axes of the squeezing ellipses,

$$\psi_n(\boldsymbol{q},\Omega) = \frac{1}{2}\arg\{U_n(\boldsymbol{q},\Omega)V_n(-\boldsymbol{q},-\Omega)\}, \qquad (11.83)$$

and on the degree of squeezing $r_n(\boldsymbol{q},\Omega)$,

$$e^{\pm r_n(\boldsymbol{q},\Omega)} = |U_n(\boldsymbol{q},\Omega)| \pm |V_n(\boldsymbol{q},\Omega)|. \qquad (11.84)$$

The phase of the amplified quadrature components is given [36] by

$$\phi_n(\boldsymbol{q},\Omega) = -\frac{1}{2}\arg\{U_n(\boldsymbol{q},\Omega)V_n^*(-\boldsymbol{q},-\Omega)\}. \qquad (11.85)$$

For the type-I phase-matched traveling-wave OPAs, the coefficients $U_n(\boldsymbol{q},\Omega)$ and $V_n(\boldsymbol{q},\Omega)$ are given by

$$U_n(\boldsymbol{q},\Omega) = \exp\Big\{i\Big[(k_z(\boldsymbol{q},\Omega)-k)l - \delta(\boldsymbol{q},\Omega)/2\Big]\Big\}\Big[\cosh\Gamma_n(\boldsymbol{q},\Omega) + \frac{i\delta(\boldsymbol{q},\Omega)}{2\Gamma_n(\boldsymbol{q},\Omega)}\sinh\Gamma_n(\boldsymbol{q},\Omega)\Big], \qquad (11.86)$$
$$V_n(\boldsymbol{q},\Omega) = \exp\Big\{i\Big[(k_z(\boldsymbol{q},\Omega)-k)l - \delta(\boldsymbol{q},\Omega)/2\Big]\Big\}\frac{g_n}{\Gamma_n(\boldsymbol{q},\Omega)}\sinh\Gamma_n(\boldsymbol{q},\Omega).$$

Here l is the length of the nonlinear crystal, $k_z(\boldsymbol{q},\Omega)$ is the longitudinal component of the wave-vector $\boldsymbol{k}(\boldsymbol{q},\Omega)$ for the wave with frequency $\omega+\Omega$ and transverse component $\boldsymbol{q}$. The dimensionless mismatch function $\delta(\boldsymbol{q},\Omega)$ is given by

$$\delta(\boldsymbol{q},\Omega) = \Big(k_z(\boldsymbol{q},\Omega) + k_z(-\boldsymbol{q},-\Omega) - k_p\Big)l \approx (2k-k_p)l + k''_\Omega l\Omega^2 - q^2l/k, \qquad (11.87)$$

where k_p is the wave number of the pump wave; $k_p - 2k = 0$ in the degenerate case. We have assumed the paraxial approximation. The parameter $\Gamma_n(\boldsymbol{q},\Omega)$ is defined as

$$\Gamma_n(\boldsymbol{q},\Omega) = \sqrt{g_n^2 - \delta^2(\boldsymbol{q},\Omega)/4}, \qquad (11.88)$$

where g_n is the dimensionless coupling strength of nonlinear interaction, taken as real for simplicity. It is proportional to the nonlinear susceptibility, the length of the crystal, and the amplitude of the pump field.

B Homodyne Detection with Spatial Resolution

In this Appendix we shall discuss the validity of Eqs. (11.10), (11.54) for the photocurrent density operators in the balanced homodyne detection scheme with spatial resolution. For definiteness we shall consider the homodyne detection of the X quadrature component. The results for the conjugate component are obtained in a similar way.

In Fig. 11.13 we present the schematic of point-by-point balanced homodyne detection with a photodetector D_x. To achieve spatial resolution, the pixels of the CCD matrices are assumed to be much smaller than the coherence area S_c of the EPR beams. We shall discuss the properties of the quantum operator for the surface density of the photocurrent and show that the definitions (11.10), (11.54) of this observable are in agreement with the standard description of photodetection with resolution in space and time, based on the Glauber field correlation functions [37]. The output signal of

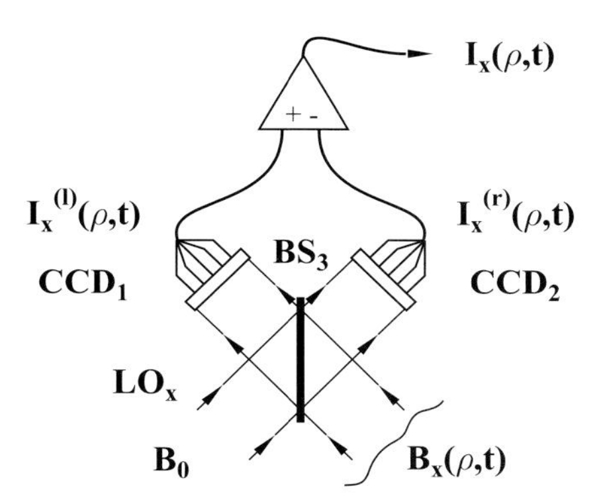

Fig. 11.13. Schematic of balanced homodyne detection with spatial resolution.

the balanced homodyne detector D_x is given by

$$\hat{I}_x(\boldsymbol{\rho}, t) = \hat{I}_x^{(l)}(\boldsymbol{\rho}, t) - \hat{I}_x^{(r)}(\boldsymbol{\rho}, t), \tag{11.89}$$

where $\hat{I}_x^{(l)}(\boldsymbol{\rho}, t)$ and $\hat{I}_x^{(r)}(\boldsymbol{\rho}, t)$ are the surface photocurrent densities, measured by the left and right CCD matrices. The local oscillator plane wave LO_x with complex amplitude $B_x^{(H)}$ is classical and strong, $|B_x^{(H)}| \gg |B_x(\boldsymbol{\rho}, t)|$, where $B_x(\boldsymbol{\rho}, t)$ is the input field of the detector. In this limit the quantum operator for the difference photocurrent surface density is introduced in analogy to the earlier-investigated spatially single-mode case; see [26, 27] and in a more general context [28]. The classical field strength is replaced by the quantum field operator $\hat{B}_x(\boldsymbol{\rho}, t)$ in the difference surface power of beatings between the object and the local oscillator waves:

$$\hat{I}_x(\boldsymbol{\rho}, t) = B_x^{(H)^*} \hat{B}_x(\boldsymbol{\rho}, t) + h.c. \tag{11.90}$$

The quantum efficiency of the CCD matrices for simplicity is assumed to be equal to unity and the additional quantum noise due to the light losses in detectors is neglected. The beamsplitter BS_3 is described by the matrix similar to (11.6).

Let us show that the operator (11.90) is in agreement with the Glauber photodetection theory [37] and, in particular, describes correctly the shot-noise of photodetection in space and time. In standard photodetection theory the second-order correlation function of the photocurrent density is related to the fourth-order correlation function of the field amplitudes. In our case of differenced photodetection this relation reads

$$\begin{aligned}\langle\frac{1}{2}\left\{\hat{I}_x(\boldsymbol{\rho},t),\hat{I}_x(\boldsymbol{\rho}',t')\right\}_+\rangle &= \langle\hat{\Phi}_x^{(l)}(\boldsymbol{\rho},t)+\hat{\Phi}_x^{(r)}(\boldsymbol{\rho},t)\rangle\delta(\boldsymbol{\rho}-\boldsymbol{\rho}')\delta(t-t')+\\ \langle T_N\left\{\left(\hat{\Phi}_x^{(l)}(\boldsymbol{\rho},t)-\hat{\Phi}_x^{(r)}(\boldsymbol{\rho},t)\right)\right.&\left.\left(\hat{\Phi}_x^{(l)}(\boldsymbol{\rho}',t')-\hat{\Phi}_x^{(r)}(\boldsymbol{\rho}',t')\right)\right\}\rangle. \qquad (11.91)\end{aligned}$$

Here the quantities

$$\begin{aligned}\hat{\Phi}_x^{(l)}(\boldsymbol{\rho},t) &= \frac{1}{2}\left(B_x^{(H)}+\hat{B}_x(\boldsymbol{\rho},t)\right)^\dagger\left(B_x^{(H)}+\hat{B}_x(\boldsymbol{\rho},t)\right),\\ \hat{\Phi}_x^{(r)}(\boldsymbol{\rho},t) &= \frac{1}{2}\left(-B_x^{(H)}+\hat{B}_x(\boldsymbol{\rho},t)\right)^\dagger\left(-B_x^{(H)}+\hat{B}_x(\boldsymbol{\rho},t)\right) \qquad (11.92)\end{aligned}$$

are the surface densities of the photon flux on the left and right CCD matrices, and $\{\ ,\ \}_+$ stands for the anticommutator. The T_N ordering of the field operators in (11.91) means: (i) the normal ordering, and (ii) the time ordering of the positive-frequency (annihilation) operators, such that the time argument grows up from right to left, and the inverse time ordering of the negative-frequency (creation) field operators.

In the limit of a strong local oscillator we can assume in (11.91) the following approximation,

$$\hat{\Phi}_x^{(l)}(\boldsymbol{\rho},t)-\hat{\Phi}_x^{(r)}(\boldsymbol{\rho},t)\approx B_x^{(H)^*}\hat{B}_x(\boldsymbol{\rho},t)+h.c. \qquad (11.93)$$

Consider now the same second-order symmetrized correlation function of the output photocurrents (see the left side of Eq. (11.91)), directly substituting into it the above-introduced surface photocurrent density operator (11.90). Bringing the field operators to the T_N order with the use of the commutation relation similar to (11.1), we arrive at

$$\begin{aligned}\langle\frac{1}{2}\left\{\hat{I}_x(\boldsymbol{\rho},t),\hat{I}_x(\boldsymbol{\rho}',t')\right\}_+\rangle &= |B_x^{(H)^*}|^2\delta(\boldsymbol{\rho}-\boldsymbol{\rho}')\delta(t-t')\\ +\langle T_N\left\{\left(B_x^{(H)^*}\hat{B}_x(\boldsymbol{\rho},t)+h.c.\right)\right.&\left.\left(B_x^{(H)^*}\hat{B}_x(\boldsymbol{\rho}',t')+h.c.\right)\right\}\rangle. \qquad (11.94)\end{aligned}$$

This expression is in agreement with the Glauber correlation function (11.91), approximated with the use of Eq. (11.93). The deltalike contribution describes the shot noise of photodetection in space and time. In Sections 11.3.3 and 11.4.2 we apply the difference photocurrent surface density operator (11.90) to the analysis of our teleportation and dense coding schemes.

References

1. *The Physics of Quantum Information*, ed. by D. Bouwmeester, A. Eckert and A. Zeilinger (Springer, Berlin, 2000).
2. *Quantum Information with Continious Variables*, ed. by S. Braunstein and A. Pati (Kluwer, Dordrecht, 2003).
3. C. H. Bennett, G. Brassard, C. Crepeau, R. Jozsa, A. Peres, and W. K. Wooters, Phys. Rev. Lett. **70**, 1895 (1993).
4. L. Vaidman, Phys. Rev. A **49**, 1473 (1994).
5. S. L. Braunstein and H. J. Kimble, Phys. Rev. Lett. **80**, 869 (1998).
6. T. C. Ralph and P. K. Lam, Phys. Rev. Lett. **81**, 5668 (1998).
7. D. Bouwmeester et al, Nature (London) **390**, 575 (1997); D. Boschi et al, Phys. Rev. Lett. **80**, 1121 (1998).
8. A. Furusawa, J. L. Sorensen, S. L. Braunstein, C. A. Fuchs, H. J. Kimble, and E. S. Polzik, Science, **282**, 706 (1998).
9. W. P. Bowen, N. Treps, B. C. Buchler, R. Schnabel, T. C. Ralph, H.-A. Bachor, T. Symul, and P. K. Lam, Phys. Rev. A **67**, 032302 (2003).
10. S. L. Braunstein, Phys. Rev. Lett. **80**, 4084 (1998); S. L. Braunstein, Nature (London) **394**, 47 (1998); S. Lloyd and J.-J. E. Slotine, Phys. Rev. Lett. **80**, 4088 (1998).
11. S. L. Braunstein and H. J. Kimble, Phys. Rev. A **61**, 042302 (2000).
12. T. S. Ralph, Phys. Rev. A **61**, 010303(R) (1999).
13. P. van Loock, S. L. Braunstein, and H. J. Kimble, Phys. Rev. A **62**, 022309 (2000).
14. I. V. Sokolov, M. I. Kolobov, A. Gatti, and L. A. Lugiato, Opt. Comm. **193**, 175 (2001).
15. A. Gatti, I. V. Sokolov, M. I. Kolobov and L. A. Lugiato, Eur. Phys. J. D **30**, 123 (2004).
16. C. H. Bennett and S. J. Wiesner, Phys. Rev. Lett. **69**, 2881 (1992).
17. K. Mattle, H. Weinfurter, P. G. Kwiat, and A. Zeilinger, Phys. Rev. Lett. **76**, 4656 (1996).
18. X. Y. Li et al, Phys. Rev. Lett. **88**, 047904 (2002).
19. T. Yu. Golubeva, Yu. M. Golubev, I. V. Sokolov, and M. I. Kolobov, J. Mod. Opt., **53**, 699 (2006).
20. R. E. Slusher, L. W. Hollberg, B. Yurke, J. C. Mertz, and J. F. Valley, Phys. Rev. Lett. **55**, 2409 (1985).
21. H.-A. Bachor, *A Guide to Experiments in Quantum Optics* (Wiley–WCH, Vienna, 1998).
22. M. I. Kolobov, I. V. Sokolov, JETP **69**, 1097 (1989); M. I. Kolobov and I. V. Sokolov, Phys. Lett. A **140**, 101 (1989).
23. M. I. Kolobov, Rev. Mod. Phys. **71**, 1539 (1999).
24. M. I. Kolobov, Phys. Rev. A **44**, 1986 (1991).
25. E. Brambilla, A. Gatti, L. A. Lugiato, Phys. Rev. A **69**, 023802 (2004).
26. B. Yurke, Phys. Rev. A **32**, 311 (1985).
27. S. Z. Ou, H. J. Kimble, Phys. Rev. A **52**, 3126 (1995).
28. S. L. Braunstein, Phys. Rev. A **42**, 474 (1990).
29. M. I. Kolobov and I. V. Sokolov, Europhys. Lett. **15**, 271 (1991).
30. A. Gatti and L. A. Lugiato, Phys. Rev. A **52**, 1675 (1995).
31. S. L. Braunstein, Optics and Photonics News, Jan. 1999, p. 10.

32. A. S. Holevo, *Probabilistic and Statistical Aspects of Quantum Theory* (North-Holland, Amsterdam, 1982).
33. N. Ph. Georgiades, E. S. Polzik, K. Edamatsu, H. J. Kimble, and A. S. Parkins, Phys. Rev. Lett. **75**, 3426 (1995).
34. B. Julsgaard, J. Sherson, J. Fiurasek, J. I. Cirac, and E. S. Polzik, Nature **432**, 482 (2004).
35. I. V. Sokolov, J. Opt. B: Quant. Semicl. Opt. **2**, 179 (2000).
36. I. V. Sokolov, M. I. Kolobov, and L. A. Lugiato, Phys. Rev. A **60**, 2420 (1999).
37. R. J. Glauber, Phys. Rev. **130**, 2529 (1963); R. J. Glauber, in *Quantum Optics and Electronics (Les Houches Summer School of Theoretical Physics, University of Grenoble)*, ed. by C. DeWitt, A. Blandin, and C. Cohen-Tannoudji (Gordon and Breach, New York, 1965), p. 53.

12 Orbital Angular Momentum of Light

Stephen M. Barnett and Roberta Zambrini

Department of Physics, University of Strathclyde, 107 Rottenrow East, G4 ONG Glasgow (UK) steve@phys.strath.ac.uk

12.1 Introduction

The study of the mechanical effects of light, including angular momentum, has a long history. The recent rapid growth of interest, however, can be traced to the observation that the Laguerre–Gaussian modes, familiar from laser physics, carry a well-defined quantity of orbital angular momentum [1]. To be specific, the mode $u_{p\ell}^{LG}$ (Eq. (12.21)) propagating in the z-direction, with azimuthal dependence $\exp(i\ell\phi)$, carries a z-component of orbital angular momentum of $\ell\hbar$ per photon. This idea is strongly suggested by the powerful analogy between paraxial optics and the Schrödinger equation, together with the operator corresponding to the z-component of orbital angular momentum $L_z = -i\hbar\partial/\partial\phi$ [2]. A convincing demonstration follows from an analysis of the Poynting vector and the associated angular momentum density [1, 3] as briefly described in Section 12.2.

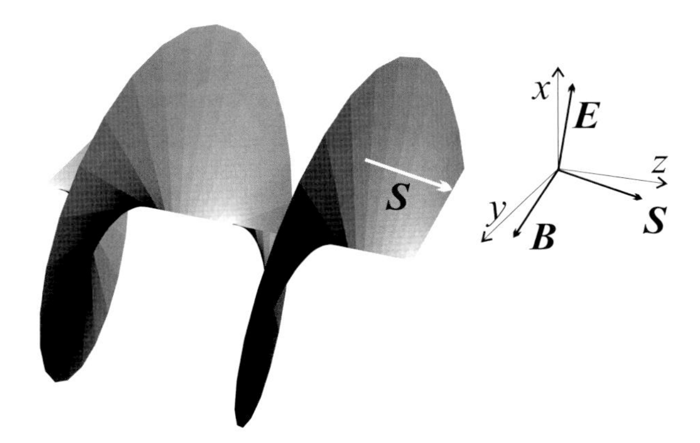

Fig. 12.1. The helical phase front for an $\ell = 1$ mode. The electric and magnetic fields lie in the plane of this phase front so that the momentum density, which is proportional to Poynting's vector, is everywhere normal to the phase front.

The physical origin of the angular momentum may be readily appreciated by reference to Fig. 12.1. The electric and magnetic fields at any point lie in the plane tangent to the helical phase front. This means that the local momentum density, $\epsilon_0 \boldsymbol{E} \times \boldsymbol{B}$, is normal to the phase front. Hence the momentum density itself follows a helical path along the beam and there is an orbital angular momentum associated with this rotation of the momentum [3]. The locus of the momentum is illustrated in Fig. 12.2 by the antlers of the male greater kudu. The angular momentum arises from the azimuthal component of the momentum density. Along the z-axis the azimuthal coordinate is undefined and the fields must tend to zero there for beams with nonzero orbital angular momentum. This is reminiscent of the eye of a hurricane (see Fig. 12.2), the windless center of the storm. For $\ell = 1$ the phase fronts have the form of a simple screw thread, but in general there are ℓ intertwined phase fronts. For $\ell = 2$ the phase fronts form a double helix, like that found for the DNA molecule and for $\ell = 3$ they have the form of the familiar fusilli pasta (see Fig. 12.2). Any beam having the azimuthal dependence $\exp(i\ell\phi)$ will carry an associated orbital angular momentum of $\ell\hbar$ per photon.

The orbital angular momentum of light is rapidly becoming a mature field of study encompassing a wide range of physical phenomena and there is only room here to touch on a few of these, with special emphasis placed on quantum and nonlinear optical phenomena. For readers interested in acquiring a broader introduction there is a review article [4], and a recent book on the topic [5] which includes reprints of many of the key papers.

12.2 Angular Momentum in Electromagnetism

The mechanical properties of light, namely energy, momentum, and angular momentum originate from the fundamental electric and magnetic fields. We can associate a local density with each of these and express a local conservation law in terms of this density and an associated flux or "current" [6]. The energy density for the free field is

$$W = \frac{1}{2}(\epsilon_0 E^2 + \mu_0^{-1} B^2). \tag{12.1}$$

This, together with Poynting's vector,

$$\boldsymbol{S} = \mu_0^{-1} \boldsymbol{E} \times \boldsymbol{B}, \tag{12.2}$$

satisfies a continuity equation corresponding to the local conservation of energy:

$$\frac{\partial W}{\partial t} + \nabla \cdot \boldsymbol{S} = 0. \tag{12.3}$$

Poynting's vector (divided by the square of the speed of light) also plays the role of the momentum density for the field:

Fig. 12.2. Left: the momentum density follows a helical locus much like the antlers of the male greater kudu (picture from *http://www.sa-venues.com/wildlife*). Right top: satellite image of a hurricane with the eye visible at its center (picture from *http://rsd.gsfc.nasa.gov/rsd/images*). Right bottom: the form of the $\ell = 3, 2$ phase fronts illustrated by the $\ell = 3$ and the less common $\ell = 2$ fusilli pasta.

$$\boldsymbol{p} = \epsilon_0 \boldsymbol{E} \times \boldsymbol{B}, \tag{12.4}$$

which is also locally conserved. The corresponding momentum flux density is

$$T_{ij} = \frac{1}{2}\delta_{ij}(\epsilon_0 E^2 + \mu_0^{-1} B^2) - \epsilon_0 E_i E_j - \mu_0 B_i B_j, \tag{12.5}$$

with momentum conservation expressed as

$$\frac{\partial S_i}{\partial t} + \frac{\partial T_{ji}}{\partial x_j} = 0, \tag{12.6}$$

where we employ the summation convention for the three spatial coordinates.

The form of the angular-momentum density follows from that for the momentum density, by analogy with mechanics, as the cross-product of the momentum density and the position:

$$\boldsymbol{j} = \epsilon_0 \boldsymbol{r} \times (\boldsymbol{E} \times \boldsymbol{B}). \tag{12.7}$$

Angular momentum is also locally conserved so there is an angular momentum flux density:

$$M_{li} = \varepsilon_{ijk} x_j T_{kl} \tag{12.8}$$

(where ε_{ijk} is the alternating or permutation symbol) and an associated conservation law:

$$\frac{\partial j_i}{\partial t} + \frac{\partial M_{li}}{\partial x_l} = 0. \tag{12.9}$$

Note that the dimensions of the angular-momentum flux density are of an angular momentum per unit area per unit time. This suggests that we can understand this as a flow of angular momentum through a surface.

All light beams carry angular momentum; we can see this from the definition of the angular momentum density (12.7). All that is required is for the total momentum to have a component perpendicular to the position relative to the axis of rotation. A more interesting situation arises when considering the z-component of angular momentum of a beam that is itself propagating in the z-direction (so that its total momentum points in this direction). Clearly this can only happen if the momentum densities in different parts of the beam are not parallel to the z-axis and if the remaining local x- and y-components of the momentum density conspire to create a nonzero z-component of angular momentum. There are essentially two ways in which this can occur: one, which is associated with spin angular momentum, has its origins in the rotation of the electric field for circularly polarized light (see Fig. 12.3) and the other, which we associate with the orbital angular momentum, arises due to the existence of helical wave-fronts as depicted in Fig. 12.1. We will be interested here only in the angular momentum of light about its propagation axis and particularly in the orbital component.

12.2.1 Spin and Orbital Angular Momentum

In mechanics it is often convenient to separate the angular momentum of a body into spin and orbital components; for a planet the former gives rise to days and nights and the latter is responsible for the annual cycle. A comparable separation for the electromagnetic field is far from straightforward. The usual approach is to introduce the vector potential $\boldsymbol{A}$ and to perform an integration by parts to give [7–9]:

$$\boldsymbol{J}_s = \epsilon_0 \int \boldsymbol{E} \times \boldsymbol{A} d^3 r, \tag{12.10}$$

$$\boldsymbol{J}_o = \epsilon_0 \int E_l (\boldsymbol{r} \times \nabla) A_l d^3 r, \tag{12.11}$$

as the total spin and orbital angular momenta. The spin part gives the difference between the total numbers of right and left circularly polarized photons [8,9], but it appears that the separation is not physically observable [7]. More serious is the observation that neither $\boldsymbol{J}_s$ nor $\boldsymbol{J}_o$ is actually an angular momentum [10, 11]. There are also aesthetic grounds for questioning the forms of $\boldsymbol{J}_s$ and $\boldsymbol{J}_o$. Maxwell's equations for the free field and the local mechanical properties described above are unchanged by the transformation

$$\begin{aligned} \boldsymbol{E} &\to \cos\theta \boldsymbol{E} + \cos\theta c \boldsymbol{B}, \\ \boldsymbol{B} &\to \cos\theta \boldsymbol{B} - \cos\theta \frac{1}{c} \boldsymbol{E}, \end{aligned} \tag{12.12}$$

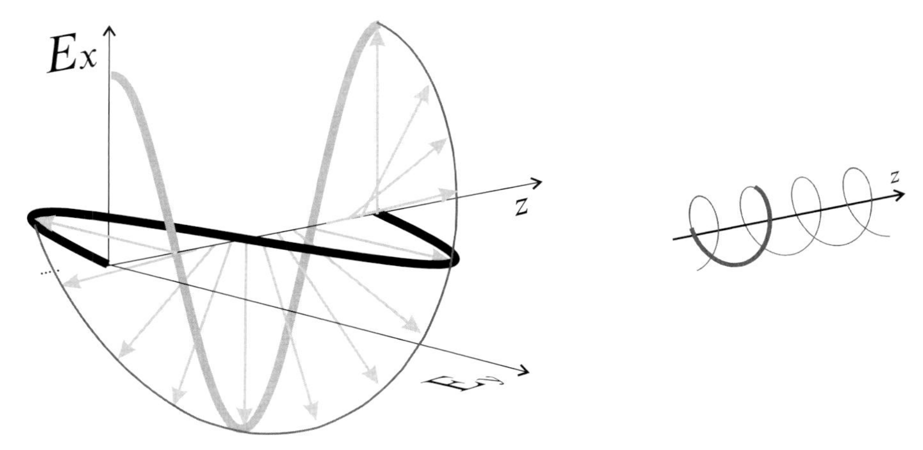

Fig. 12.3. The electric field vector for a circularly polarized beam. The rotation of this and of the associated magnetic field give rise to a spin angular momentum.

for any angle θ. This is not true, however, for the densities associated with $\boldsymbol{J}_s$ and $\boldsymbol{J}_o$. The situation is simplified somewhat if we consider the angular-momentum flux rather than the density of the angular momentum [12]. We find that for a monochromatic beam of light, a separation into a spin and an orbital component is both meaningful and physically reasonable. It should be noted, however, that a different form exists for the spin density and flux [13] and it remains to be seen whether the difference is significant.

The form of the angular momentum greatly simplifies, however, within paraxial optics and it is in this regime that most of the work has been done. In order to understand the angular momentum of a light beam, therefore, it is reasonable to examine it within the paraxial regime.

12.2.2 Angular Momentum in Paraxial Optics

For most practical purposes it suffices to consider beams of light, propagating in the z-direction, for which the transverse beam profile changes only slowly with z. We consider a monochromatic field with angular frequency $\omega = kc$. For the amplitude distribution $u(x, y, z)e^{ikz}$, the associated paraxial approximation amounts to ignoring $\partial^2 u/\partial^2 z$ in comparison with $k\partial u/\partial z$ and $\partial u/\partial z$ compared with ku in the scalar wave equation. The resulting paraxial wave equation is noteworthy for its similarity to the Schrödinger equation:

$$i\frac{\partial u}{\partial z} = -\frac{1}{2k}\left(\frac{\partial^2}{\partial x^2} + \frac{\partial^2}{\partial y^2}\right)u. \tag{12.13}$$

In order to make a connection with the fields we need a method to construct them from u. The simplest procedure is to work in the Lorentz gauge and to write the vector potential as [14]

$$\boldsymbol{A} = (\alpha\hat{\boldsymbol{x}} + \beta\hat{\boldsymbol{y}})ue^{ikz}. \tag{12.14}$$

The Lorentz gauge is used, in preference to the more usual Coulomb gauge, so that we do not need to force $\boldsymbol{A}$ to be transverse. Here α and β are a pair of complex numbers with $|\alpha|^2 + |\beta|^2 = 1$ and $\hat{\boldsymbol{x}}$, $\hat{\boldsymbol{y}}$ are unit vectors in the x- and y-directions, respectively. Within the paraxial approximation, the corresponding positive-frequency components of the electric and magnetic fields are

$$\begin{aligned}\boldsymbol{E} &= i\omega\boldsymbol{A} - \nabla\left(\frac{c^2}{i\omega}\nabla\cdot\boldsymbol{A}\right)\\ &= \left[i\omega\alpha u\hat{\boldsymbol{x}} + i\omega\beta u\hat{\boldsymbol{y}} - c\left(\alpha\frac{\partial u}{\partial x} + \beta\frac{\partial u}{\partial y}\right)\hat{\boldsymbol{z}}\right]e^{ikz}, \end{aligned} \tag{12.15}$$

$$\boldsymbol{B} = \left[-i\beta ku\hat{\boldsymbol{x}} + i\alpha ku\hat{\boldsymbol{y}} + \left(\beta\frac{\partial u}{\partial x} - \alpha\frac{\partial u}{\partial y}\right)\hat{\boldsymbol{z}}\right]e^{ikz}. \tag{12.16}$$

From these, it is straightforward to calculate the (time-averaged) momentum density [1,3]

$$\begin{aligned}\bar{\boldsymbol{p}} &= \frac{\epsilon_0}{2}\left(\boldsymbol{E}^* \times \boldsymbol{B} + \boldsymbol{E}^* \times \boldsymbol{B}\right)\\ &= \frac{\epsilon_0}{2}[i\omega(u\nabla u^* - u^*\nabla u) + 2\omega k|u|^2\hat{\boldsymbol{z}} + \omega\sigma\nabla|u|^2 \times \hat{\boldsymbol{z}}],\end{aligned} \tag{12.17}$$

where $\sigma = i(\alpha\beta^* - \alpha^*\beta)$. It simplifies matters considerably to work in cylindrical polar coordinates (ρ, ϕ, z). The second term represents the linear momentum in the direction of the beam and is consistent with assigning a z-component of momentum equivalent to $\hbar k$ per photon. (Consistency with the paraxial approximation requires us to ignore the z-component of the first term.) The ϕ-component may be associated with a twisting of the light about its propagation direction and is responsible for the angular momentum. There are two terms in Eq. (12.17) that can give rise to a ϕ-component of the momentum density: the first term through $(u\nabla u^* - u^*\nabla u)_\phi$ and the last term through $(\nabla|u|^2 \times \hat{\boldsymbol{z}})_\phi$. The first of these arises naturally when the amplitude has the form

$$u = v(\rho, z)e^{i\ell\phi}, \tag{12.18}$$

and leads to a momentum density that is proportional to ℓ. The second occurs when $\alpha\beta^*$ has an imaginary part and is clearly associated with a circular or elliptical polarization of the beam. The remaining ρ-component is associated with diffraction of the light.

If we specialize to a beam with amplitude of the form (12.18) then we find that the angular momentum density has the simple form [1]

$$j_z = \rho\bar{p}_\phi = \epsilon_0\omega l|u|^2 - \frac{\epsilon_0}{2}\omega\sigma\rho\frac{\partial|u|^2}{\partial\rho}. \tag{12.19}$$

The flux of angular momentum in the paraxial approximation is simply cj_z [12] and hence the ratio of the total flux of angular momentum to the total flux of energy (given by Poynting's vector) is

$$\frac{\int\int \rho d\rho d\phi c j_z}{\int\int \rho d\rho d\phi c^2 \bar{p}_z} = \frac{\ell + \sigma}{\omega}. \tag{12.20}$$

This simple result has an intriguing interpretation that emerges if we multiply and divide this expression by $\hbar$ to give $(\hbar\ell + \hbar\sigma)/\hbar\omega$. We have associated an energy $\hbar\omega$ with each photon and the result we have derived suggests that we should also associate an orbital angular momentum $\hbar\ell$ and a spin angular momentum $\hbar\sigma$ with each photon. This is indeed the case and an analysis based on the angular momentum flux reveals that this is also true for nonparaxial beams [12].

12.2.3 Mechanical Effects

It is important to realize that the spin and orbital angular momenta carried by our light beam are true mechanical angular momenta. This means that the light can exert a torque on a material body and this has been confirmed in a sequence of careful experiments [15–19].

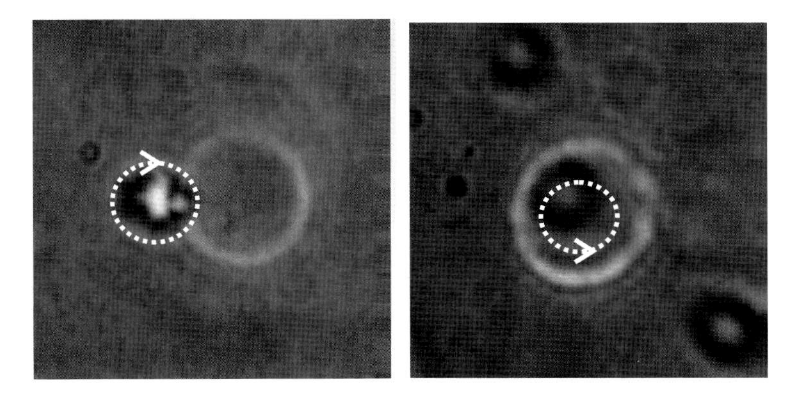

Fig. 12.4. Left: a birefringent particle spinning around its axis when illuminated by a circularly polarized beam. Right: an absorbing particle orbiting around the axis of a linearly polarized beam carrying orbital angular momentum $\ell = 8$. Further details of this experiment may be found in [19].

The effects of a light beam carrying angular momentum on a particle are illustrated in Fig. 12.4. A beam carrying spin but no orbital angular momentum $\sigma = \pm 1, \ell = 0$ induces a particle to spin about its own center of mass. A beam carrying orbital angular momentum but no spin angular momentum induces a particle to orbit about the center of the beam. The polarization-induced rotation of a particle trapped at the radial intensity maximum seems, at first sight, to be at odds with our formula (12.19), which suggests that the optical spin angular momentua should be zero there (as $\partial|u|^2/\partial\rho = 0$). The resolution of this paradox lies in the fact that we should be calculating the torque in terms of the change in the total flux of optical angular momenta on passing through the material. This procedure yields results that are fully in accord with experimental observations [20]

12.3 Beams Carrying Orbital Angular Momentum

12.3.1 Phase Singularities and Spatial Properties

We have seen that a wave with an amplitude having azimuthal dependence $\exp(i\ell\phi)$ carries $\ell\hbar$ units of orbital angular momentum for each photon. The presence of this phase factor tells us that the orbital angular momentum is fundamentally a property of the form of the wave in the plane perpendicular to the direction of propagation. In particular, on moving around the z-axis in a closed loop, the phase will change by $2\pi\ell$. The amplitude (12.18) has an undefined phase or phase singularity on the z-axis, that is, at $\rho = 0$. It necessarily follows that the field must be zero there and we will see that this is a common feature for fields carrying orbital angular momentum.

We should note that phase singularities and their associated lines of darkness are topological features of the field. They commonly occur for a variety of waves and have been studied widely [21,22]. In the lower part of Fig. 12.5 we have plotted the intensity and phase for a three-slit interference experiment. The phase clearly contains points, two of which are labeled, at which all the colors meet and the associated phase is undefined. Examination of the intensity reveals that these correspond to dark regions. The more familiar two-slit pattern exhibits the less common phenomenon of planes of darkness on which the phase is undefined.

Here we are interested in fields carrying orbital angular momentum and so will limit our attention to phase singularities lying on the z-axis with azimuthal dependence $\exp(i\ell\phi)$. Such fields have zero intensity along the z-axis.

12.3.2 Laguerre–Gaussian and Bessel Beams

The paraxial wave equation (12.13) has simple solutions with the required $\exp(i\ell\phi)$ dependence. The most important and most widely used of these are the Laguerre–Gaussian modes which have the form [24]

$$u_{p\ell}^{LG} = \frac{C_{p\ell}^{LG}}{w(z)}\left(\frac{\rho\sqrt{2}}{w(z)}\right)^{|\ell|} L_p^{|\ell|}\left(\frac{2\rho^2}{w^2(z)}\right)\exp\left(\frac{-\rho^2}{w^2(z)}\right)\exp(i\ell\phi)$$

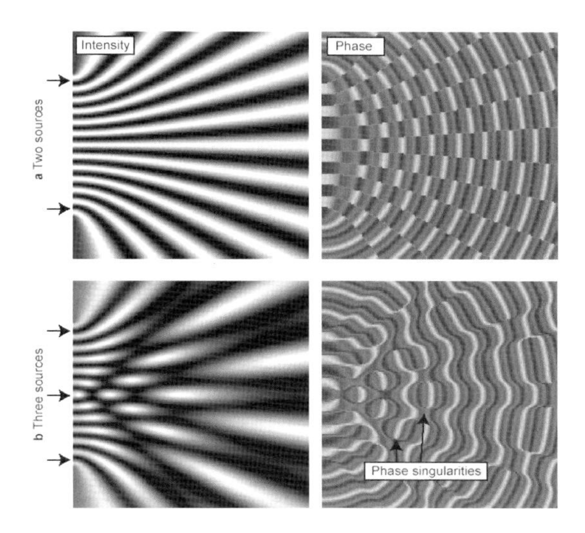

Fig. 12.5. The intensity and phase for a Young two-slit experiment and its less familiar three-slit counterpart. The three-slit system exhibits phase singularities along lines piercing the plane of the figure. Reproduced from [23].

$$\times \exp\left[\frac{ik\rho^2 z}{2(z^2+z_R^2)}\right]\exp\left[-i(2p+|\ell|+1)\tan^{-1}(z/z_R)\right], \quad (12.21)$$

where z_R is the Rayleigh range, $w(z) = [2(z^2+z_R^2)/kz_R]^{1/2}$ is the radius of the beam, and where $L_p^{|\ell|}$ is an associated Laguerre polynomial, obtained from the more familiar Laguerre polynomials by differentiation:

$$L_p^{|\ell|} = (-1)^{|\ell|}\frac{d^{|\ell|}}{dx^{|\ell|}}L_{p+|\ell|}(x). \quad (12.22)$$

The constant

$$C_{p\ell}^{LG} = \left(\frac{2p!}{\pi(p+|\ell|)!}\right)^{1/2} \quad (12.23)$$

is chosen so that the modes are normalized in the transverse plane $\iint dxdy|u|^2 = 1$. The focal plane of the Laguerre–Gaussian mode is at $z=0$ and it is there that it has its smallest width. The Rayleigh range, z_R, is the characteristic length scale associated with diffraction. Parts (a) to (c) of Fig. 12.6 present the intensity, real part, and phase of the Laguerre–Gaussian mode with $\ell = 3$ and $p = 2$ in its focal plane. The intensity has the dark central spot characteristic of nonzero orbital angular momentum, and $p+1$ bright rings. The phase increases by $2\pi\ell = 6\pi$ as we traverse a closed circuit around the central phase singularity. The term $(2p+|\ell|+1)\tan^{-1}(z/z_R)$ is the Gouy phase, which describes the change in the phase, in addition to kz, on propagating along the beam. Its value depends only on the mode order $N = 2p + |\ell|$. This makes it relatively easy to deal with superpositions of modes having the same order, but combining modes of different order can result in a complicated evolution.

A second important class of solutions are the Bessel beams:

$$u^B_{\kappa\ell} = J_{|\ell|}(\kappa\rho)\exp(i\ell\phi)\exp\left(\frac{-i\kappa^2 z}{2k}\right). \tag{12.24}$$

Unlike the Laguerre–Gaussian beams, they are not welllocalized at small values of ρ and cannot be normalized. They are of interest as they do not diffract and also because they are also solutions of the full, that is, nonparaxial, with the only requirement being to replace $-\kappa^2/2k$ by $\sqrt{k^2 - \kappa^2} - k$. We should note that practical realizations of Bessel beams always have finite transverse extent and hence are nondiffracting only for a finite propagation length [25].

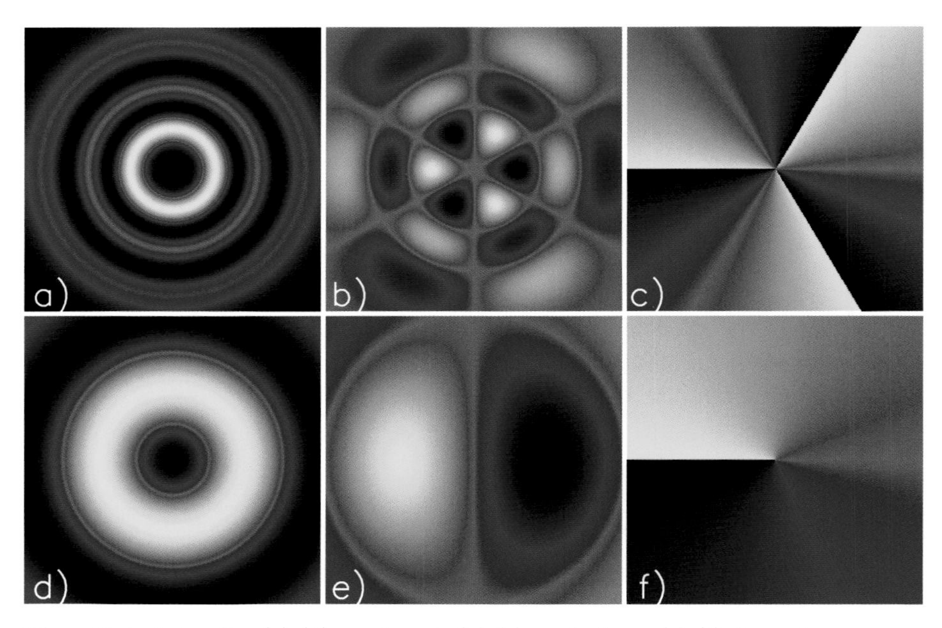

Fig. 12.6. Intensity (a),(d), real part (b),(e) and phase (c),(f) for a Laguerre–Gaussian (a)–(c) beam with $\ell = 3, p = 2$ in its focal plane and for a Bessel (d)–(f) beam with $\ell = 1$. The color scale can be clearly recognized from the last plot.

12.3.3 Generation and Conversion

We have not yet explained how beams carrying orbital angular momentum are generated. There are essentially three methods that have been applied to date and we briefly describe each in turn.

Cylindrical Lens Mode Converter

It is simplest to start by recalling the manner in which polarization, or spin angular momentum, is routinely manipulated in the laboratory. Birefringent waveplates have different refractive indices for two characteristic orthogonal linear polarizations and this means that the two polarizations acquire different phase shifts on propagation through the waveplate. There is, therefore, an associated transformation of the polarization that depends on the thickness and orientation of the waveplate. The most often-used waveplates are the quarter- and half-waveplates, which induce a relative phase shift of $\pi/2$ or π, respectively, between the two characteristic linear polarizations. They are commonly used to convert linear polarization to circular polarization and to rotate the linear polarization, but a combination of them can be used to generate any desired change in polarization.

The cylindrical lens mode-converter for orbital angular momentum works by direct analogy with the waveplate for polarization [26–28]. To see this we must first introduce the Hermite–Gaussian modes [24]:

$$\begin{aligned} u_{nm}^{H} &= \frac{C_{nm}^{HG}}{1+z^2/z_R^2} \exp\left(\frac{-(x^2+y^2)}{w^2(z)}\right) \exp\left(\frac{-ik(x^2+y^2)z}{2(z^2+z_R^2)}\right) \\ &\times \exp\left(i(2p+|\ell|+1)\tan^{-1}(z/z_R)\right) \\ &\times H_n\left(\frac{x\sqrt{2}}{w(z)}\right) H_m\left(\frac{y\sqrt{2}}{w(z)}\right), \end{aligned} \tag{12.25}$$

where H_n is the nth Hermite polynomial and the constant C_{nm}^{HG} has been chosen to enforce normalization. These are the analogues of linear polarizations along the x- and y-directions and may be generated by breaking the rotational symmetry in a laser cavity to suppress the fundamental TEM_{00} mode. We can transform these modes into the required Laguerre–Gaussian modes, or into rotated Hermite–Gaussian modes, by means of cylindrical lenses, which act to focus the beam along one transverse direction only. A pair of such lenses, of focal length f, separated by $f/\sqrt{2}$ acts as a $\pi/2$-converter. This is the analogue of the quarter-waveplate and converts suitably oriented Hermite–Gaussian modes into simply related Laguerre–Gaussian modes with $\ell = \pm|m-n|$ and $p = \min(m,n)$, so that the mode order is unchanged ($N = m+n = 2p+|\ell|$). If the lenses are separated by $2f$ then the lenses act as a π-converter, which allows us to rotate the Hermite–Gaussian mode. Combining $\pi/2$- and π-converters provides considerable freedom in manipulating the mode and its orbital angular momentum [29].

Spiral Phase Plate

The essential feature of modes carrying an orbital angular momentum of $\ell\hbar$ is that they have an azimuthal phase dependence of $\exp(i\ell\phi)$. The simplest Au: ok?

method for achieving this, at least conceptually, is to use a transparent material, the thickness of which increases linearly with ϕ, so that it resembles a spiral staircase (see Fig. 12.7a). If the height of the step corresponds to a phase difference of $2\pi\ell$, then the device will imprint an azimuthal phase profile of $\exp(i\ell\phi)$ on an incoming wave [30, 31]. If we start with a Gaussian beam, which has zero orbital angular momentum, then a beam carrying $\ell\hbar$ units of orbital angular momentum per photon will be generated. Naturally, the initial radial profile of the beam is largely unaffected so that a superposition of many Laguerre–Gaussian modes is generated with a range of values of p but a single value of ℓ. The superposition is dominated, however, by the $p = 0$ contribution. Generation of beams carrying orbital angular momentum has been demonstrated both in the microwave [31] and optical [30, 32] regions of the spectrum. In the latter, a high degree of precision in fabrication is required owing to the small value of the wavelength.

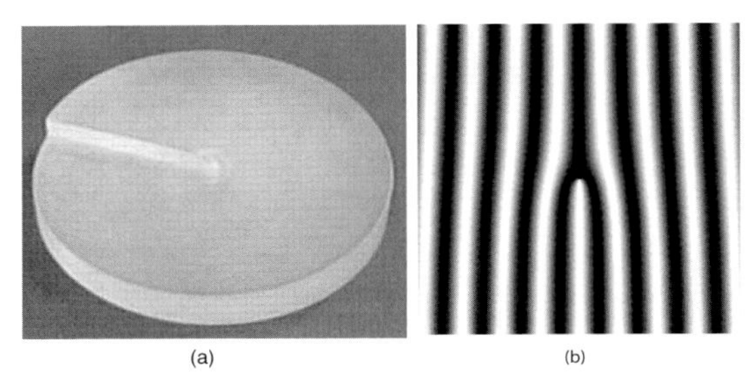

Fig. 12.7. (a) A spiral phase plate and (b) a hologram designed for generating light-carrying orbital angular momentum.

Hologram

Holographic methods for generating orbital angular momentum are closely related to the spiral phase plate. The essential idea is to use a phase hologram to impose an $\exp(i\ell\phi)$ phase shift on the incoming beam in much the same way as the spiral phase plate. Inevitably, however, there will remain a non-diffracted component of the initial beam carrying no angular momentum. In order to separate the desired beam we superpose a phase diffraction grating onto the desired phase pattern. The resulting hologram has a "pitchfork" dislocation with the ℓ-value imposed corresponding to the difference between the number of lines above and below the dislocation [33, 34]. A hologram designed to create a beam with $\ell = 1$ is given in Fig. 12.7b. As with the spiral phase plate, the diffracted beam will be a superposition of Laguerre–Gaussian beams with a range of different p values, but it can be arranged for the value $p = 0$ to dominate.

12.3.4 Other Field Spatial Profiles

We have seen that beams with azimuthal phase dependence $\exp(i\ell\phi)$ and an associated phase singularity or vortex along the z-axis carry an orbital angular momentum of $\ell\hbar$ per photon. It would be quite wrong, however, to infer that the angular momentum is a property of the vortex core itself. There is no light at the vortex core and so the on-axis densities and fluxes of energy, momentum, and angular momentum are all zero. The angular momentum is, of course, a property of the beam as a whole. It is quite possible to form a superposition of Laguerre–Gaussian modes such that the total angular momentum of the beam differs from the topological charge (or ℓ value) associated with the vortex. A dramatic example of this can be seen in an experiment that demonstrated the inversion of an optical vortex under free-space propagation [35]. Naturally, angular momentum is conserved in this situation. A further example is given by the astigmatic modes (see Fig. 12.8). These are characterized in the transverse plane by an elliptical intensity profile and elliptical equiphase contours with an angle between the major axes of the two ellipses. Such modes have been shown to carry large quantities of orbital angular momentum but do not display a vortex at all [36, 37].

This somewhat perplexing situation can be demystified by appealing to quantum theory and considering a single photon. We have seen that each photon in a Laguerre–Gaussian beam carries an orbital angular momentum of $\ell\hbar$. We can also expand any mode in terms of the complete set of Laguerre–Gaussian modes. This suggests that we can consider each photon in an astigmatic mode as being in a superposition of different angular momentum states. The large orbital angular momentum observed for highly astigmatic modes is then the average angular momentum $\langle\ell\rangle\hbar$ multiplied by the number of photons.

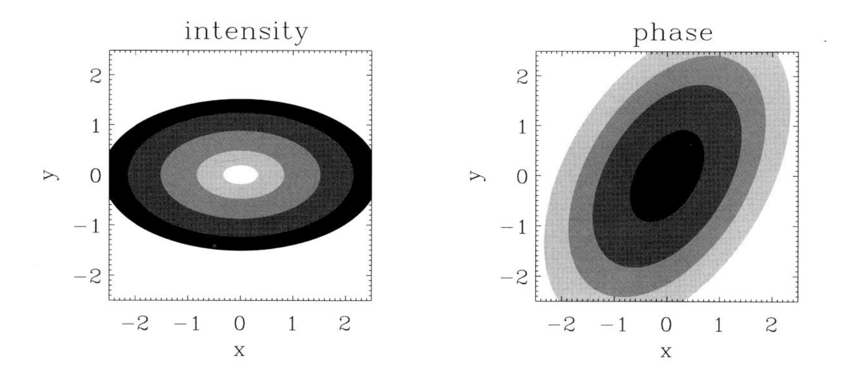

Fig. 12.8. Contours of equal intensity and phase in the transverse plane for a typical astigmatic mode.

12.3.5 Fractional Orbital Angular Momentum

We have seen that a superposition of Laguerre–Gaussian modes can be characterized by a well-defined mean orbital angular momentum and this can, of course, take on any desired value, integer or noninteger. By fractional orbital angular momentum, however, we refer to a beam specifically prepared to have an azimuthal phase dependence of the from $\exp(i\ell\phi)$ where ℓ is not an integer. The first study of such modes that we are aware of [98] showed that a degenerate optical parametric oscillator pumped by an $\ell = 1$ Laguerre–Gaussian beam should produce a field with half-integer orbital angular momentum and an azimuthal dependence $\exp(i\phi/2)$. Such a field necessarily displays a radial discontinuity at some given angle, that is, a radial line of zero intensity, so that the field can be single-valued. We can view this line as a discontinuity or domain wall between equally stable states in which the generated subharmonic field differs in overall sign, that is, by a phase of π.

Fractional orbital angular momentum has been generated by means of spiral phase plates with a phase step not equal to an integer multiple of 2π and by analogous holograms. The latter exhibit a radial discontinuity between the fringes. The form of the mode is characterized by both the value of ℓ and the angular position of the discontinuity [32, 38, 39]. This contrasts with the integer orbital angular momentum modes, of course, for which there is no discontinuity.

The fractional angular momentum states are not eigenmodes of the paraxial wave equation and so do not preserve their form under spatial propagation. We can view this process as interference between the Gouy phases associated with different mode orders. Propagation produces a rich structure of vortices that has been both derived analytically and observed experimentally [38, 39].

12.4 Quantum Optical Angular Momentum

12.4.1 States of Spin and Orbital Angular Momentum

The optics described in the preceding section, together with the familiar polarization-dependent wave plates and polarizing beamsplitters, give us a high degree of control over both the spin and orbital components of the optical angular momentum. This control extends into the quantum regime, where experiments have been performed on optical angular momentum at the single photon level [40–43, 96]. For a single photon, at least within the paraxial approximation, it is permissible to write down a state vector for the angular momentum in the form $|\ell\rangle \otimes |\sigma\rangle$ as eigenstates of the total angular momentum:

$$\hat{J}_z|\ell\rangle \otimes |\sigma\rangle = (\ell + \sigma)\hbar|\ell\rangle \otimes |\sigma\rangle. \tag{12.26}$$

Here we readily identify the spin ($\sigma\hbar$) and orbital ($\ell\hbar$) contributions to the total angular momentum.

The polarization is described by two orthonormal state vectors with $\sigma = \pm 1$ or by any superposition of these. This has been widely employed as a physical implementation of the *qubit* or quantum bit of information [47] and underlies many implementations of quantum key distribution [48]. Using the orbital angular momentum offers the prospect of a much larger, and in principle unbounded, space of states, each associated with a different integer value of ℓ. There has been some progress in this direction including a demonstration of entanglement between a pair of photons with a well-defined total angular momentum but with the angular momentum of each individual photon unspecified (see Section 12.6.3) [40]. There has also been a demonstration of a free-space communications system in which information is encoded on the value of ℓ (see Section 12.5.3) [49].

The orbital angular momentum is intimately connected with the phase profile of the electromagnetic field in the plane perpendicular to the propagation axis. For this reason there is a natural link between the study of orbital angular momentum and transverse effects. In the quantum theory of transverse effects it is convenient to introduce continuum annihilation and creation operators $\hat{A}_\sigma(\boldsymbol{x})$ and $\hat{A}_\sigma^\dagger(\boldsymbol{x})$, where $\boldsymbol{x} = x\boldsymbol{i} + y\boldsymbol{j}$ is any position in the plane perpendicular to the propagation axis (chosen to correspond to the z-axis) and σ denotes the polarization [50]. We include quantum effects by means of the commutation relation:

$$\left[\hat{A}_\sigma(\boldsymbol{x}), \hat{A}_{\sigma'}^\dagger(\boldsymbol{x}')\right] = \delta_{\sigma\sigma'}\delta(\boldsymbol{x} - \boldsymbol{x}'). \tag{12.27}$$

The Laguerre–Gaussian modes form a complete orthonormal set in the x, y plane and it may be useful to introduce a complete set of annihilation operators to describe these modes:

$$\hat{a}_{p\ell\sigma} = \int_{-\infty}^{\infty}\int_{-\infty}^{\infty} u_{p\ell}^{LG}(\boldsymbol{x})\hat{A}_\sigma(\boldsymbol{x})dxdy. \tag{12.28}$$

It is straightforward to show that these operators and their creation operator counterparts satisfy the required commutation relations for a set of independent boson modes:

$$\begin{aligned} \left[\hat{a}_{p\ell\sigma}, \hat{a}_{p'\ell'\sigma'}^\dagger\right] &= \delta_{pp'}\delta_{\ell\ell'}\delta_{\sigma\sigma'}, \\ \left[\hat{a}_{p\ell\sigma}, \hat{a}_{p'\ell'\sigma'}\right] = 0 &= \left[\hat{a}_{p\ell\sigma}^\dagger, \hat{a}_{p'\ell'\sigma'}^\dagger\right]. \end{aligned} \tag{12.29}$$

Hence we can describe transverse effects in quantum optics using both the continuum mode approach and, where it is suitable, operators for modes with well-defined orbital angular momentum.

12.4.2 Measuring Orbital Angular Momentum

If we are to explore and exploit quantum-optical orbital angular momentum then it is important to have an efficient method for measuring it. Figure 12.9

illustrates some of the methods that have been employed successfully for this purpose. Perhaps the most conceptually simple method is depicted in Fig. 12.9a. A Mach–Zehnder interferometer is employed in which a Dove prism is inserted in one arm. This prism acts to reflect, by means of total internal reflection, the beam about its vertical axis. The resulting interference pattern exhibits $2|\ell|$ fringes and allows us to determine ℓ by counting fringes [27, 44]. This method is not suitable, however, at the single photon level as we will then only obtain a single spot on our detector, rather than the required $2|\ell|$ fringes.

For single photons we can employ a hologram or a spiral phase plate to change the value of ℓ by a selected amount, $\ell \rightarrow \ell - \Delta\ell$. If and only if the change induced results in an $\ell = 0$ beam will the resulting beam have an intensity on the axis. Hence focusing the light onto a pinhole will produce transmitted light only if the original beam had $\ell = \Delta\ell$ (see Fig. 12.9b) [45]. Detecting a photon, therefore, corresponds to determining that the value of ℓ was the single value $\Delta\ell$; indeed failure to find the photon does not even allow us to conclude that $\ell \neq \Delta\ell$ as most of an $\ell = 0$ mode will be focused outside the pinhole. More complicated computer-generated holograms have been used to detect several different ℓ states, but the efficiency of this process cannot exceed the reciprocal of the number of different ℓ values to be detected [46].

One method, depicted in Fig. 12.9c, is, at least in principle, 100% efficient for a known set of ℓ values. The interferometer from Fig. 12.9a is modified by the inclusion of a second Dove prism. The net effect of these is to introduce an ℓ-dependent phase shift, equal to $\ell\alpha$, between the fields in the two interferometer arms, where α is twice the angle between the orientations of the two prisms. If, for example, we choose $\alpha = \pi$, then the fields reaching the output beamsplitter will be in phase for even values of ℓ but out of phase for odd values and we can therefore determine whether ℓ is even or odd, for even a single photon, by monitoring the direction in which it exits the interferometer [41]. Placing additional similar interferometers at the output allows us to further separate ℓ-values. Figure 12.9d presents the results of two stages of ℓ-sorting, with three interferometers in total, allowing measurement of ℓ modulo 4. A practical improvement of this design is described in [51]. The spin angular momentum can easily be transformed and measured using polarization-sensitive elements such as waveplates and polarizing beamsplitters. These can be incorporated into the orbital angular momentum sorting interferometer to allow us to measure the total angular momentum $\ell + \sigma$ [52].

12.5 Angle and Angular Momentum

12.5.1 Uncertainty Relation for Angle and Angular Momentum

The modes corresponding to precisely defined orbital angular momentum have a dependence on the azimuthal coordinate in the form $\exp(i\ell\phi)$ and

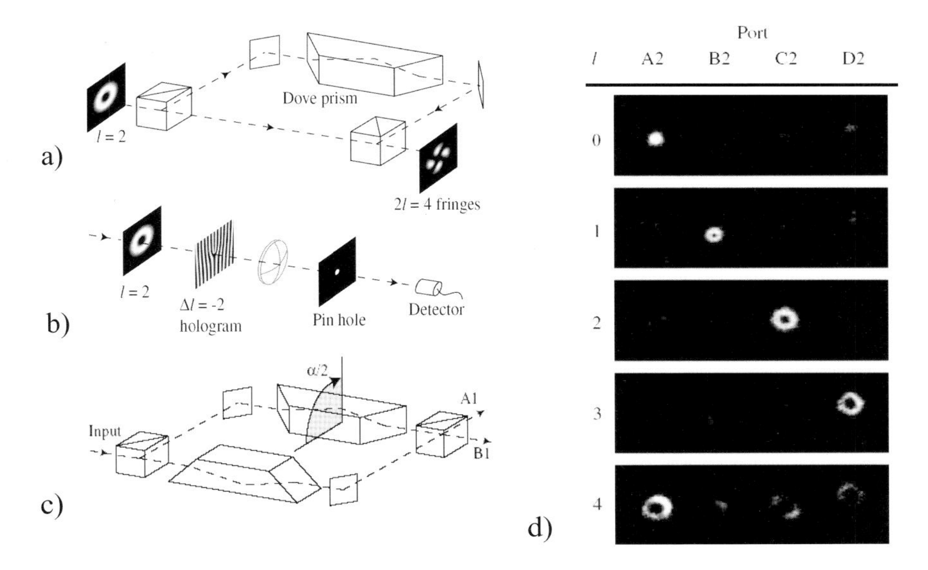

Fig. 12.9. Different methods to detect orbital angular momentum (a)–(c) and experimental observation of ℓ modulo 4 with three interferometers (d). Reproduced from [41].

hence the associated intensity for the mode has cylindrical symmetry. This is analogous to the behavior found in quantum mechanics where the eigenstates of $\hat{L}_z$ are associated with cylindrically symmetric probability distributions. It follows that any attempt to localize the angular coordinate will inevitably introduce a spread of angular momentum values. This situation is a reflection of the conjugate relationship between the angular position ϕ and the corresponding angular momentum L_z.

We can express the complementarity of ϕ and L_z by means of an uncertainty principle. In doing so, however, we need to take account of the fact that all physical properties are periodic functions of the angular position. For this reason we must restrict the values of the angle observable to lie within a 2π radian range. Thus the angle operator $\hat{\phi}_\theta$ will have eigenvalues ϕ lying in the range θ to $\theta + 2\pi$, with a common choice being $-\pi \leq \phi < \pi$. This dependence on the choice of angular range is denoted by the subscript θ on the angle operator. Deriving the correct properties of $\hat{\phi}_\theta$ requires some care, including the use of a specific limiting procedure [53], and is analogous to the derivation of the phase operator for a field mode [54, 55]. The result is that for states with a finite uncertainty in angular momentum we find [53] Au: ok?

$$\Delta\phi_\theta \Delta L_z \geq \frac{\hbar}{2} \left|1 - 2\pi P(\theta)\right| , \tag{12.30}$$

where $\Delta L_z = \hbar \Delta \ell$ and $P(\theta)$ is the angular probability density at the boundary of the chosen angular range. We can see that for the angular momentum eigenstates, $\Delta L_z = 0$ and so too does the right-hand side of the inequality as $P(\theta) = 1/2\pi$, which reflects the rotational symmetry of the angular momentum eigenstates. For states with well-localized angles, however, we can have $P(\theta) \approx 0$ so that $\Delta\phi_\theta \Delta L_z \geq \hbar/2$.

12.5.2 Intelligent and Minimum Uncertainty Product States

The uncertainty relation (12.30) limits the precision with which we simultaneously fix both the angular coordinate and the angular momentum. It is important to determine, therefore, the form of the states that minimize the uncertainty product. The lower bound, given by the right-hand side of Eq. (12.30), depends on the angle probability density $P(\theta)$ and hence the lower bound is itself state dependent. For this reason there are at least two distinct ways in which the uncertainty product can be said to be minimized. These are the intelligent states [56] and the minimum uncertainty product states [57]. We can obtain the form of these states by using Lagrange's method of undetermined multipliers [59, 60].

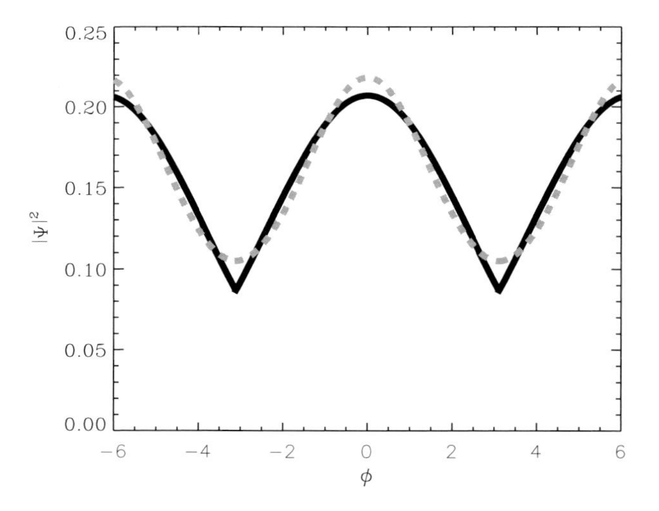

Fig. 12.10. Intelligent and minimum uncertainty product states with angle uncertainty $\Delta\phi = 1.6$.

Intelligent States

The states that saturate the uncertainty relation, that is, realize the equality in the uncertainty relation, are the intelligent states [58]. Such states satisfy a simple eigenvalue equation that, for angular momentum and angle takes the form:

$$\left[\hat{L}_z - \langle \hat{L}_z \rangle - i\hbar\lambda \left(\hat{\phi}_\theta - \langle \hat{\phi}_\theta \rangle\right)\right] |\psi\rangle = 0. \tag{12.31}$$

We can solve this equation by working in the angle representation to find the angle wavefunction $\Psi(\phi)$:

$$\left[i\frac{d}{d\phi} + \bar{\ell} + i\lambda(\phi - \bar{\phi}_\theta)\right] \Psi(\phi) = 0, \tag{12.32}$$

where $\bar{\ell} = \langle \hat{L}_z \rangle / \hbar$ and $\bar{\phi}_\theta = \langle \hat{\phi}_\theta \rangle$. In seeking the solution we note that $\Psi(\phi)$ needs to be a continuous function of ϕ so that the derivative $d\Psi(\phi)/d\phi$ is well behaved. If we choose $\theta = -\pi$, so that $-\pi \leq \phi < \pi$, we are led to the solution, normalized in the range $-\pi$ to π [56]:

$$\Psi(\phi) = \frac{(\lambda/\pi)^{1/4}}{\sqrt{\operatorname{erf}(\pi\sqrt{\lambda})}} e^{i\bar{\ell}\phi} e^{-\lambda\phi^2/2}. \tag{12.33}$$

The form of this wavefunction is given by the solid curve in Fig. 12.10. The expectation value of the angle for this state is zero and the requirement that that wavefunction should be continuous tells us that $\bar{\ell}$ must be an integer. The uncertainties in the angle and angular momentum for this state are

$$\begin{aligned} \Delta\phi &= (2\lambda)^{-1/2}\sqrt{1 - \frac{2\sqrt{\pi\lambda}e^{-\pi^2\lambda}}{\operatorname{erf}(\pi\sqrt{\lambda})}}, \\ \Delta\ell &= \lambda\Delta\phi, \end{aligned} \tag{12.34}$$

where we have written $\Delta\phi_{(-\pi)}$ as $\Delta\phi$. From these it is easy to show that

$$\Delta\phi\Delta\ell = \frac{1}{2}\left|1 - 2\pi|\Psi(\pi)|^2\right|, \tag{12.35}$$

confirming that (12.33) is indeed the intelligent state.

Minimum Uncertainty Product States

For any given value of $\Delta\phi$ the minimum value of the uncertainty product $\Delta\phi\Delta\ell$ need not be that given by the corresponding intelligent state. This is because the states minimizing the angle–angular momentum uncertainty product may have a value of $|1 - 2\pi P(\theta)|/2$ that is considerably smaller than that found for the corresponding intelligent state. The states that minimize the uncertainty product $\Delta\phi\Delta\ell$ either for given $\Delta\phi$ or for $\Delta\ell$ are the

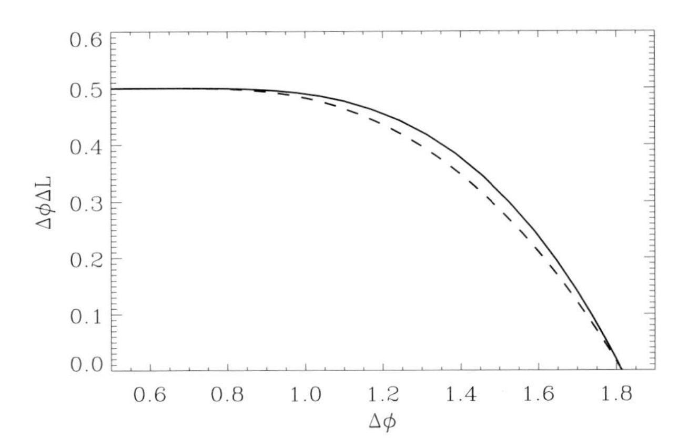

Fig. 12.11. Uncertainty product for the intelligent states (solid line) and for the minimum uncertainty product states (dashed line) plotted as a function of the angle uncertainty.

minimum uncertainty product states (or constrained minimum uncertainty product states) [57]. This procedure leads us to the states $|f\rangle$ that satisfy the eigenvalue equation

$$\left(\frac{\hat{L}_z^2}{\hbar^2} + \lambda\hat{\phi}^2\right)|f\rangle = \mu|f\rangle, \tag{12.36}$$

where λ and μ are real constants. We can solve this equation by again using the angle representation to obtain the angle wavefunction $f(\phi)$. In doing so, we need to ensure that both the wavefunction and its first derivative are continuous so that the second derivative arising from the square of the angular momentum, $\hat{L}_z^2 \to -\hbar^2\partial^2/\partial\phi^2$, is well behaved. The resulting wavefunction can be written in terms of a confluent hypergeometric function:

$$f(\phi) = \exp\left(-\frac{\sqrt{\lambda}\phi^2}{2}\right) M\left(\frac{\sqrt{\lambda}-\mu}{4\sqrt{\lambda}}, \frac{1}{2}, \sqrt{\lambda}\phi^2\right), \tag{12.37}$$

where the relationship between μ and λ is such as to ensure the required continuity of $f(\phi)$. The form of this wavefunction is given by the dashed curve in Fig. 12.10. If the angular uncertainty is not too large then this state is well approximated by a overlapping sequence of Gaussians:

$$f(\phi) \approx \kappa \sum_{n=-\infty}^{\infty} \exp\left[-\frac{\sqrt{\lambda}(\phi+2n\pi)^2}{2}\right], \tag{12.38}$$

with $\mu = \sqrt{\lambda}$, and κ is a normalization constant. The uncertainty products for the intelligent and minimum uncertainty product states are compared in Fig. 12.11.

Experimental Tests

The form of the uncertainty relation (12.30) and of the associated intelligent states (12.33) have been confirmed by an experimental test [56]. A schematic representation of the experiment is given in Fig. 12.12. The required intelligent state was prepared by means of an aperture that imposed the required Gaussian angular dependence on an initial $\ell = 0$ Gaussian mode. The angular momentum content of the resulting mode was then analyzed using the hologram and pinhole technique described in Section 12.4. The intensity measured for the different values of ℓ, together with the value of $\Delta\phi$ imposed by the aperture, was then used to calculate an experimental value for the uncertainty product $\Delta\phi\Delta\ell$.

The results obtained for the uncertainty product are plotted against $\Delta\phi$ in Fig. 12.13 and compared with the theoretical value obtained for the intelligent states from the uncertainty relation. We see that there is very good agreement for most values of $\Delta\phi$. The slightly poorer fit visible for small values of $\Delta\phi$ are due to the difficulty in accurately measuring the very large angular momenta that contribute to $\Delta\ell$. At large values of $\Delta\phi$ we are very close to an eigenstate of angular momentum and the observed angular-momentum uncertainty is dominated by experimental noise.

12.5.3 Communications

The orbital angular momentum of light can be used to carry information, with different values of ℓ corresponding to different possible "letters" in a signal. Communication using different states of orbital angular momentum is not suitable for fiber-based communications as fibers can only admit a restricted range of spatial modes and will also tend to scramble the relative phase between these. It is well suited, however, to free-space communications. A recent experiment has demonstrated the feasibility of free-space communications based on orbital angular momentum [49]. The communication device consists of transmitting and receiving telescopes. In the transmitting telescope, light with a chosen value of ℓ is created by means of a spatial light modulator acting as a hologram. The ℓ value is analyzed in the receiving telescope by means of a second hologram.

The uncertainty relation between the angle and angular momentum provides a measure of security against eavesdropping. This is because an eavesdropper will typically only be able to access a small portion of the beam and the uncertainty principle then ensures that this light will have a broadened spectrum of ℓ-values [49, 62]. The theoretical and experimental effects of cutting out part of the beam are presented in Fig. 12.14.

The orbital angular momentum is a well-behaved quantum property of light. Exploiting this degree of freedom will allow us to engineer states of so-called quNits for quantum communications [61].

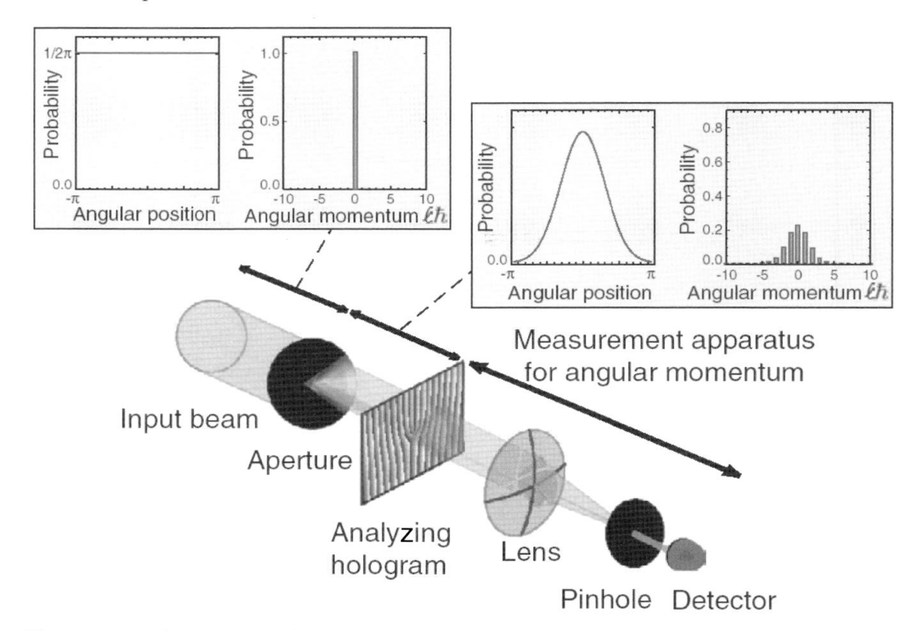

Fig. 12.12. Schematic of the experiment used to observe uncertainties in angular position and angular momentum. Passing a light beam through an aperture restricts the angular position, leading to a corresponding broadening of the light's angular momentum states. A spatial light modulator is used to produce both the aperture function and the angular-momentum-analyzing hologram. The probability of each angular momentum component is deduced from the fraction of the resulting light transmitted through a pinhole. Reproduced from [56].

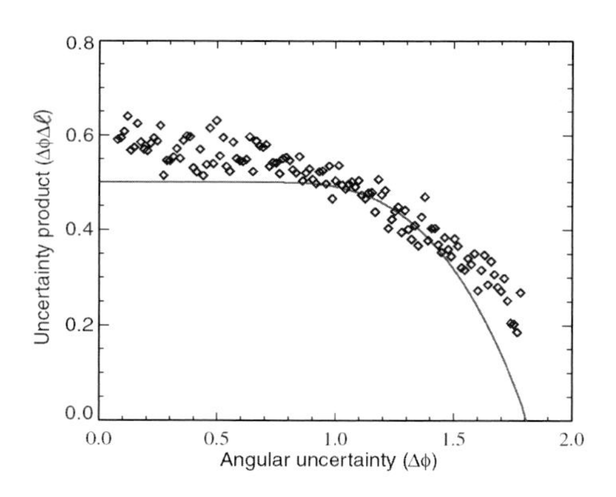

Fig. 12.13. Experimental values for the uncertainty product $\Delta\phi\Delta\ell$, compared with the theoretical value given by the solid line. Reproduced from [56].

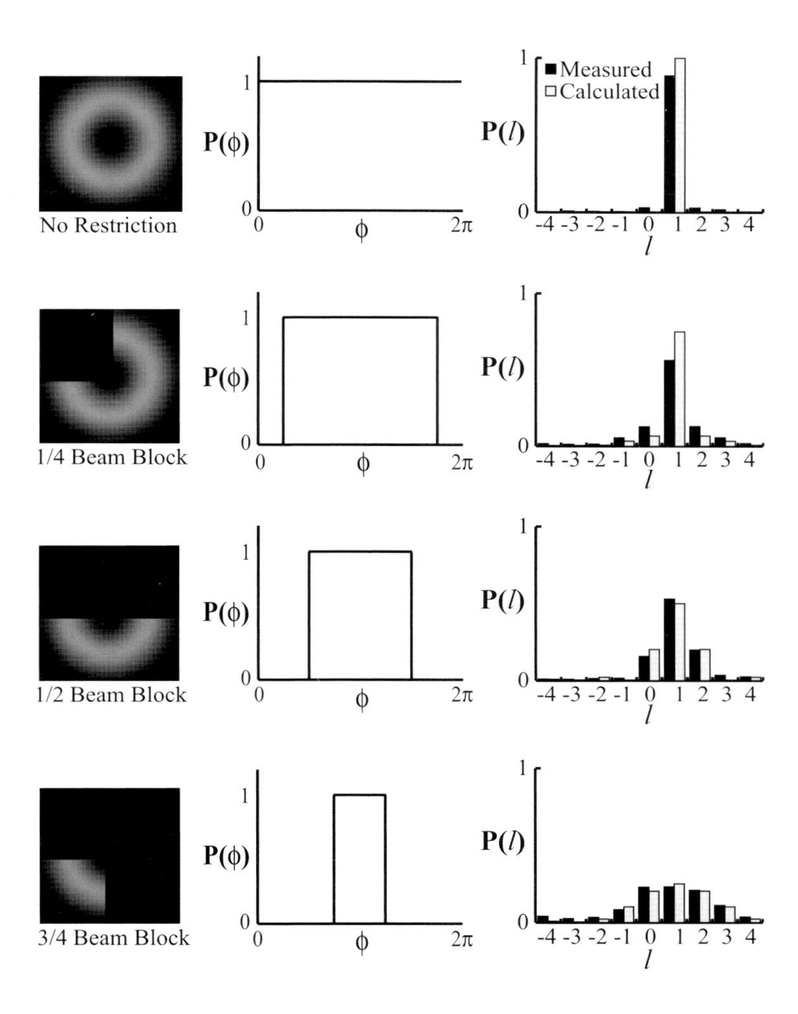

Fig. 12.14. Figure of effects on the angular momentum content of aperturing a beam with a well-defined angular momentum. Reproduced from [49].

12.5.4 Rotation Measurements

High-resolution displacement measurement of a light beam has been recently demonstrated using a proper spatial mode, a novel "flipped" mode [63]. Indeed the achieved precision depends not only on the the number of photons in the beam quantum state, as is generally true in interferometry [64,65], but also on its transverse spatial distribution. The choice of the spatial mode is also important to increase the precision in measuring the rotation of a beam of light about an optical axis. The limiting resolution in this rotation measurement does not depend specifically on the number of photons used but rather on the total number of quanta of orbital angular momentum carried by the light beam [66].

We consider a light beam propagating in the z-direction through an *image rotator*, such as a rotating Dove prism, or a pair of stationary Dove prisms with a fixed relative orientation. The output beam will be azimuthally rotated by an angle $\delta\phi$ around the z-axis. A natural way to measure small angles $\delta\phi$ is to observe the azimuthal displacement of a spot of light impinging on the edges of the rotator. The achievable experimental precision is then limited by the finite size of the optical elements used. For a device with radial aperture R, the azimuthal resolution is [66]

$$\delta\phi \propto \frac{1}{\sqrt{\ell_M}} f(N), \tag{12.39}$$

where $\ell_M = (R/w_0 - 1)^2$ is the maximum angular momentum index transmitted by the device. The function $f(N)$ has the form $N^{-1/2}, N^{-3/4}, N^{-1}$ for coherent, strongly squeezed, or number states, respectively, where N is the average number of photons [67].

The resolution of rotation measurements can be improved by choosing the appropriate spatial mode. If the incoming beam is an angular momentum eigenstate then the only effect of the rotator is to add a constant phase shift. This suggests the use of an interferometer with the rotator placed along one of the paths and input beams in angular momentum eigenstates. The difference in the intensities of the two output beams depends both on the phase shift, here $\ell\delta\phi$, and on the quantum state of the incoming beams, leading to the smallest detectable phase shift [66]

$$\delta\phi \propto \frac{1}{\ell} f(N). \tag{12.40}$$

For Fock states ($f(N) = 1/N$) the minimum detectable rotation is set by the total number of quanta of orbital angular momentum, $N\ell$. Therefore we can increase the sensitivity in rotation measurements not only by the use of nonclassical states of light (dependence on N), but also by using easily accessible eigenmodes of orbital angular momentum (geometrical function of ℓ_M). Moreover, the use of these modes to enhance the angular resolution is relatively robust, as no matter how many photons are lost, each of the remaining photons still carries $\ell\hbar$ units of angular momentum.

12.6 Orbital Angular Momentum in Quantum Nonlinear Optics

Vortex beams carrying orbital angular momentum made an early appearance in nonlinear classical optics, but were mostly studied as a peculiar case of *singularities*, namely phase singularities or screw dislocations [68,69]. Even after the identification of a well-defined amount of orbital angular momentum per photon in Laguerre–Gaussian beams in 1992 [1], most of the literature in singular optics was not explicitly concerned with this mechanical property of light [70]. This is perhaps not too surprising as the link between vortices and orbital angular momentum is not generally as simple as that for the Laguerre–Gaussian modes. In this section we focus on orbital angular momenta in nonlinear optics, and for general studies on singular optics and vortex solitons we refer to [71, 94, 72].

In Section 12.3.3 we have summarized some of the techniques used to generate beams with orbital angular momentum based on mode converters, spiral phase plates, and holograms. These are devices with a linear response to the input beam. Beams carrying orbital angular momentum can also be generated spontaneously in nonlinear devices. We mention here in particular the example of lasers, in which optical elements with circular symmetry are embedded in the resonator allowing for the oscillation of higher-order LG modes [73,74].

An active area of investigations about angular momentum in nonlinear processes focuses on the effects observed when LG or Bessel beams are pumped in optical devices as quadratic crystals [40,75], nonlinear cavities [81], lasers [83,84], or cold atoms [85]. An important question arising when waves mix in nonlinear devices is the conservation of their mechanical properties (see Section 12.6.1). Several experiments have confirmed that the orbital angular momentum of photons is conserved in up- and down-conversion as well as in four-wave mixing. These are discussed in Sections 12.6.2 and 12.6.3. Another question of interest is the stability of beams carrying orbital angular momentum during propagation in nonlinear media: a subject that challenged several theoretical studies in the last decade. An example of a stable ring soliton has been recently observed in two-dimensional optically induced photonic lattices [86,87]. On the other hand, the instability of vortex solitons in propagation allows us to observe a clear and dynamical manifestation of the orbital angular momentum of light, as discussed in Section 12.6.4. We conclude this section reporting on some observations of spontaneously formed spatial structures that carry a fractional angular momentum per photon.

12.6.1 Phase Matching

Wave-mixing processes in nonlinear materials take place efficiently only if the phase of the waves is properly matched to produce constructive interference. If we consider three-wave mixing, as observed in quadratic crystals, then the frequency and phase-matching conditions are $\omega_0 = \omega_1 + \omega_2$ and $\boldsymbol{k}_0 = \boldsymbol{k}_1 + \boldsymbol{k}_2$. Considering the nonlinear interaction at the quantum level, we see that the frequencies and wave-vectors of the created photons sum to the same values as for the destroyed photons. We can interpret these conditions as expressions of the conservation of energy and linear momentum, respectively. The interactions between light and matter considered here do indeed assume that neither energy nor momentum is transferred to matter during the interaction.

Recent experiments have shown that orbital angular momentum is also a conserved quantity within nonlinear processes [40, 75], as detailed in the following sections. In the case of mixing of three waves this is expressed by $\ell_0 = \ell_1 + \ell_2$, being $\ell_{0,1,2}$ the angular momentum of pump, signal, and idler photons. In [77] it was shown that the conservation of orbital angular momentum is also a consequence of phase-matching conditions. More complicated situations occur in the nonparaxial regime [78–80] or for noncollinear interactions [76].

Interestingly, the spin angular momentum is usually not conserved in wave-mixing processes: in type-I phase matching, for instance, the lower-frequency waves are generally polarized orthogonally to the higher-frequency wave. As a matter of fact, in order to phase match three-wave mixing, it is necessary to use birefringent materials. As we have seen in Section 12.2.3, the interaction is then anisotropic and spin angular momentum in the light involved in the process is not conserved, with the outstanding angular momentum transferred to the medium.

12.6.2 Second-Harmonic Generation of Laguerre–Gaussian Beams

Second-harmonic generation results from the mixing of three waves in quadratic media, when the frequency of a pump wave ($\omega = \omega_1 = \omega_2$) is doubled ($2\omega = \omega_0$). It has been shown that if the pump carries $\ell\hbar$ units of orbital angular momentum per photon, the second-harmonic carries $2\ell\hbar$ units [88]. A simple explanation of this phenomenon is obtained assuming that the generated wave is proportional to the square of the pump wave: for a wave in a Laguerre–Gaussian mode $L_{0,\ell}$ the second harmonic will be $L_{0,2\ell}$ with a beam waist reduced by a factor $\sqrt{2}$. A consistent picture was also obtained in [88] by considering that the interacting waves must have collinear Poynting vectors. As the Poynting vector of a pump $u_{0\ell}^{LG}$ beam spirals at the rate ℓ/kr [89], and the generated beam has a doubled k, ℓ must also be doubled in order to have the same rotation rates in both waves.

The experimental demonstration of the conservation of angular momentum is achieved through the observation of the spatial structure of both fundamental and second-harmonic beams. In [88] this was achieved by the use of cylindrical mode converters, transforming Hermite–Gaussian in Laguerre–Gaussian modes with $n - m = \ell$ and $m = p = 0$ (Section 12.3.3). Given the fundamental relation between orbital momentum and azimuthal phase distribution of the beam, interferometric techniques are also particularly useful. In [75] the beam was interfered with its mirror image generated with a Dove prism, giving the characteristic fork diagrams shown in Fig. 12.15. The ℓ value is given by one half of the number of the bright fringes in the fork. The doubling of ℓ can then be easily recognized in the experimental images of the fundamental and second-harmonic.

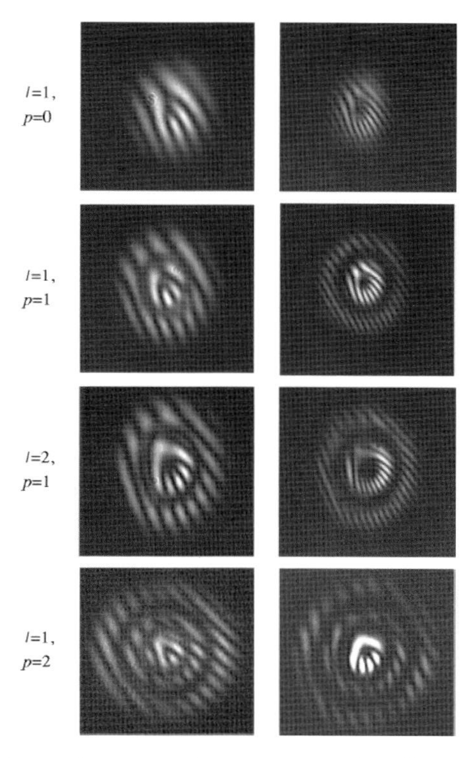

Fig. 12.15. Forked interferograms for a variety of $u_{p\ell}^{LG}$ modes (left) and their second-harmonic counterpart (right). Both $p = 0$ and $p \neq 0$ are studied. Reproduced from [75].

The conservation of orbital angular momentum in second-harmonic generation has also been demonstrated for Laguerre–Gaussian modes with $p \neq 0$. For a pump beam in a multiringed mode $p \neq 0$ the second harmonic will be in a superposition of modes with different p indices, instead of in a single mode as in the case $p = 0$. These modes have different Gouy phases and interfere, giving a spatial distribution that changes during propagation, but with a fixed azimuthal index 2ℓ. The experimental data displayed in Fig. 12.15 show the conservation of orbital angular momentum for $p = 1, 2$. Equivalent results can be found when Bessel instead of Laguerre–Gaussian beams are considered [92].

Second-harmonic generation is a particular (degenerate) case of sum-frequency processes, in which two waves with frequencies ω_1 and ω_2 are injected in a quadratic crystal and sum in a third wave $\omega_3 = \omega_1 + \omega_2$. The conservation of orbital angular momentum has been demonstrated also in this more general case [90]. This allows us to produce the arithmetical sum and difference of topological charges, and equivalently to the orbital angular momenta of single $u_{0\ell}^{LG}$ modes.

We have focused here on the conservation of angular momenta for nonlinear interactions of vortex beams under conditions for stable propagation. A more accurate analysis reveals that ring-shaped beams in a nonlinear optical material can be azimuthally unstable [93]. This problem has been extensively studied analytically and numerically in the literature, as reviewed in [94]. We will describe a recent experiment on this subject in the context of high-order nonlinearities in Section 12.6.4.

12.6.3 Down-Conversion and Entanglement

In the previous section we discussed the conservation of orbital angular momentum in sum-frequency processes, where the spatial distribution of two input waves dictates the shape of the sum wave through phase matching. In the case of parametric down-conversion a pump wave of frequency ω_0 is converted in two waves $\omega_1 + \omega_2 = \omega_0$, known as the signal and idler. Phase matching dictates in this case only the complementarity of the phases of the down-converted beams, but not their individual values. Indeed either the signal or idler is generally in a superposition of orbital angular momentum eigenstates and known to be individually incoherent. However, coincidence measurement allows us to observe that the orbital angular momenta of each down-converted photon pair do sum to the value of a pump photon [40].

The lack of a well-defined phase profile and angular momentum in the signal as well as in the idler beam prevent immediate observation of the conservation of this quantity from the detection of the spatial profile of the beams [95]. Indeed the conservation in these spontaneous processes is observed only when looking at cross-correlations between different spatial modes in the signal and idler [40, 96]. Alternatively well-defined spatial modes in the down-converted fields can be achieved in stimulated parametric down-conversion [97], by strong transverse selection of the spontaneously down-converted field obtained with Bessel beams [82], or in optical parametric oscillators (OPO) [81]. In [97], in addition to the pump, the idler is also injected on the crystal inducing the emission of a well-defined mode. Detecting the three interacting fields with a CCD camera it was confirmed that the idler angular momentum was $\ell_2 = \ell_0 - \ell_1$. The conservation can also be observed in parametric oscillators operating above threshold, but only under certain resonance and mode degeneracy conditions, ultimately depending on the crystal birefringence [81].

A degenerate OPO pumped by a c.w. LG mode was studied in [98] in the context of the stabilization of domain walls. It was shown that for a pump $u_{0\ell}^{LG}$ with even ℓ there is a stable signal with phase profile $e^{i\ell\phi/2}$. The signal would be in a discontinuous state with fractional angular momentum for odd ℓ. This ill-defined phase induces one (or more) domain wall with vanishing intensity in the radial direction. Signals with a number of domain walls with the same parity of the pump index ℓ are shown in Fig. 12.16. These light distributions with one or several lines of darkness were called "optical sprinklers," as they actually rotate in time. The peculiarity of this state is that it is a mode with fractional orbital angular momentum given by the noninteger azimuthal index $\ell/2$, for ℓ odd (Section 12.3.5). We note that some authors use this same definition of beams carrying fractional angular momentum referring to the average angular momentum per photon. This quantity is generally noninteger for any superposition of few $u_{0\ell_i}^{LG}$ modes with an odd value of $\sum \ell_i$, without any discontinuities in the phase. Nice examples of structures with fractional average angular momentum are self-trapped necklaces in the nonlinear Schrödinger equation [99].

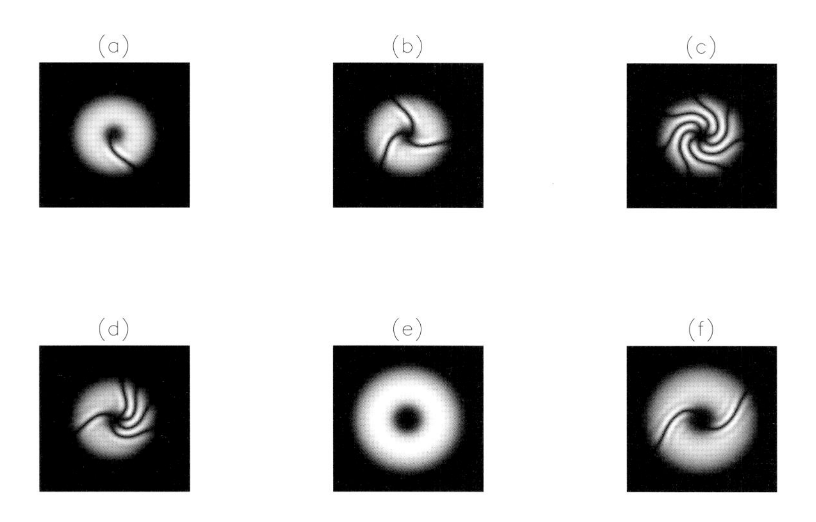

Fig. 12.16. Numerically calculated "optical sprinklers" in the signal intensity of an OPO pumped by a $u_{0\ell}^{LG}$ mode of azimuthal index $\ell = 1$ (a)–(d) and $\ell = 2$ (e)–(f) for different pump intensities. Reproduced from [98].

The identification of a conserved quantity in the simultaneous generation of photons suggests the possibility of generating orbital angular momentum entanglement in a Hilbert space with increased dimensionality with respect to the spin case. This entanglement was already shown in [40] and discussed in [77]. The two-photon state generated in the paraxial regime for a thin crystal can then be written as

$$|\psi\rangle = \sum_{\ell_1=-\infty}^{+\infty} C_{\ell_1} |\ell_1\rangle |\ell_0 - \ell_1\rangle, \tag{12.41}$$

with $|\ell\rangle$ denoting an angular momentum eigenstate. The angular distribution of coincidence counts has a more complicated structure. For a Gaussian ($\ell = 0$) pump we expect these to be well localized on opposite sides of the down-conversion cone. It has recently been suggested that this simultaneous correlation between the orbital angular momentum and the azimuthal coordinate could form the basis of a demonstration of the famous EPR paradox [91]. For a Laguerre–Gaussian pump, however, the "image" produced in the idler, in coincidence with detecting the signal photon in a small region, is given by the Fourier transform of the pump field inside the nonlinear crystal [43]. It has recently been demonstrated that photons are also entangled in fractional angular momenta [32].

12.6.4 High-Order Nonlinearity

In Sections 12.6.2 and 12.6.3 we have reported on experiments focusing on the conservation of orbital angular momentum in three-wave mixing processes. Higher-order nonlinearities allow us to increase the number of coupled beams. Experiments in noncollinear configurations both in cold cesium atoms [85] and in a doped color glass [100] have recently confirmed angular momentum conservation for Kerr nonlinear processes mixing four waves in LG modes. However, the question that is attracting the attention of several groups is the possibility of obtaining stable propagation of beams carrying angular momentum in nonlinear media [102].

A novel class of self-trapped beams characterized by the rotation of the field phase and more resistant to whole-beam collapse was proposed in 1985 [103]. These beams are indeed spatial optical solitons, carrying a nonzero angular momentum, and known as *vortex solitons*. After their first experimental observation in 1992 [104], different strategies to obtain vortex solitons have been studied, including competing nonlinearities, multimode vector solitons (in which a vortex is stabilized by the potential introduced by another mode), nonlocal coupling, and photonic crystals. Extensive numerical and theoretical investigations, however, have shown that ring beams suffer strong azimuthal instabilities in both a saturable Kerr medium and in a material with a competing quadratic and cubic nonlinearity [107]. Due to the azimuthal symmetry-breaking instability, beams with a phase profile $\exp(i\ell\phi)$ usually decay into $2|\ell|$ (for the self-focusing Kerr-like medium) or $2|\ell| + 1$ (for the quadratic media) fundamental optical solitons [107].

The instability of these beams offers the possibility of observing a nice dynamical effect of the conservation of angular momentum, as seen both in Kerr-like [105] and quadratic [106] media. In particular, the propagating ring decays into a number of filaments that travel off tangentially to the unstable ring, conserving the total orbital angular momentum, as shown in Fig. 12.17(i). The results of a recent experiment with a saturable nonlinearity [101] are shown in Fig. 12.17(ii). When a laser pulse in an LG mode with

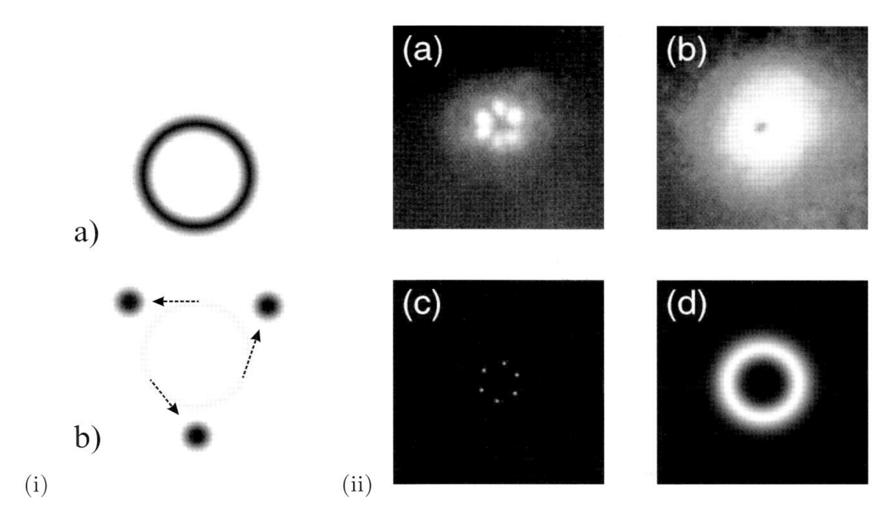

Fig. 12.17. (i) Graphical evolution of an $\ell = 3$-ring mode (a) decaying into three solitons flying away tangentially to the initial ring (b). (ii) Experimental output (a) for an $\ell = 3$ beam breaking into six filaments at resonance, and (b) tuned far from resonance. In (c) and (d) the equivalent results from numerical simulations of the nonlinear propagation equation modeling for the sodium vapor. Reproduced from [101].

orbital angular momentum $\ell = 3$ passes through a dense sodium vapor it breaks up into six filaments, as predicted in [107].

12.7 Conclusion

The orbital angular momentum of light is fundamentally associated with the phase properties of the field in the plane perpendicular to the propagation direction. This contrasts with the more familiar spin angular momentum that is associated with the polarization or vector character of the electromagnetic field. It is clear that the orbital angular momentum is fundamentally a spatial property of the field and so is intimately connected with imaging and spatial correlations, which are the principal topics of this book.

We should emphasize that optical angular momentum is a genuine mechanical property and this has been amply demonstrated in the experiments described in Section 12.2.3. It is also clear that the nonlinear instabilities described in Section 12.6.4 can most easily be understood in terms of a conserved mechanical angular momentum.

The angular momentum is conjugate to the azimuthal angular coordinate and these are constrained by an uncertainty relation, or analogue of the Fourier bandwidth theorem. We showed in Section 12.5 that this uncertainty principle leads to well-defined intelligent and minimum uncertainty product states and that these have been demonstrated experimentally.

The study of orbital angular momentum in the quantum domain is only in its infancy. It is already clear, however, that entanglement can be produced and that angular momentum can be manipulated and measured at the single-photon level. There is a great deal of potential for the realization of quantum communications and information-processing-based orbital angular momentum.

Acknowledgments

We are grateful to the friends and colleagues with whom we have enjoyed investigating orbital angular momentum. In particular we thank Les Allen, Johannes Courtial, Sonja Franke-Arnold, Graham Gibson, Jörge Götte, Jonathan Leach, Rodney Loudon, Miles Padgett, David Pegg, Ken Sheldon, Laura Thomson, and Eric Yao. This work was supported by the UK Engineering and Physical Sciences Research Council.

References

1. L. Allen, M. W. Beijersbergen, R. J. C. Spreeuw, and J. P. Woerdman, Phys. Rev. A **45**, 8185 (1992).
2. D. Marcuse, *Light Transmission Optics* (Van Nostrand, New York, 1972).
3. L. Allen, M. J. Padgett, Opt. Commun. **184**, 67 (2000).
4. L. Allen, M. J. Padgett, and M. Babiker, in *Progress in Optics*, vol. XXXIX, 291 (1999).
5. L. Allen, S. M. Barnett, and M. J. Padgett, *Optical Angular Momentum* (Institute of Physics Publishing, Bristol, 2003).
6. J. D. Jackson, *Classical Electrodynamics*, 3rd edn. (Wiley, New York, 1999).
7. C. Cohen-Tannoudji, J. Dupont-Roc, and G. Grynberg, *Photons and Atoms: Introduction to Quantum Electrodynamics* (Wiley, New York, 1997).
8. S. J. van Enk and G. Nienhuis, Opt. Commun. **94**, 147 (1992).
9. L. Mandel and E. Wolf, *Optical Coherence and Quantum Optics* (Cambridge University Press, Cambridge, 1995).
10. S. J. van Enk and G. Nienhuis, Europhys. Lett. **25**, 497 (1994).
11. S. J. van Enk and G. Nienhuis, J. Mod. Opt. **41**, 963 (1994).
12. S. M. Barnett, J. Opt. B: Quantum Semiclass. Opt. **4**, S7 (2002).
13. C. N. Alexeyev, Y. A. Fridman, and A. N. Alexeyev: Ukr. Phys. J. **46**, 43 (2001).
14. H. A. Haus, *Waves and Fields in Optoelectronics* (Prentice-Hall, Englewood Cliffs, NJ, 1984).
15. R. A. Beth, Phys. Rev. **50**, 115 (1936).
16. N. He, M. E. J. Friese, N. R. Hechenberg, and H. Rubinstein-Dunlop, Phys. Rev. Lett. **75**, 826 (1995).

17. M. E. J. Friese, J. Enger, H. Rubinstein-Dunlop, and N. R. Heckenberg, Phys. Rev. A **54**, 1593 (1996).
18. N. B. Simpson, K. Dholakia, L. Allen, and M. J. Padgett, Opt. Lett. **22**, 52 (1997).
19. A. T. O'Neil, I. MacVicar, L. Allen, and M. J. Padgett, Phys. Rev. Lett. **88**, 053601 (2002).
20. R. Zambrini and S. M. Barnett, J. Mod. Opt. **52**, 1045 (2005).
21. J. F. Nye, *Natural Focusing and Fine Structure of Light* (Institute of Physics Publishing, Bristol, 1999).
22. M. S. Soskin and M. V. Vasnetsov, in *Progress in Optics*, vol. 42, ed. by E. Wolf (Elsevier, Amsterdam, 2001).
23. J. Leach, PhD Thesis, University of Glasgow, Glasgow (2005).
24. A E. Siegman, *Lasers* (University Science Books, Mill Valley, CA, 1986).
25. J. Durnin, J. J. Miceli, and J. H. Eberly, Phys. Rev. Lett. **58**, 1499 (1987).
26. L. Allen, H. E. L. O. van der Veen, and J. P. Woerdman, Opt. Commun. **96**, 123 (1993).
27. M. J. Padgett, J. Arlt, N. Simpson, and L. Allen, Am. J. Phys. **64**, 77 (1996).
28. M. J. Padgett and L. Allen, J. Opt. B: Quantum Semiclass. Opt. **4**, S17 (2002).
29. L. Allen, J. Courtial, and M. J. Padgett, Phys. Rev. E **60**, 7497 (1999).
30. M. W. Beijersbergen, R. P. C. Coerwinkel, M. Kristensen, and J. P. Woerdman, Opt. Commun. **112** 321 (1994).
31. G. A. Turnbull, D. A. Robertson, G. M. Smith, L. Allen, and M. J. Padgett, Opt. Commun. **127**, 183 (1996).
32. S. S. R. Oemrawsingh, A. Aiello, E. R. Eliel, G. Nienhuis, and J. P. Woerdman, Phys. Rev. Lett. **92**, 217901 (2004).
33. V. Y. Bazhenov, M. V. Vasnetsov, and M. S. Soskin, JEPT Lett. **52**, 429 (1990).
34. N. R. Heckenberg, R. McDuff, C. P. Smith, H. Rubinsztein-Dunlop, and M. J. Wegener, Opt. Quantum Electron. **24**, S951 (1992).
35. G. Molina-Terriza, J. Recolons, J. P. Torres, L. Torner, and E. M. Wright, Phys. Rev. Lett. **87**, 023902 (2001).
36. J. Courtial, K. Dholakia, L. Allen, and M. J. Padgett, Opt. Commun. **144**, 210 (1997).
37. J. Visser and G. Nienhuis, Phys. Rev. A **70**, 013809 (2004).
38. M. V. Berry, J. Opt. A **6**, 259 (2004).
39. J. Leach, E. Yao, and M. J. Padgett, New J. Phys. **6**, 71 (2004).
40. A. Mair, A. Vaziri, G. Weihs, and A. Zeilinger, Nature **412**, 313 (2001).
41. J. Leach, M. J. Padgett, S. M. Barnett, S. Franke-Arnold, and J. Courtial, Phys. Rev. Lett. **65**, 033823 (2002).
42. N. K. Langford, R. B. Dalton, M. D. Harvey, J. L. O'Brien, G. J. Pryde, A. Gilchrist, S. D. Bartlett, and A. G. White, Phys. Rev. Lett. **93**, 053601 (2004).
43. A. R. Altman, K. G. Köprülü, E. Corndorf, P. Kumar, and G. A. Barbosa, Phys. Rev. Lett. **94**, 123601 (2005).
44. M. Harris, C. A. Hill, P. R. Tapster, and J. M. Vaughan, Phys. Rev. A **49**, 3119 (1994).
45. G. Weihs and A. Zeilinger, Nature **412**, 313 (2001).
46. V. V. Kotlyar, V. A. Soifer, and S. N. Khonina, J. Mod. Opt. **44**, 1409 (1997).
47. M. A. Nielsen and I. L. Chuang, *Quantum Computation and Quantum Information* (Cambridge University Press, Cambridge, 2000).

48. S. J. D. Phoenix and P. D. Townsend, Contemp. Phys. **36**, 165 (1995).
49. G. Gibson, J. Courtial, M. J. Padgett, M. Vasnetsov, V. Pas'ko, S. M. Barnett, and S. Franke-Arnold, Opt. Express **12**, 5448 (2004).
50. A. Gatti, H. Wiedemann, L. A. Lugiato, I. Marzoli, G. L. Oppo, and S. M. Barnett, Phys. Rev. A **56**, 877 (1997).
51. H. Wei, X. Xue, J. Leach, M. J. Padgett, S. M. Barnett, S. Franke-Arnold, E. Yao, and J. Courtial, Opt. Commun. **223**, 117 (2003).
52. J. Leach, J. Courtial, K. Skeldon, S. M. Barnett, S. Franke-Arnold, and M. J. Padgett, Phys. Rev. Lett. **92**, 013601 (2004).
53. S. M. Barnett and D. T. Pegg, Phys. Rev. A **41**, 3427 (1990).
54. D. T. Pegg and S. M. Barnett, Phys. Rev. A **39**, 1665 (1989).
55. S. M. Barnett and D. T. Pegg, J. Mod. Opt. **36**, 7 (1989).
56. S. Franke-Arnold, S. M. Barnett, E. Yao, J. Leach, J. Courtial, and M. Padgett, New J. Phys. **6**, 103 (2004).
57. D. T. Pegg, S. M. Barnett, R. Zambrini, S. Franke-Arnold, and M. Padgett, New J. Phys. **7**, 62 (2005).
58. C. Aragone, E. Chalbaud, S. Salam, J. Phys. A: Math. Gen. **7**, L149 (1974).
59. R. Jackiw, J. Math. Phys. **9**, 339 (1968).
60. G. S. Summy and D. T. Pegg, Opt. Commun. **77**, 75 (1990).
61. G. Molina-Terriza, J. P. Torres, and L. Torner, Phys. Rev. Lett. **88**, 013601 (2002).
62. M. V. Vasnetsov, V. A. Pas'ko, and M. S. Soskin, New J. Phys. **7**, 46 (2005).
63. N. Treps, U. Andersen, B. Buchler, P. K. Lam, A. Maître, H.-A. Bachor, and C. Fabre, Phys. Rev. Lett. **88**, 203601 (2002).
64. C. M. Caves, Phys. Rev. D **23**, 1693 (1981).
65. M. J. Holland and K. Burnett, Phys. Rev. Lett. **71**, 1355 (1993).
66. S. M. Barnett and R. Zambrini, J. Mod. Opt., **53**, 613 (2006).
67. S. M. Barnett, C. Fabre, and A. Maître, Eur. Phys. J. D **22**, 513 (2003).
68. V. L. Ginzburg and L. P. Pitaevskii, Sov. Phys. JETP **34**, 858 (1958).
69. J F. Nye and M. V. Berry, Proc. R. Soc. London Ser. **336**, 165 (1974).
70. L. Allen, J. Opt. B, Qunantum Semiclass. Opt. **4**, S1 (2002).
71. Y. S. Kivshar and B. Luther-Davies, Phys. Rep. **298**, 81 (1998).
72. M. Berry, M. Dennis, and M. Soskin (Guest Editors), J. Opt. A **6**, S155 (2004).
73. J. M. Vaughan and D. V. Willetts, Opt. Commun. **30**, 263 (1979).
74. P. Coullet, L. Gil, and F. Rocca, Opt. Commun. **73**, 403 (1989).
75. J. Courtial, K. Dholakia, L. Allen, and M. J. Padgett, Phys. Rev. A **56**, 4193 (1997).
76. G. Molina-Terriza, J. P. Torres, and L. Torner, Opt. Commun. **228**, 155 (2003).
77. S. Franke-Arnold, S. M. Barnett, M. J. Padgett, and L. Allen, Phys. Rev. A **65**, 033823 (2002).
78. H. H. Arnaut and G. A. Barbosa, Phys. Rev. Lett. **85**, 286 (2000).
79. E. R. Eliel, S. M. Dutra, G. Nienhuis, and J. P. Woerdman, Phys. Rev. Lett. **86**, 5208 (2001).
80. H. H. Arnaut and G. A. Barbosa, Phys. Rev. Lett. **86**, 5209 (2001).
81. M. Martinelli, J. A. O. Huguenin, P. Nussenzveig, and A. Z. Khoury, Phys. Rev. A **70**, 013812 (2004).
82. V. Pyragaite, A. Piskarskas, K. Regelskis, V. Smilgevicius, A. Stabinis, S. Mikalauskas, Opt. Commun. **240**, 191 (2004).

83. S. F. Pereira, M. B. Willemsen, M. P. van Exter, and J. P. Woerdman, Appl. Phys. Lett. **73**, 2239 (1998).
84. Y. F. Chen and Y. P. Lan, Phys. Rev. A, **63**, 063807 (2001).
85. S. Barreiro and J. W. R. Tabosa, Phys. Rev. Lett. **90**, 133001 (2003).
86. D. N. Neshev, T. J. Alexander, E. A. Ostrovskaya, Y. S. Kivshar, H. Martin, I. Makasyuk, and Z. Chen, Phys. Rev. Lett. **92**, 123903 (2004).
87. J. W. Fleischer, G. Bartal, O. Cohen, O. Manela, M. Segev, J. Hudock, and D. N. Christodoulides, Phys. Rev. Lett. **92**, 123904 (2004).
88. K. Dholakia, N. B. Simpson, M. J. Padgett, and L. Allen, Phys. Rev. A **54**, R3742 (1996).
89. M. J. Padgett and L. Allen, Opt. Commun. **121**, 36 (1995).
90. A. Berzanskis, A. Matijosius, A. Piskarskas, V. Smilgevicius, and A. Stabinis, Opt. Commun. **140**, 273 (1997).
91. J. Götte, S. Franke-Arnold, and S. M. Barnett, J. Mod. Opt., **53**, 627 (2006).
92. D. McGloin and K. Dholakia, Cont. Phys. **46**, 15 (2005).
93. L. Torner and D. V. Petrov, J. Opt. Soc. Am. B **14**, 2017 (1997).
94. A. V. Buryak, P. Di Trapani, D. V. Skryabin, and S. Trillo, Phys. Rep. **370**, 63 (2002).
95. J. Arlt, K. Dholakia, L. Allen, and M. J. Padgett, Phys. Rev. A 59, **3950** (1999).
96. S. P. Walborn, A. N. de Oliveira, R. S. Thebaldi, and C. H. Monken, Phys. Rev. A **69**, 023811 (2004).
97. D. P. Caetano, M. P. Almeida, P. H. Souto Ribeiro, J. A. O. Huguenin, B. Coutinho dos Santos, and A. Z. Khoury, Phys. Rev. A **66**, 041801(R) (2002).
98. G-L. Oppo, A. J. Scroggie, and W. J. Firth, Phys. Rev. E **63**, 066209 (2001).
99. M. Soljacic and M. Segev, Phys. Rev. Lett. **86**, 420 (2001).
100. V. Pyragaite, K. Regelskis, V. Smilgevicius, and A. Stabinis, Opt. Commun. **198**, 459 (2001).
101. M. S. Bigelow, P. Zerom, and R. W. Boyd, Phys. Rev. Lett. **92**, 083902 (2004).
102. Y. S. Kivshar and G. P. Agrawal *Optical Solitons: From Fibers to Photonics Crystals* (San Diego, CA, Academic, 2003).
103. V. I. Kruglov and R. A. Vlasov, Phys. Lett. A **111**, 401 (1985).
104. G. A. Swartzlander and C. T. Law, Phys. Rev. Lett. **69**, 2503 (1992).
105. V. Tikhonenko, J. Christou, and B. Luther-Davies, Phys. Rev. Lett. **76** 2698 (1996).
106. D. V. Petrov, L. Torner, J. Martorell, R. Vilaseca, J. P. Torres, and C. Cojocaru, Opt. Lett. **23**, 1444 (1998).
107. W. J. Firth and D. V. Skryabin, Phys. Rev. Lett. **79**, 2450 (1997).

Index